T0198105

SAE AND THE EVOLVED PACKET CORE: DRIVING THE MOBILE BROADBAND REVOLUTION

SAE and the Evolved Packet Core

Driving The Mobile Broadband Revolution

Magnus Olsson
Shabnam Sultana
Stefan Rommer
Lars Frid
Catherine Mulligan

AMSTERDAM • BOSTON • HEIDELBERG • LONDON • NEW YORK • OXFORD
PARIS • SAN DIEGO • SAN FRANCISCO • SINGAPORE SYDNEY • TOKYO
Academic Press is an imprint of Elsevier

Academic Press is an imprint of Elsevier
Linacre House, Jordan Hill, Oxford OX2 8DP, UK
30 Corporate Drive, Suite 400, Burlington, MA 01803, USA

First edition 2009

Notice
Knowledge and best practice in this field are constantly changing. As new research and experience broaden our understanding, changes in research methods, professional practices, or medical treatement may become necessary.

Practitioners and researchers must always rely on their own experience and knowledge in evaluating and using any information, methods, compounds, or experiments described herein. In using such information or methods they should be mindful of their own safety and the safety of others, including parties for whom they have a professional responsibility.

To the fullest extent of the law, neither the Publisher nor the authors, contributors, or editors, assume any liability for any injury and/or damage to persons or property as a matter of products liability, negligence or otherwise, or from any use or operation of any methods, products, instructions, or ideas contained in the material herein.

British Library Cataloguing in Publication Data
SAE and the evolved packet core.
   1. Wireless communication systems.
   I. Olsson, Magnus.
   621.3'8215-dc22

Library of Congress Control Number: 2009927258

ISBN: 978-0-12-374826-3

For information on all Academic Press publications
visit our website at elsevierdirect.com

Printed and bound in Great Britain

09 10 11   12 11 10 9 8 7 6 5 4 3 2 1

Contents

Foreword by Dr. Ulf Nilsson

The history of modern mobile telephony, which is about 30 years by now, has certainly been fascinating. The first analogue systems deployed in the early 1980s followed by GSM in the early 1990s provided users a basic voice service with mobility support. The addition of GPRS subsequently introduced support for packet-oriented mobile services. After about another 10 years or so, the third generation mobile system UMTS appeared with better capacity and higher throughput for packet services. For a long time, however, voice services dominated the operators' service offerings and the mobile network traffic. But just as the Internet changed the nature of fixed-access networks, it finally changed the usage of the mobile networks as well. In front of their computers at home, more and more users were realizing what a great source for information, entertainment, interactivity and productivity the Internet was. They also discovered new ways of communicating with others through, for example, chat. The wish, or in many cases probably the need, to bring the Internet along to wherever you happened to go, led to the mobile broadband revolution of recent years. In many markets, the best selling mobile device is no longer a phone but rather a mobile broadband modem for laptops and computers. This is a paradigm change that the whole mobile industry needs to understand and come to grips with.

For a mobile operator, the mobile broadband revolution with its rapidly increasing traffic volumes has resulted in a number of challenges. Our customers want ubiquitous network coverage, high bandwidths and reliable services for reasonable price, while investors and owners require constant efficiency improvements, reduced operational costs and higher profits. In order to cope with such diverse requirements, operators rely as always on the mobile communications industry to continuously improve already deployed networks but also, when the evolutionary tracks finally come to an end, to define new network solutions.

Currently we clearly see that if we rely on only enhancements to the GSM/GPRS/UMTS core and access networks, it will be impossible to cope with the foreseen future demands. In fact, they might not be enough even in the near-time. Therefore the SAE/LTE network developed by 3GPP is extremely important, not only for an operator like TeliaSonera, but also for the whole mobile industry. It is what we shall deploy and live with for a number of years in the new mobile broadband-dominated market place.

As the SAE/LTE network is important for the mobile industry, it will be absolutely necessary for everybody working in the area, or aiming to work in the area, to have a solid understanding of what the new network is capable of and what possibilities it provides. There is no doubt that this book, which appears just when the mobile industry starts its transition away from legacy GSM/GPRS and UMTS networks into the future, will become the reference work on SAE/LTE. There are no better-qualified persons than the authors of this book to provide both communication professionals and an interested general public with insights into the inner workings of SAE/LTE. Not only are they associated with one of the largest mobile network equipment vendors in the world, they have all actively contributed to and, in some cases, been the driving forces behind the development of SAE/LTE within 3GPP.

Dr. Ulf Nilsson
TeliaSonera R&D
Mobility Core and Connectivity

Foreword by Dr. Kalyani Bogineni

There are billions of mobile devices operating on various types of 2G and 3G wireless networks. Projections are for several billion more devices in the next few years on newer technologies with expectations of simultaneous services with high throughput and low-latency requirements. There will be multiple wireless devices for each user and there will be wireless devices embedded in machines supporting automation of many functions. In short, the users will be 'any-where any-time on any-device'. This is heralding an era of communication and information exchange that will test the limits of many existing telecommunications and data technologies. Hence there is a need for implementing concepts born out of disruptive thinking combined with pragmatic application of innovations.

From a service provider point of view, this is a time for laying the foundation for many of the features needed in future generation networks in order to meet the above expectations. For example, the networks need to enable signalling and low-latency media paths across segments of different technologies to support real-time applications like voice and gaming. Fundamentals like mobility and roaming, the pillars of global cellular technologies, need services management based on availability of resources, offered via policy-peering mechanisms between the home and visited networks. Simplicity and ease of using devices and services is enabled through unified authentication and subscription validation mechanisms across various access networks and application platforms. Coexistence and cooperation is needed between end-user-driven intelligent devices and intelligent network elements.

The 3GPP has specified a core network based on the Internet Protocol (IP) that provides numerous operational benefits in addition to meeting the above-mentioned expectations. The specification

– Allows evolution of any deployed wireless or wired access technology network towards a common architecture with benefits of seamless mobility between various generations of access networks and global roaming capabilities on different technologies.

- Enables network designs based on high availability, reliability, scalability and manageability paradigms as well as efficient bandwidth usage on access, backhaul and core networks.
- Supports delivery of combinations of advanced telephony and Internet services that can be hosted by any access network or application provider.
- Provides the user security functions like privacy and confidentiality while protecting the network through functions like mutual authentication, firewalls, etc.
- Minimizes the number of services databases and the number of services controllers which reduces the number of provisioning points in the network.
- Provides an efficient charging architecture that reduces the number of network elements sending billing records and minimizes the number of billing records formats.

The scope of the 3GPP specifications is ambitious but essential. The authors have done an excellent job in writing this book. Their familiarity with the requirements, concepts and solution alternatives, as well as the standardization work allows them to present the material in a way that provides easy communication between Architecture and Standards groups and Planning/Operational groups within service provider organizations.

Dr. Kalyani Bogineni
Principal Architect, Verizon

Preface

The outcome of the 3GPP SAE (system architecture evolution) technical study and specification work is a set of standards that specifies the evolution of the packet core network for GSM/GPRS and WCDMA/HSPA to an all-IP architecture and enables a feature-rich 'common packet core' for radio accesses developed within 3GPP and also by other standardization fora. This common core is referred to as EPC (evolved packet core) and the full system is known as EPS (evolved packet system) which includes support for 3GPP radio access technologies (LTE, GSM and WCDMA/HSPA) as well as support for non-3GPP access technologies. Unlike its predecessor, EPC provides support for multiple access technologies and provides for mobility between them, allowing end-users to move between, for example LTE, WLAN and other 3GPP and non-3GPP accesses. The architecture, in comparison to the one used for 2G/3G packet core is also optimized for efficient payload handling; a so-called 'flatter' architecture. In addition to these benefits, EPC provides updates to all of the already established parts of the 2G/3G packet core network, for example security, connectivity management and so on. In short, the SAE work has prepared the core network for the mobile broadband revolution, through the specification of EPC.

The standards produced by the SAE work item can be perceived as complex, as they span several thousands of pages. This makes it difficult for any individual not involved in the development of the standard to find time to examine these specifications in detail. Many people in the industry have often mentioned to us that the system spans a multitude of specifications and it would be beneficial for the readers within the telecom industry to have a single source description of the new packet core. This book is an answer to this demand; a concise and comprehensive description of the different aspects of the SAE standards for several different reader groups with interests for mobile communications industry.

Our goal is that reading this book will improve the overall understanding of the network architecture and protocols included in the EPC system. It is, however, significantly more than just annotated 3GPP specifications. It provides a detailed analysis of the network architecture, nodes and protocols involved in EPC. In addition we have described the main reasons why certain decisions were taken in the standards bodies; the context of many of these technical decisions is often imperative to a full understanding of how the architecture fits together. This is extra knowledge that we have tried to capture for the reader through our experiences in the standardization process of the SAE work item.

This book provides a thorough grounding for anyone wishing to learn about how operators and other actors in the industry may implement the EPS and also the different migration paths that may be taken. It also provides an overview of the services that will be utilizing LTE and EPC.

Readers who are already familiar with EPC, LTE or IMS will hopefully also benefit substantially from this book as it identifies how these concepts fit together in order to deliver the promise of mobile broadband. For example, readers familiar with IMS will gain a new depth of insight into how voice services will fit together with the new network architecture and protocols. Appendix A covers the different specifications that are relevant for SAE. It should be noted, however, that this book is not just for readers interested in 3GPP specifications, but it also covers the implementation scenarios for 3GPP2 and also interconnection with non-3GPP accesses such as WLAN, WiMAX or fixed access. Readers interested in only one access technology, or indeed interested in only one protocol, will also gain a good depth of understanding of how their part fits in with the overall network architecture.

We have divided this book into five different parts, each of which contains several chapters.

Part I: Introduction – Background and Vision of EPC

These chapters put SAE and EPC in the correct context with regard to other technologies that affect the evolution of telecommunications networks, specifically mobile broadband and non-3GPP access technologies. These chapters also give a description of the history behind SAE and why the core network needed to be evolved.

Chapter 1

This chapter provides the 'outside view' of telecommunications networks as they stand today and where EPC sits in relation to this, covering the following points:

- Why evolve the core network?
- Technologies connected to EPC.
- Standards bodies involved in SAE work.

Chapter 2

This chapter provides the reasoning within the industry for evolving the core network and the role of different players in the standards bodies.

- Why SAE was started and what the initial targets were
- How did these initial targets evolve during the process?
- Description of the different aspects of the standardization process and the impact they had on the architecture.

Part II: Overview of EPS

This section provides technical descriptions of EPS, including functional descriptions of the different components of EPC. This section also covers different migration and introduction scenarios as well as illustrates how the concepts and standards described in previous chapters are connected together to create services in an operator's network.

Chapter 3

Chapter 3 provides a high-level introduction to the main concepts of the EPS system designed to give a basic understanding of SAE/LTE services.

- A brief description of the EPS services.
- Simplified network diagrams to give the reader an initial understanding of the EPS network and where EPC is placed in the overall network.
- Introductory information on the fundamental choices in LTE.
- Terminal perspective.
- Short LTE overview and its relation to EPC.

Chapter 4

Chapter 4 provides descriptions of how EPC may be deployed based on the situation of the market where it is being deployed as well as its relation to LTE deployment.

- Brief description of the overall NW when deploying EPC/LTE in different operator configurations.

Chapter 5

This chapter provides a description of the data and voice services that will be used on an EPC network, aiming to bring the whole EPS and its concepts together, analysing it from several different potential evolution paths for the services.

- Description of the predicted target services:
    - Data services and applications
    - Voice services
    - Messaging services

Part III: Key Concepts

Chapter 6

This chapter provides a description of key concepts within EPS. Owing to the nature of EPC compared with previous core network architectures, this chapter will provide a clear description of these new concepts and compare them to the previous core networks. This aims to provide readers with a clear point of reference for the key concepts after the evolution of the core network.

Chapter 7

This chapter provides details on security including user authentication/authorization as well as network security mechanism for both 3GPP and non-3GPP accesses connecting to EPS.

Chapter 8

This chapter provides readers with in-depth view of quality of service and policies to control and manage services and to differentiate charging. This chapter also includes a high-level overview of the 3GPP charging models and mechanisms.

Chapter 9

This chapter provides an in-depth view of the usage of DNS as well as 3GPP developed mechanisms as tools for the operations of the EPS network efficiently by selecting the 'right entity' for the right user in an operator's network.

Part IV: The Nuts and Bolts of EPC

Chapters 10, 11 and 12 together illustrate in detail how the EPS system is built end-to-end by using the network entities, the interfaces connecting them together and protocols that provide the 'meat' for the 'backbone' of the system carrying the information between these entities and then some high-level procedures illustrating key scenarios such as attaching to the EPC, detaching from the EPC, handover of various kinds between 3GPP and non-3GPP access technologies as well as between 3GPP access technologies.

Part V: Conclusion and Future of EPS

Chapter 13

This chapter includes observations and conclusions regarding the EPC and some discussion on what may lie ahead for the future evolution.

Acknowledgements

A work of this nature is not possible without others' support. The authors would like to gratefully acknowledge the contribution of our colleagues at Ericsson, in particular Ralf Keller, György Miklós, Mats Näslund, Reiner Ludwig, John Stenfelt, Louis Segura, Maurizio Iovieno, Erik Dahlman, Per Beming, Peter Malm, Anki Sander, Göran Hall, Anders Lundström, Paco Cortes, Jesús De Gregorio, Lars Lövsen and Patrik Teppo.

We would also like to thank our families. Writing this book would not have been possible without their generosity and support throughout the process.

Part I
Introduction – Background and Vision of EPC

1

Mobile broadband and the core network evolution

The telecommunications industry is in a period of radical change with the advent of mobile broadband radio access and the convergence of Internet and mobile services. Part of this radical change is enabled by a fundamental shift in the underlying technologies; mobile telephony is moving towards an all Internet Protocol (IP) network architecture after several decades of circuit switched technology. The evolution of the core network to support the new high-bandwidth services promised by mobile broadband is a monumental breakthrough. This book covers that evolution.

The phenomenal success of GSM (Global System for Mobile Communications) was built on the foundation of circuit switching. Services, meanwhile, were built by developers specialized in telecommunication applications. During the early 1990s, usage of the Internet also took off, in later years leading to a demand for 'Mobile Internet'; Internet services that could be accessed from an end-user's mobile device. The first such services had limitations due to the processing capacity of terminals and also a very limited bandwidth on the radio interface. This has now changed with the evolution of radio access networks (RANs) with high data rates delivered by High Speed Packet Access (HSPA) and Long Term Evolution (LTE). The speed of this change is set to increase dramatically as a number of other developments emerge in addition to the new high-speed radio accesses; the rapid advances in the processing capacity of semiconductors for mobile terminals and also in the software that developers can use to create services. IP and packet-switched technology are soon expected to be the base for data and voice services on both the Internet and mobile communications networks.

The core network is the part that links these worlds together, combining the power of high-speed radio access technologies with the power of the innovative application development enabled by the Internet. The evolution of the core network, or Evolved Packet Core (EPC), is a fundamental cornerstone of the mobile broadband revolution; without it neither the RANs nor mobile Internet services would realize their full potential. The new core network was developed with high-bandwidth services in mind from the outset, combining the best of IP infrastructure with mobility. It is designed to truly enable mobile broadband services and applications and to ensure a smooth experience for both operators and end-users as it also connects multiple radio access technologies.

This chapter introduces the reasoning behind the evolution of the core network and a brief introduction to the technologies related to EPC.

System Architecture Evolution (SAE) is the name of the Third Generation Partnership Project (3GPP) standardization work item which is responsible for the evolution of the packet core network, more commonly referred to as EPC. This work item is closely related to the LTE work item, covering the evolution of the radio network. Evolved Packet System (EPS) covers the radio access, the core network and the terminals that comprise the overall mobile system. Also provides support for other high-speed RANs that are not based on 3GPP standards, for example, WLAN, WiMAX or fixed access. This book is all about EPC and EPS – the evolution of the core network in order to support the mobile broadband vision and an evolution to IP-based core networks for all services.

The broad aims of the SAE work item were to evolve the packet core networks defined by 3GPP in order to create a simplified all-IP architecture, providing support for multiple radio accesses, including mobility between the different radio standards. So, what drove the requirement for evolving the core network and why did it need to be a globally agreed standard? We will start with looking into this.

1.1 The need for global standards

There are many discussions today regarding the evolution of standards for the communications industries, in particular when it comes to convergence between IT and telecommunications services. A common question that pops up occasionally is why is a global standard needed at all? Why does the cellular industry follow a rigorous standards process, rather than, say, the de-facto standardization process that the computer industry often uses? There is a lot of interest in the standardization process for work items like LTE and SAE, so there is obviously a commercial reason for this, or very few companies would see value in participating in the work.

The necessity for a global standard is driven by many factors, but there are two main points. First of all, the creation of a standard is important for interoperability in a truly global, multi-vendor operating environment. Operators wish to ensure that they are able to purchase network equipment from several vendors, ensuring competition. For this to be possible nodes from different vendors must inter-work with one another; this is achieved by specifying a set of 'interface descriptions', through which the different nodes on a network are able to communicate with one another. A global standard therefore ensures that an operator can select whichever network equipment vendor they like and that end-users are able to select whichever handset that they like; a handset from vendor A is able to connect to a base station from vendor B and vice versa. This ensures competition which in itself attracts operators and drive deployments by ensuring a sound financial case through avoiding dependencies on specific vendors.

Secondly, the creation of a global standard is about reducing fragmentation in the market for all the actors involved in delivering network services to end-users; operators, chip manufacturers, equipment vendors, etc. A global standard ensures that there will be a certain market for the products that, for example,. an equipment vendor develops. The larger the volume of production for a product, the greater the volume there is to spread the cost of production across the end-users that will use the products. Essentially, with increased volumes a vendor should be able to produce each node at a cheaper per unit cost. Vendors can then achieve profitability at lower price levels, which ultimately leads to a more cost-effective solution for both operators and end-users. Global standards are therefore a foundation stone of the ability to provide inexpensive, reliable communications networks and the aims behind the development of EPC were no different.

There are several different standards bodies that have been directly involved in the standardization processes for the SAE work. These standards bodies include the 3GPP, the lead organization initiating the work, the Third Generation Partnership Project 2 (3GPP2), the Internet Engineering Task Force (IETF), WiMAX Forum and Open Mobile Alliance (OMA). 3GPP 'owns' the EPS specifications and refers to IETF and occasionally OMA specifications where necessary, while 3GPP2 complements these EPS specifications with their own documents that cover the impact on EPS and 3GPP2-based systems. WiMAX forum also refers to 3GPP documentation where appropriate for their specification work.

The readers who are not familiar with the standardization process are referred to Appendix 1, where we provide a brief description of the different bodies involved and the processes that are followed during the development of these specifications.

1.2 Origins of the EPC

Over the years, many different radio standards that have been created worldwide, the most commonly recognized ones are GSM, CDMA and WCDMA/HSPA.

The GSM/WCDMA/HSPA and CDMA radio access technologies were defined in different standards bodies and also had different core networks associated with each one as we describe below.

In order to understand why evolution was needed for 3GPP's existing packet core, we therefore also need to consider where and how the various existing core network technologies fit together in the currently deployed systems. Chapter 2 provides more details on the background and history of the evolution towards EPC from the perspective of the standardization bodies. The following section presents a discussion around why the evolution was necessary. While the number of acronyms may appear daunting in this section for anyone new to 3GPP standards, the rest of the book explains the technology in great detail. This section highlights only some of the main technical reasons for the evolution.

1.2.1 3GPP radio access technologies

GSM was originally developed within the European Telecommunication Standards Institute (ETSI), which covered both the RAN and the core network supplying Circuit Switched telephony. The main components of the core network for GSM were the Mobile Switching Centre (MSC) and the Home Location Register (HLR). The interface between the GSM BSC (Base Station Controller) and the MSC was referred to as the 'A' interface. It is common practice for interfaces in 3GPP to be given a letter as a name, in later releases of the standards there are often two letters, for example, 'Gb' interface. Using letters is just an easy shorthand method of referring to a particular functional connection between two nodes.

Over time, the need to support IP traffic was identified within the mobile industry and the General Packet Radio Services (GPRS) system was created as an add-on to the existing GSM system. With the development of GPRS, the concept of a packet-switched core network was needed within the specifications. The existing GSM radio network was evolved, while two new logical network entities or nodes were introduced into the core network – the SGSN (Serving GPRS Support Node) and the GGSN (Gateway GPRS Support Node).

GPRS was developed during the period of time when PPP, X.25 and Frame Relay were state-of-the-art technologies (mid to late 1990s) for packet data transmission on data communications networks. This naturally had some influence on the standardization of certain interfaces, for example, the Gb interface, which connects the BSC in the GSM radio network with GPRS packet core.

During the move from GSM EDGE Radio Access Network (GERAN) to UMTS Terrestrial Radio Access Network (UTRAN), an industry initiative was launched to handle the standardization of radio and core network technologies in a global forum, rather than ETSI, which was solely for European standards. This initiative became known as the 3GPP and took the lead for the standardization of the core network for UTRAN/WCDMA, in addition to UTRAN radio access itself. 3GPP later also took the lead for the creation of the Common IMS specifications. IMS is short for IP Multimedia Subsystem, and targets network support for IP-based multimedia services. We discuss the IMS more in Chapter 5.

The core network for UTRAN reused much of the core network from GERAN, with a few updates. The main difference between was the addition of the interface between the UTRAN Radio Network Controller (RNC), the MSC and the SGSN, the Iu-CS and the Iu-PS, respectively. Both of these interfaces were based on the A interface, but the Iu-CS was for circuit-switched access, while the Iu-PS was for the packet-switched connections. This represented a fundamental change in thinking for the interface between the mobile terminal and the core network. For GSM, the interface handling the circuit-switched calls and the interface handling the packet-switched access were very different. For UTRAN, it was decided to have one common way to access the core network, with only small differences for the circuit-switched and packet-switched connections. A high-level view of the architecture of this date, around 1999, is shown in Figure 1.2.1.1 (to be completely accurate, the Iu-CS interface was split into two parts, but we will disregard that for now in order not to make this description too complex).

The packet core network for GSM/GPRS and WCDMA/HSPA forms the basis for the evolution towards EPC. As a result, it is worthwhile taking the time for a

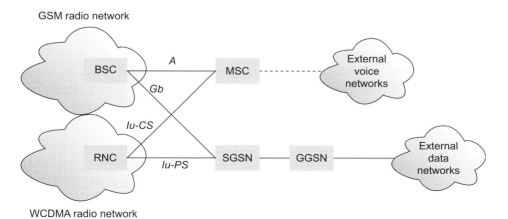

Figure 1.2.1.1 *High-level architecture WCDMA and GSM radio networks.*

brief review of the technology. Again, do not be put off by the number of acronyms, Parts II and III provide more details.

The packet core architecture was designed around a tunnelling protocol named GTP (GPRS Tunnelling Protocol) developed within ETSI and then continued within 3GPP after its creation. GTP is a fundamental part of 3GPP packet core, running between the two core network entities the SGSN and the GGSN. GTP runs over IP and provides mobility, Quality of Service (QoS) and policy control within the protocol itself. As GTP was created for use by the mobile community use, it has inherent properties that make it suitable for the robust and time-critical systems such as mobile networks. Since GTP is developed and maintained within 3GPP, it also readily facilitates the addition of the special requirements of a 3GPP network such as the use of the Protocol Configuration Option (PCO) field between the terminal and the core network. PCO carries special information between the terminal and the core network, allowing for flexible, efficient running and management of the mobile networks.

GTP has from time to time faced criticism, however, from parts of the communication industry outside 3GPP. This has mainly been due to the fact that it was not developed in the IETF community, the traditional forum for Internet and IP. GTP is instead a unique solution for 3GPP packet data and is therefore not automatically a good choice for other access technologies. GTP was instead tailor-made to suit the needs of 3GPP mobile networks. Whether the criticism is justified or not, is largely dependent on the viewpoint of each individual person.

Regardless, GTP is today a globally deployed protocol for 3GPP packet access technologies such as HSPA, which has emerged as the leading mobile broadband access technologies deployed prior to LTE. Due to the number of subscribers using GSM and WCDMA packet data networks, now counting in billions in total for both circuit and packet- switched systems, GTP has been proven to scale very well and to fulfil the purposes for which it has been designed.

Another significant aspect of GPRS is that it uses SS7-based signalling protocols such as MAP (Mobile Application Part) and CAP (CAMEL application part), both inherited from the circuit-switched core network. MAP is used for user data management and authentication and authorization procedures, and CAP is used for CAMEL-based on line charging purposes. Further details on CAMEL (Customized Applications for Mobile networks Enhanced Logic) are far beyond the scope of this book. For our purposes, it is enough to understand that CAMEL is a concept designed to develop non-IP-based services in mobile networks. The use of SS7-based protocols can be seen a drawback for a packet network created for delivering Internet connections and IP-based services.

3GPP packet core uses a network-based mobility scheme for handling user and terminal mobility. Another aspect that was to become a target for optimization at a later date was the fact that it has two entities (i.e. SGSN and GGSN) through which user data traffic is carried. With the increased data volumes experienced as a result of WCDMA/HSPA, an optimization became necessary and was addressed in 3GPP Rel-7, completed in early 2007 with the enhancement of the packet core architecture to support a mode of operation known as 'Direct tunnel' where the SGSN is not used for the user plane traffic. Instead, the radio network controller connects directly to the GGSN via Iu-user plane (based on GTP). This solution, however, only applies to non-roaming cases, and also requires packet data charging functions to reside in the GGSN instead of the SGSN.

For further details on the packet core domain prior to SAE/EPC, please refer to 3GPP Technical Specification TS 23.060 [23.060].

1.2.2   3GPP2 radio access technologies

In North America another set of radio access technology standards was developed. This was developed within the standards body called 3GPP2, under the umbrella of ANSI/TIA/EIA-41 which includes North American and Asian interests towards developing global standards for those RAN technologies supported by ANSI/TIA/EIA-41.

3GPP2 developed the radio access technologies cdma2000®, providing 1xRTT and HRPD (High Rate Packet Data) services. cdma2000 1xRTT is an evolution of the older IS-95 cdma technology, increasing the capacity and supporting higher data speeds. HRPD defines a packet-only architecture with capabilities similar to the 3GPP WCDMA technology. The set of standards for the packet core network developed within 3GPP2 followed a different track to 3GPP, namely the reuse of protocols directly from the IETF, such as the Mobile IP family of protocols as well as a simpler version of IP connectivity known as Simple IP, over a PPP link. The main packet core entities in this system are known as PDSN (Packet Data Serving Node) and HA (Home Agent), where terminal-based mobility concepts from IETF are used, in conjunction with 3GPP2 developed own mechanisms. It also uses Radius-based AAA infrastructure for its user data management, authentication and authorization and accounting.

1.2.3   Other forums involved in SAE

WiMAX forum indirectly participated in the development of SAE through companies and participants who took part in both WiMAX forum and 3GPP work.

A number of key WiMAX Forum contributing members were also dominant 3GPP2 members and thus the packet core network developed within WiMAX forum bears strong similarity with 3GPP2 packet core in its usage of the same IETF protocols like MIPv4, MIPv6 and Radius for the same purposes as 3GPP2.

1.2.4 Dawn of EPC

At the beginning of the work on EPC there were therefore two main variants of core network architectures designed with several different RANs in mind. With the evolution to mobile broadband, in particular LTE, the opportunity to evolve the core network to better utilize IP technologies in order to bring significant cost savings for operators and end-users, became evident. It was also immediately apparent that the new core network would need to support legacy access networks in order to ensure that existing radio network installations would be able to connect to EPC.

At the same time, many operators within 3GPP2 became interested in the evolution of the core network ongoing in 3GPP as they wished to join the LTE ecosystem and development of common packet core work under the umbrella of SAE. As a result, work in both 3GPP and 3GPP2 was established to ensure that the EPS could support interworking towards 3GPP2 networks. EPS then needed to support the evolution of two very different types of core network and that created the framework of SAE work in 3GPP.

1.2.5 SAE – building bridges between different networks

This meant that SAE was to target both improving and building a bridge between two very distinct packet core networks.

The existing packet core networks were developed to serve certain market and operator requirements. These requirements have not changed with the evolution to EPS. Rather, with the evolution towards new radio networks and also the need to deliver new types of services across the core network, the EPS is instead required to support extra requirements on top of the old ones.

The added common requirements towards the system are as follows in no particular order of importance or priority:

1. Support for non-3GPP access networks
2. Support handovers between 3GPP and non-3GPP accesses

3. Network-based mobility mechanisms were preferred
4. Common Security framework
5. Common User management and Authentication and Authorization framework
6. Common Policy and Charging support
7. Common framework for On and Off line Charging and Accounting
8. Provide Optimized handover to/from existing deployed Radio access and Packet Core networks: 3GPP's GERAN, UTRAN and HSPA and 3GPP2's HRPD networks
9. Common Evolved Packet Core for access to Common IMS and Applications and Service framework
10. Common operations and management of Terminals.

Additional requirements identified from the implementation and use of 3GPP packet core were related to:

1. Improving the performance of the existing system
2. Securing high performance packet handovers between 3GPP accesses
3. Clear separation of control and user plane operation
4. The option to have a single user plane entity during non-roaming mode of operation
5. Easy migration to EPS
6. Manageable impacts on roaming infrastructure upgrades, implying the ability to gradually migrate from the existing GRX (roaming) networks and the ability to support inbound/outbound roamers from LTE and to/from existing 3GPP networks belonging to operator partners that have not upgraded to EPS.

In order to support the key components described above, EPS developed multiple options of the architecture where different protocol suites were tailored to fulfil the majority of the requirements.

It was clear from early on that IETF-based protocols would play a key role in EPS. 3GPP had been very deeply involved in continuing the development of the IMS and PCC (Policy and Charging Control) Systems, where all the protocols are built on IETF-developed base protocols and then enhanced within the IETF as per 3GPP's requirements. This was not new or unchartered territory for 3GPP member companies since 3GPP already had contributed extensively to the development within IETF of SIP, AAA, Diameter and various security related protocols.

As such, once the EPS architecture and protocol choices were settled, the work progressed quite smoothly. The most contentious area of protocol selection was related to mobility management, where there were a few competing proposals in the IETF and progress was slow. Once the IETF settled on developing PMIP (Proxy Mobile IP) as the network-based mobility protocol, however, 3GPP EPS development followed suit.

Additional standard bodies that 3GPP collaborated with in order to develop the necessary components for EPS are 3GPP2 and OMA. 3GPP2 helped develop the aspects related to CDMA technologies in relation to supporting optimized handover between HRPD and LTE as well as Single Radio Voice Call Continuity and CS Fallback solutions. OMA cooperation involved the areas of device management; focusing on aspects of network discovery and selection procedures for non-3GPP accesses.

Even though 3GPP collaborates with various other standardization fora in order to develop the final system, the actual details are worked out within 3GPP protocol working groups according to the architectural requirements set by architecture (SA) and RAN working groups. As necessary, various IETF drafts and other specifications are then prepared and progressed as per 3GPP specifications to finalize the 3GPP specification development process.

1.2.6   Introducing EPC – an operator's and end-user's perspective

What, then, are some of the key changes that a network operator and an end-user will experience due to the evolution of the packet core? This section will focus on the EPS, which is composed of the EPC, End-User Equipment (more commonly known as the UE), Access Network (including 3GPP access such as GERAN, UTRAN, E-UTRAN and non 3GPP accesses such as HRPD, WiMAX, WLAN, etc). The combination of these enables access to an operator's services and also to the IMS, which provides voice services.

Before getting into the technical aspects of EPS, let us look at the system from the end-user's and network operator's perspective. What do they need to do before their subscribers are able to connect to the new network? This depends on whether an operator is migrating to the new radio access technology or utilizing the existing 3GPP radio or other non-3GPP access technologies with EPC. For the sake of simplicity we will for the moment assume that LTE is the operator's chosen radio access technology.

From an end-user's perspective, the actions are quite simple; if an end-user wishes to use mobile broadband, then they need to select an operator which supports it and purchase a device capable of supporting the highest bandwidth that is provided through using EPS via E-UTRAN access. While the UICC card that the user currently owns may be reused, the enhanced security functions and data-only functions may encourage the purchase of SIM cards designed specifically for EPS.

From an operator's perspective, in order to provide a data-only mobile broadband service, the infrastructure must be upgraded to EPS. EPS provides components that allow existing 2G/3G access networks to utilize EPC components. For those incumbent 2G/3G operators the existing CS network can provide access to voice calls in the short term, but the deployment of IMS in conjunction with EPS would provide an All-IP network with access to speech services.

In the following chapters, we will focus initially on a mobile broadband deployment scenario where access to data services is the predominant mode of operation. We will later also describe how EPS is supporting voice services.

2

SAE history and background

In Chapter 1, we mentioned that SAE is a work item developed within 3GPP and it also incorporates the use of protocols from other standards bodies where necessary in order to prevent overlapping work as well as benefit from expertise of work performed outside of 3GPP, in particular protocols developed by the IETF and other standards bodies. This may sound like a relatively easy task; but the process can become quite difficult due to both political as well as completion timing issues. Sometimes, the process of standardization within the IETF can take longer than expected, or different companies within 3GPP itself can have different priorities regarding the content and functions for the selection of protocols or the functionality required from the protocols. All of these aspects can affect the resulting standards developed within 3GPP and the SAE work item was no exception.

2.1 Impact of standardization processes on SAE

Chapter 1 outlined the process of standardization including the organization of the standards bodies and the approval processes involved within those organizations. This is only one part of the standards process; the other aspect of the process not captured in any such description is the human aspect and its impact on the standardization process. This section attempts to outline the types of discussions that affected the final set of standards known as SAE today. Some concepts and terms used in this chapter have been described more in detail in later chapters of this book. Note that any such analysis usually is subjective in nature, but we have made every attempt to provide an unbiased and objective overview of the process and events.

As with most standardization efforts within 3GPP, the SAE work item builds upon existing technologies; in the case of SAE/EPC, the base was the existing 3GPP packet core system used for GSM/GPRS and WCDMA/HSPA. As outlined in Chapter 1, the progression of the LTE is a closely related work item to SAE. As the development of LTE progressed within the RAN groups of 3GPP, the opportunities to develop improvements to Packet Core were identified by several different interested parties.

One of the key aspects to understand about any standardization activity in any standards forum worldwide is the role of conflicting and diverging goals among the participating companies/stakeholders. These are often commercially driven goals, but are also the result of technical visions and the desire for their realization into 'real systems'. The purpose of the standards bodies is to help resolve difficult technical and political issues leading ultimately to achieving the common goal of a strong architecture designed to serve the needs of the entire community.

The standardization process is dynamic and exciting; an analysis of these aspects is a good basis for understanding the development of the work items involved. Firstly, with the introduction of a work item that is designed to create an 'all-IP' network naturally meant that there was a tremendous amount of interest in 3GPP's work in this area from new participants. In fact, the number of participants within 3GPP SA WG2 dealing with Architecture and Overall System aspects rose dramatically from around 100 people to about 180–225 participants during the peak period; approximately 75% of these participants were actively involved in attempting to shape the SAE work and EPS architecture. Naturally, with such a large increase in the number of participants, there was a period of adjustment for everyone involved. While the new participants learnt the 3GPP working procedures and also adjusted to having to handle the implications of defining an architecture that included providing a clear migration path from an existing system to a new one. Whereas the incumbents needed to adapt to a more dynamic exchange of ideas from various new sources and find a common ground to move forward.

What were the key driving forces behind some of these groups and companies just on the 3GPP accesses and its associated core network evolution? One major factor was the mix of incumbent vendors and newcomers to the 3GPP system. The enormous interest in LTE/SAE work drew a large number of companies who had previously not participated in the 3GPP standardization process, many had indeed spent time working on other systems in various other standardization fora. These new entrants to 3GPP forged alliances and joined the SAE work. This created quite a contradictory vision for the future which was initially hard to reconcile since it is the vision between 'continuity' and a completely 'new beginning of sorts'.

Some vendors and operators, experienced with the existing 3GPP systems, considered it important that continuity would be maintained with the 3GPP packet core when evolving the system, whereas others initially wanted to explore new avenues based on technology used by other standardization fora rather than 3GPP. There were also a few existing 3GPP operators and vendors who solely focused on creating an architecture that was not based on the existing 3GPP architecture. All these different inflection angles and viewpoints were fed into the work on SAE, and resulting in a very dynamic and diverse working environment. Initial investigations as documented in the 3GPP Technical report 23.882 [23.882] reflect various options with one common theme; that it makes sense to separate control and user plane entities. Some of the key architecture options that were discussed during the initial stages of the standardization process may be viewed as follows:

1. Evolution of the existing 3GPP packet core architecture, i.e. evolving the GTP protocol, but not necessarily reusing the architecture, with a single User Plane Gateway (GW) for the non-roaming cases, a local anchor and a GW for roaming and using the network-based mobility protocol developed in 3GPP.
2. In principle, follow a very similar architecture as outlined in the point 1 above, but where the two GW entities; thus the roaming and non-roaming architecture is exactly the same which is slightly different than 1, the protocols between these GWs would be developed in IETF.
3. An overlay model where a control plane entity and a GW/Home Agent are used in the architecture with client-based mobility protocols.

As will be seen during the rest of the book, a suitable compromise was reached through the standardization development process, essentially an architecture that is a combination of the above proposals.

Another key element to this discussion related to the architecture of the RAN for LTE. Though not discussed in detail in this book, it played a crucial role in creating additional ambiguity and delays since there were two very distinct views. One followed the principle of a need for a central radio network controller such as an RNC as is part of the architecture of WCDMA/HSPA, while the other view argued that a flat radio network architecture without RNC is better for the future and that the drawbacks of such an RNC-like entity were higher than the advantages. A decision on this matter was crucial since the functional division between the core and radio network for the 3GPP system depended on that point. Here the progress of SAE and LTE were closely interlinked for a period of time.

A significant amount of time and effort were placed in discussing the functional division between core and radio networks and the pros and cons of the different approaches. In the end, the 3GPP community made the decision to go with the architecture option without RNC-like entity for the Radio Access Network and the community then focused on settling the functions belonging to Radio and Core aspects and move forward with the investigations of developing the new architecture. Great efforts were put into the work on arriving at the preferred functional split between the LTE RAN (now only consisting of base stations) and the packet core network as defined by EPC. 3GPP finally arrived at a decision where the functional split between RAN and CN for LTE are similar in nature as for WCDMA with few exceptions.

Another key aspect that influenced the work on EPC was what role WiMAX as a technology would have. WiMAX was at the time promoted by some parties as a potentially important and preferred next generation mobile radio access technology, partly in competition with LTE. A key question here was the ambition level for supporting connection of WiMAX access networks (e.g. interworking and handover) to the 3GPP family of access technologies. The 3GPP community here took decisive action to postpone this tight coupling for the time being, clearing the path for completion of the SAE work. The main rationale behind this decision was that the most important interworking cases with rigid performance demands are between LTE and legacy access networks which are already widely deployed, meaning GSM, WCDMA/HSPA and CDMA/HRPD. Interworking with all other access technologies including WiMAX were considered to have less stringent performance demands.

Another difficult architecture decision was the selection of Policy Control and Charging mechanism. In the 3GPP systems so far, the GTP protocol carries not only mobility information but also QoS and Charging and policy control information. This way of transferring information that supports the PCC infrastructure is also known as the 'On-Path model'. Since two options for handling mobility within EPC was decided to be supported using GTP and PMIP respectively, and the IETF PMIP protocol is unable to carry QoS and charging information itself the On-Path model used for GTP could not be supported for PMIP. The two options then considered in case of PMIP PCC was a form of On-Path model where a protocol like for example Diameter may be used directly between the two involved entities (see subsequent section for PCC for more details on the two models) to carry the necessary data and the other model also known as Off-Path model where the data is then carried via the PCC infrastructure and thus taking some extra hops on its way. There were some additional subtleties between the two models where QoS enforcement and management are not handled by the same entities for GTP vs. the PMIP variant of the architecture. In the end, the community selected to use Off-Path model for the PMIP variant, and thus the major hurdles for the architecture work were considered removed.

As usual, any new area of work brings a lot of research ideas to the forefront and the same happened for the SAE work. This has special significance when the issues affect crucial aspects for operators, e.g. migration, interworking with existing infrastructure, and the overall cost of implementation and deployment. These issues were not necessarily considered as a priority until the end of the architecture development phase.

With such an approach the direction of work risked becoming in some sense decoupled from the reality of the every day business of running the networks.

The positive aspect of this was that the development progressed with a completely new vision for the future though still considering that 3GPP needed to maintain some, if not most, of the key functions of the current system.

As investigations of various alternatives progressed, some new enthusiasm in the form of a few incumbent CDMA operators came into the 3GPP community about mid way through the evaluation/development process. These CDMA operators made strong commitments to help develop SAE/LTE and took steps to commit to the SAE/LTE track as an alternative next step for the migration of existing HRPD network.

As work progressed, it became clear that the SAE work would not emerge with one set of protocols and design choices for the EPS. With somewhat divergent operator requirements and migration/evolution strategies, 3GPP needed to take a hard decision; either following one architecture alternative (which was rather impossible to achieve), or allowing for multiple alternatives. At the end of the day, 3GPP emerged with multiple protocol variants used within one overall architecture framework satisfying these requirements. An aspect worth noting is that not only did 3GPP select two protocol options with slightly different architectural variants for network-based mobility, including the GTP evolution and PMIP-track based on the IETF-developed PMIP protocol suite and 3GPP specific extensions to this mobility mechanism, 3GPP also continued to develop a terminal-based mobility option though with rather limited interest from operator community.

A final twist came in the conclusion phase when it was decided that the existing packet core architecture should be maintained in parallel with EPC, while the original assumption and understanding of the work was that EPC would replace the existing packet core architecture. The consequence of this decision was the creation of two variants of how to interconnect to GERAN and WCDMA/HSPA access networks towards packet data networks (e.g. GPRS and EPC).

One milestone accomplishment during the architecture work is that regardless of choice of network-based mobility model (GTP or PMIP), the user device acts in the same way. There is no dependency between the mobility protocol used by the network entities and how the terminal connects to the network. This transparency should definitely help future development of EPS as a whole.

Figure 2.1.1 provides a timeline for some of key decisions taken during the SAE development phases primarily in 3GPP Architecture Working Group also known as SA WG2. Though other working groups were working in parallel in areas such as Security, Charging, Lawful Intercept, the main emphasis in this section has been centred around SA WG2 and its relation with other WGs including RAN and other fora.

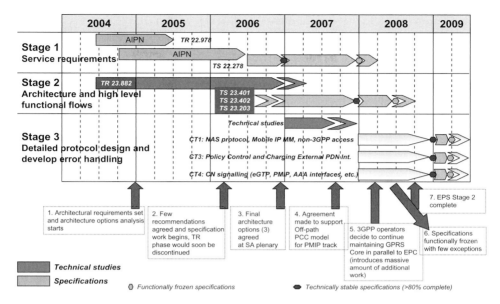

Figure 2.1.1 *3GPP rough Time line for major mile stones achieved for SAE.*

2.2 Terminologies used in this book

As you progress through the chapters in this book, you will notice that there are several different acronyms used to describe different aspects of the core network evolution. You will notice that these acronyms are being used extensively in the industry as well, so here we have included a brief description of the meanings of the different acronyms and how we have selected to use these terms in this book.

The common terminology used in the industry is not necessarily the same as the terms that have been used in standardization. On the contrary there is somewhat of a mismatch between the mostly used terms in the mobile industry and the terms actually used in 3GPP specification work. SAE is a prime example of a term that has one meaning inside the 3GPP standardization community and a different meaning outside. Formally, in 3GPP the term SAE is the name of the work item developing the overall architecture for the EPS. The term SAE does not appear as a name on any part of the system in the 3GPP specifications but in the industry and the general public it is probably more familiar than the term EPS, which would be the formally correct way to refer to the overall system that implements the LTE radio.

Below follows a list of terms describing some of the most common acronyms in this book.

EPC: The new Packet Core architecture itself as defined in 3GPP Rel-8.

SAE: The work item, or standardization activity, within 3GPP that was responsible for defining the EPC specifications.

EPS: A 3GPP term which refers to a complete end-to-end system, that is, the User Equipment, E-UTRAN (and UTRAN and GERAN connected via EPC) and Core Network.

SAE/LTE: A term often used to refer to the complete NW; it is more commonly used outside of 3GPP instead of EPS. In this book we have used the terms EPS instead of the term SAE/LTE.

E-UTRAN: Evolved UTRAN; the 3GPP term denoting the RAN that implements the LTE radio interface technology.

UTRAN: The RAN for WCDMA/HSPA.

GERAN: The GSM RAN.

LTE: Formally the name of the 3GPP work item (Long Term Evolution) that developed the radio access technology and E-UTRAN, but in daily talk, it is used more commonly instead of E-UTRAN itself. In the book we use LTE for the radio interface technology. In the overall descriptions we have allowed ourselves to use LTE for both the RAN and the radio interface technology. In the more technical detailed chapters (Parts III and IV) of the book we strictly use the terms E-UTRAN for the RAN and LTE for the radio interface technology.

2G/3G: A common term for both the GSM and WCDMA/HSPA radio access and the core networks. In a 3GPP2-based network 2G/3G refers to the complete network supporting CDMA/HRPD.

GSM: 2G RAN. In this book, the term does not include the core network.

GSM/GPRS: 2G RAN and the GPRS core network for packet data.

WCDMA: The air interface technology selected for the 3G UMTS standard. Commonly also used to refer to the RAN formally known as UTRAN.

WCDMA/HSPA: 3G RAN and the enhancements of the 3G RAN to high-speed packet services. Commonly also used to refer to a UTRAN that is upgraded to support HSPA.

GSM/WCDMA: Both the second and third-generation radio access technologies and RAN.

HSPA: A term which covers both HSDPA (High Speed Downlink Packet Access) and Enhanced Uplink together. HSPA introduces several concepts into WCDMA allowing for the provision of downlink and uplink data rates up to 42 and 12 Mb/s respectively (February 2009).

CDMA: For the purposes of this book, CDMA refers to the system and standards defined by 3GPP2; in the context of this book, it is used as a short form for cdma2000®, referring to the access and core networks for both circuit switched services and packet data.

HRPD: High Rate Packet Data; the high-speed CDMA-based wireless data technology. For EPC, HRPD has been enhanced further to connect to EPS and support handover to and from LTE. Thus we also refer to as eHRPD; evolved HRPD network which supports interworking with EPS.

We also want to bring the attention to the use of **UE, Terminal, and Mobile Device** in this book. These terms are used interchangeably in the book and all refer to the device that an end-user accesses the network with.

Also, we use the word 'interface' to refer to both the reference points and the actual interfaces.

cdma2000® is the trademark for the technical nomenclature for certain specifications and standards of the Organizational Partners (OPs) of 3GPP2. Geographically (and as of the date of publication), cdma2000® is a registered trademark of the Telecommunications Industry Association (TIA-USA) in the United States.

Part II
Overview of EPS

3
Architecture overview

This chapter introduces the EPS Architecture, firstly presenting a high-level perspective of the complete system as defined in the SAE work item. In subsequent sections, we introduce the logical nodes and functions in the network. By the end of this chapter, the main parts of the EPS architecture should be understandable and readers will be prepared for the full discussion about each function and interface, as well as all applicable signalling flows that follow in Parts III and IV.

3.1 EPS Architecture

There are several domains in EPS, each one a grouping of logical nodes that interwork to provide a specific set of functions in the network.

A network implementing 3GPP specifications is illustrated in Figure 3.1.1.

On the left of the diagram are four clouds that represent different RAN domains that can connect to the EPC, including the second and third generations of mobile access networks specified by 3GPP, more commonly known as GSM and WCDMA respectively. LTE is of course the latest mobile broadband radio access

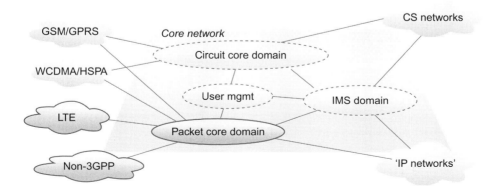

Figure 3.1.1 *3GPP architecture domains.*

as defined by 3GPP. Finally, there is the domain called 'non-3GPP access networks'. This denotes any packet data access network that is not defined by 3GPP standardization processes, for example eHRPD, WLAN, fixed network accesses or some combination of these. This also means that 3GPP does not specify the details about these access technologies – these specifications are instead handled by other standardization fora, such as 3GPP2, IEEE or Broadband Forum. Interworking with these access domains will be covered in more detail in Section 6.3.

The Core Network is divided into multiple domains (Circuit Core, Packet Core and IMS), as illustrated above. As can also be seen, these domains interwork with each other over a number of well-defined interfaces. The user management domain provides coordinated subscriber information and supports roaming and mobility between and within the different domains.

The Circuit Core domain consists of nodes and functions that provide support for circuit-switched services over GSM and WCDMA.

Correspondingly, the Packet Core domain consists of nodes and functions that provide support for packet-switched services (primarily IP connectivity) over GSM, WCDMA and HSPA. Furthermore, the Packet Core domain also provides support for packet-switched services over LTE and non-3GPP access networks which in general have no relation to the Circuit Core (except for some specific features needed for voice handovers in relation to LTE). The packet core domain also provides functions for management and enforcement of service and bearer level policies such as QoS.

The IMS domain consists of nodes and functions that provide support for multimedia sessions based on SIP (Session Initiation Protocol), and utilizes the IP connectivity provided by the functions in the Packet Core domain.

In the middle of all of this, there is also a user management domain, where resides the handling of the data related to the subscribers utilizing the services of the other domains. Formally, in the 3GPP specifications, it is not a separate domain in and of itself. Instead, there are user management functions in the Circuit Core, Packet Core and IMS domains interacting with subscriber data bases defined by 3GPP. However, for the purposes of clarity, we have selected to show this as a domain in and of itself.

The main emphasis here is the EPC architecture (defined within SAE work) which means the evolution of the Packet Core domain and user management domain. The development of LTE as a new 3GPP access technology is of course closely related to the design of EPC. Due to the importance of LTE in relation to EPC (since LTE only connects via Packet Core domain) we also provide a brief description of LTE on a high level. For a deeper insight into the interesting area of advanced radio communications called 3G Evolution, we recommend [Dahlman, 2008]. The Circuit Core and the IMS domains are described in Chapter 5 where we look further on the topic of voice services.

We will now leave the high-level view of the 3GPP network architecture and turn our attention to the evolution of the Packet Core domain, or EPC.

While the logical architecture may look quite complex to anyone not familiar with the detailed functions of EPC, do not be put off. The EPS architecture consists of a few extra new functional entities in comparison to the previous core network architectures, with a large number of additional new functions and many new interfaces where common protocols are used. We will address the need for these additions and the perceived complexity by investigating all of these step by step.

Illustrated in Figure 3.1.2 is the logical architecture developed for EPS, together with the Packet Core domain defined prior to EPC. It also shows how the connection to this 'old' 3GPP packet core is designed (in fact, this specific connection comes in two flavours itself, a fact that adds to the complexity of the diagram, but more about that later).

Note here that Figure 3.1.2 illustrates the complete architecture diagram, including support for interconnection of just about any packet data access network one can think of. It is unlikely that any single network operator would make use of all these logical nodes and interfaces; this means that deployment options and interconnect options are somewhat simplified.

What is not visible in the above diagram is the 'pure' IP infrastructure supporting the logical nodes as physical components of a real network. These functions are contained in the underlying transport network supporting the functions needed to run IP networks, specifically IP connectivity and routing between the entities, DNS functions supporting selection and discovery of different network elements within and between operators networks, support for both IPv4 and IPv6 in the transport and application layer (the layers are more clearly visible when we go into the details in Parts III and IV.

Be aware that all nodes and interfaces described in this chapter (and in fact, throughout the complete book) are logical nodes and interfaces, that is, in a

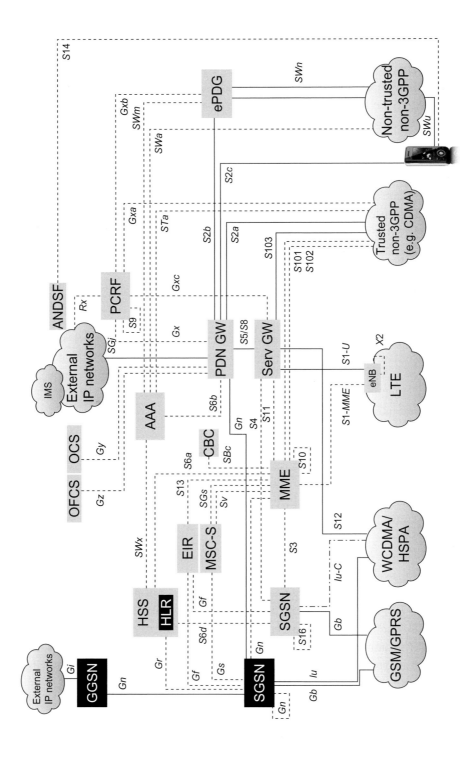

Figure 3.1.2 *Architecture overview.*

real network implementation, some of these functions may reside on the same physical piece of infrastructure equipment; different vendors may have different implementations. In essence, different functions may be implemented in soft-ware and connect with one another via an internal interface, rather than via an actual cable. Also, the physical implementation of a particular interface may not run directly between two nodes; it may be routed via another physical site. Naturally, interfaces may also share transmission links.

One example is that the X2 interface connecting two eNodeBs (which will be explained in more detail later) may physically be routed from eNodeB A together with the S1 interface (which connects an eNodeB to an MME in the core network) to a site in the network with core network equipment. From this site, it would be routed back onto the radio access and finally to eNodeB B. This is illustrated in Figure 3.1.3.

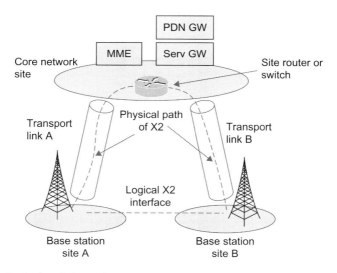

Figure 3.1.3 *Logical and physical interfaces.*

3.1.1 *Basic IP connectivity over LTE access*

At the core of the EPC architecture is the function required to support basic IP connectivity over LTE access. The plain vanilla EPC architecture, which one cannot live without when deploying LTE, appears as in Figure 3.1.1.1.

Two main principles have been guiding the design of the architecture. First of all, the strong wish to optimize the handling of the user data traffic itself, through designing a 'flat' architecture. A flat architecture in this context means that as few nodes as possible are involved in processing the user data traffic.

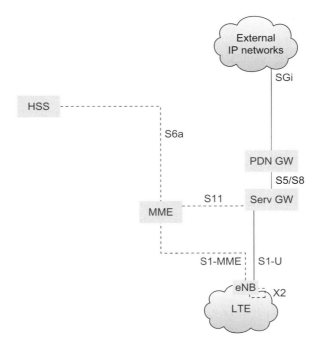

Figure 3.1.1.1   *Basic EPC architecture for LTE.*

The primary motivation for this was to allow a cost efficient scaling of the infrastructure operating on the user data traffic itself, an argument increasingly important as mobile data traffic volumes are growing fast and are expected to grow even faster in the future with the introduction of new services relying on IP as well as new powerful access technologies such as LTE.

The second guiding principle was to separate the handling of the control signalling (shown as dotted lines) from the user data traffic. This was motivated by several factors. The need to allow independent scaling of control and user plane functions were seen as important since control data signalling tends to scale with the number of users, while user data volumes may scale more dependent on new services and applications as well as the capabilities in terms of the device (screen size, coder, etc). Allowing for both the control signalling functionality and the user data functionality to be implemented in optimized ways was another rationale for separating these functions in the logical architecture. A third important factor guiding the decision was that the split between control signalling and user data functions allowed for more flexibility in terms of network deployment as it allowed for the freedom of locating infrastructure equipment handling user data functions in a more distributed way in the networks, while at the same time allowing for a centralized deployment of the equipment handling the control signalling. This was in order to save valuable transmission resources and minimize delays between two parties connected and utilizing a real-time service such as voice or gaming. In addition to this, the split between the control signalling and user data functions allows for optimized operational costs through having these functions at separate physical locations in the network; through separating this functionality, the network nodes are more scalable in particular when it comes to supporting high-bandwidth traffic. Only those nodes that are associated with end-user traffic need to be scaled for the high throughput, rather than both the traffic and signalling nodes as would have been the case previously. Finally, the chosen architecture was similar to the existing packet core architecture for evolved HSPA, allowing the possibility of a smooth migration and co-location of functionality supporting both LTE and HSPA.

Looking at the architecture itself, let us start with the radio network. First of all, in the LTE radio network there is at least one eNodeB (the LTE base station). The functionality of the eNodeB includes all features needed to realize the actual wireless connections between user devices and the network. The features of the LTE eNodeB will be described in Chapter 10.

In a reasonably sized network scenario, there may be several thousand eNodeBs in the network; many of these may be interconnected via the X2 interface in order to allow for efficient handovers.

All eNodeBs are connected to at least one MME (short for the 'Mobility Management Entity') over the S1-MME logical interface. The MME handles all LTE-related control plane signalling, including mobility and security functions for devices and terminals attaching over the LTE RAN. The MME also manages all terminals being in *idle mode*, including support for Tracking Area management and paging. Idle mode will be further described in Section 6.3.

The MME relies on the existence of subscription-related user data for all users trying to establish IP connectivity over the LTE RAN. For this purpose, the MME is connected to the HSS (the Home Subscriber Server) over the S6a interface. The HSS manages user data and related user management logic for users accessing over the LTE RAN. Subscription data includes credentials for authentication and access authorization, and the HSS also supports mobility management within LTE as well as between LTE and other access networks (more about this later).

The user data payload – the IP packets flowing to and from the mobile devices are handled by two logical nodes called the Serving Gateway (Serving GW) and the PDN Gateway (PDN GW) where PDN is short for 'Packet Data Network'.

The Serving GW terminates the S1-U user plane interface towards the base stations (eNodeBs), and constitutes the anchor point for intra-LTE mobility, as well as (optionally) for mobility between GSM/GPRS, WCDMA/HSPA and LTE. The Serving GW also buffers downlink IP packets destined for terminals that happen to be in idle mode, as well as supports transport level QoS through marking IP packets with appropriate DiffServ code points based on the parameters associated with the corresponding packet bearer. (An in-depth description of how QoS works in the EPS architecture can be found in Section 8.1.) For roaming users, the Serving GW always resides in the visited network, and supports accounting functions for inter-operator charging and billing settlements.

The PDN GW is the point of interconnect to external IP networks through the SGi interface. The PDN GW includes functionality for IP address allocation, charging, packet filtering and policy-based control of user-specific IP flows. The PDN GW also has a key role in supporting QoS for end-user IP services.

The Serving GW and PDN GW are connected over an interface called either S5 (if the user is not roaming, i.e. the user is attaching to his home network) or S8 (if the user is roaming, i.e. attaching to a visited LTE network).

Something quite unique with the EPC architecture is that one of the interfaces is specified in two different variants (you guessed correct – it is S5/S8). One of these variants utilizes the GTP protocol over S5/S8 (more about this in Section 10.3), which is also used to provide IP connectivity over GSM/GPRS and WCDMA/HSPA networks. The other variant utilizes the IETF PMIPv6 protocol over S5/S8 (more about this in Section 10.4).

Since PMIPv6 and GTP do not have exactly the same feature set, this means that the functional split between the Serving GW and the PDN GW is somewhat different depending on what protocol is deployed over S5/S8. In fact, nothing prevents both variants from being used simultaneously in one and the same network. This is not an unlikely situation since any operator deploying S5 based on the PMIPv6 variant may at least in the early years of LTE deployment have to rely on GTP over S8 since GTP is today the de-facto protocol used to interconnect hundreds of mobile networks world-wide, allowing for IP connectivity to be established just about anywhere in the world where there is GSM/GPRS or WCDMA/HSPA coverage.

Also note that S5 in itself may not be in use at all in most non-roaming traffic cases. It is a quite possible scenario that many operators will choose to deploy equipment that can combine Serving GW and PDN GW functionalities whenever needed, in practice reducing the amount of hardware needed to process the user data plane by 50%.

In some traffic cases, the S5 interface is however very much needed, resulting in a division of the Serving GW and PDN GW functionality between two physical pieces of infrastructure equipment (two 'Gateways'). Note that for a single user/terminal point of view, there can only be a single Serving GW active at any given time.

The split GW deployment using S5 may happen in three cases:

1. When a user wants to connect to more than one external data network at the same time, and not all of these can be served from the same PDN GW. All user data related to the specific user will then always pass the same Serving GW, but more than one PDN GW
2. When an operator's deployment scenario causes the operator to have their PDN GWs in a central location whereas the Serving GWs are distributed closer to the LTE radio base stations (eNodeBs)
3. When a user moves between two LTE radio base stations that does not belong to the same *service area*, the Serving GW need to be changed, while the PDN GW shall be retained in order not to break the IP connectivity. (The concepts of service areas and pooling will be described in detail in Section 6.6)

Control plane signalling between the MME and the Serving GW is exchanged over the S11 interface which is one of the key interfaces in the EPC architecture. Among other things, this interface is used to establish IP connectivity for LTE users through connecting Gateways and radio base stations, as well as to provide support for mobility when users and their devices move between LTE radio base stations.

3.1.2   Adding more advanced functionality for LTE access

Expanding somewhat on the basic architecture described in above means introducing some more interfaces and some additional advanced features targeting the control of end-user IP flows; these additional features are covered in the following section.

For the purposes of this section, an 'IP flow' can normally be thought of as all IP packets flowing through the network associated with a specific application in use, e g a web browsing session or a TV stream.

See the architecture diagram in Figure 3.1.2.1, which shows a bit more detail than the previous illustration.

Three new logical nodes and associated interfaces are added – the PCRF, the OCS and the OFCS.

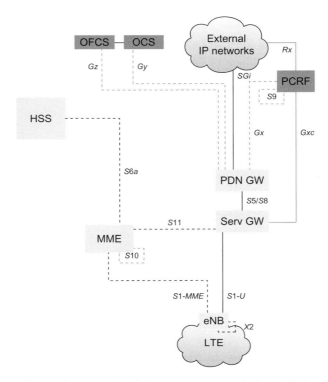

Figure 3.1.2.1 *Adding policy control and charging support to the basic EPC architcture.*

The PCRF (Policy and Charging Rules Function) make up a key part of a concept in the EPC architecture (and in the 3GPP packet core architecture in general) called PCC (Policy and Charging Control). The PCC concept is designed to enable flow-based charging, including, for example, online credit control, as well as policy control, which includes support for service authorization and QoS management.

What, then, is a 'policy' in the 3GPP architecture context? Think of it as a rule for what treatment a specific IP flow shall receive in the network, for example, how the data shall be charged for or what QoS that shall be awarded to this service. Both the charging and the policy control functions rely on that all IP flows are classified (in the PDN GW/Serving GW) using unique packet filters that operate in real time on the IP data flows.

The PCRF contains policy control decision and flow-based charging control functionalities. It terminates an interface called Rx, over which external application servers can send service information, including resource requirements and IP flow related parameters, to the PCRF. The PCRF interfaces the PDN GW over the Gx interface and for the case when PMIPv6 and not GTP is used on S5, the PCRF also interfaces the Serving GW over an interface called Gxc.

In the roaming case, a PCRF in the home network controls the policies to be applied. This is done via a PCRF in the visited network over the S9 interface which hence is a roaming interface between PCRFs.

OFCS is short for Offline Charging System while OCS is short for Online Charging System. Both logical entities interface the PDN GW (through the Gz and Gy interfaces respectively) and support various features related to charging of end-users based on a number of different parameters such as time, volume, event, etc. Section 8.3 contains a description of the charging support in the EPC architecture.

Also shown in the diagram above is an interface called S10. It connects MMEs together, and is used when the MME that is serving a user has to be changed for one reason or another, either due to maintenance, to a node failure, or the most obvious usage, when a terminal moves between two *pools*. Do not worry if this is not obvious to you – the pooling concept will be described in detail in Section 6.7.

3.1.3   Interworking between LTE and GSM/GPRS or WCDMA/HSPA

Given the fact that any new radio network is normally brought into service well before a complete radio coverage is achieved (if that ever happens), the ability to allow for a continuous service coverage through interworking with other radio networks is a key feature in any mobile network architecture. In many markets, LTE will be deployed in frequency bands around 2 GHz or higher. While the data capacity normally increases as one moves into higher frequency band (as there is more spectrum available), the ability to cover a given geographical area with a given base station output power quickly decreases with higher frequencies. Simply put, the gain in increased data capacity is unfortunately paid for by much less coverage, but that is simply the rule of physics.

For LTE deployment, interworking with existing access networks supporting IP connectivity hence becomes crucial. The EPS architecture is addressing this need with two different solutions. One is addressing GSM/GPRS and WCDMA/HSPA network operators, while the other solution is designed to allow LTE interworking with CDMA access technologies (1xRTT and eHRPD). Section 6.4 includes more details on inter-system mobility support in EPC.

First of all, be aware of that there are several different ambition levels related to 'interworking'.

The simplest way of interworking would be to allow for a portable device to connect to either LTE or, say HSPA, depending on specific preferences but also on the available radio coverage at the geographical location the user is at for the moment (assuming the user is carrying this 'device' which may be a portable laptop computer or similar). Let's call this network A. When the user moves to another location, he or she is able to connect to another access network (B) if available. The change of access network from A to B is supported through one single service subscription, and may be more or less automated or hidden for the user. This is of course a quite basic level of support which means that services and applications running on the user device may not be usable when the device has moved; some of these will require an application restart. This is due to that when the user has moved and wants to use the new access network B, the network view this as a completely new attach request. The device is normally given a new IP address from the network, which then may or may not cause problems for the applications in use in the device. Furthermore, there is normally a quite long service interruption between loss of coverage of network A and the establishment of IP connectivity to network B. This is illustrated in figure 3.1.3.1.

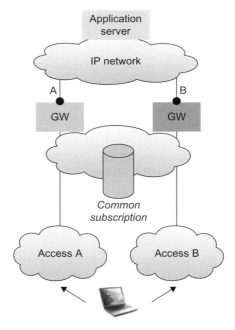

Figure 3.1.3.1   *Inter-access mobility without session continuity.*

This way of interworking basically only requires a single subscription allowing the end-user to attach over network A or network B, as well as that the user carries a portable device which includes support for both radio technologies. There is in fact no specific support required in the network itself, apart from the ability for the network operator to ensure the consistency of the user's subscription data related to networks A and B respectively.

The EPC architecture however also allows for 'session continuity', that is that an IP connectivity session which is established over any of the allowed access networks (A or B) actually *will* survive movements between the different access networks due to loss of radio coverage. This is handled through retaining a stable 'IP anchor point' in the network which allows for not having to change the IP address of the device at all. Applications and services will, in theory, then not be dependent on the access network that is in use or on any possible movements between these. This is of course only partly true. Some services may rely on really high data speeds or very low network delay, criteria that may not be possible to meet with all of networks due to limited radio coverage or by limitations in the access technologies themselves. Remember that the wireless data performance and capabilities offered by LTE may be superior to HSPA which in its turn is vastly superior to GPRS, the IP connectivity service offered on GSM networks (Figure 3.1.3.2).

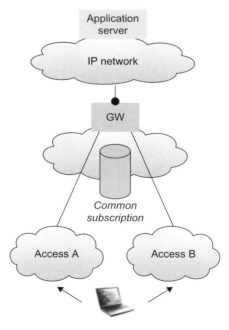

Figure 3.1.3.2   *Inter-access mobility with session continuity.*

This is however a point at which 3GPP has made the solution a bit more complex than what you could think was needed. 3GPP has in fact defined two *different* options for how to interconnect LTE and WCDMA/HSPA or GSM/GPRS. We will describe both below.

An important piece of understanding here is to note that when a terminal attaches over LTE it is served by the MME (as described above), whereas when the terminal attaches over a WCDMA/HSPA/GSM/GPRS network it is instead served by an SGSN.

The SGSN has been part of the packet core architecture since the first GSM/GPRS specification release in 1997. Back in those days, it was introduced to support a brand new service called GPRS which was and still is the packet data connectivity service of GSM. In 1999, the ability to also serve IP connectivity over WCDMA networks was added. Note that the IP connectivity service over WCDMA was greatly enhanced in 2005 through the definition of HSPA and the Direct Tunnel concept, but that has no real impact on the packet core architecture itself. HSPA is mainly an enhancement of the WCDMA radio access technology.

An SGSN connects to a GGSN which acts as the point of interconnect to external IP networks for all packet data sessions over GSM/GPRS and WCDMA/HSPA. In fact, it is the SGSN that selects which GGSN to use for a specific terminal.

When a user is moving between two networks that happen to be served by two different SGSNs, these SGSNs interact over an interface (quite illogically also called Gn) to support IP session continuity, that is, that the IP address, and all other data associated with the IP session itself, is maintained through keeping the GGSN unchanged when changing from one access network to the other.

If we disregard physical packet data equipment which may or may not have smooth migration paths to the EPC architecture, the *logical* SGSN node has a key role to play also for LTE, but that is however not the case for the *logical* GGSN node. Below we will show you why this is the case.

The legacy packet core architecture and control signalling procedures form the base for the first solution for interworking between LTE and GSM/GPRS or WCDMA/HSPA described here. It was actually the second one defined but it is the most straightforward to understand.

This solution includes the SGSN attaching to GSM and WCDMA radio networks as today, but then includes the MME and the PDN GW acting as an SGSN and a GGSN respectively. The MME and PDN GW are in fact replicating the signalling needed for movements between networks GSM/GPRS and WCDMA/HSPA to also apply for mobility with LTE (Figure 3.1.3.3).

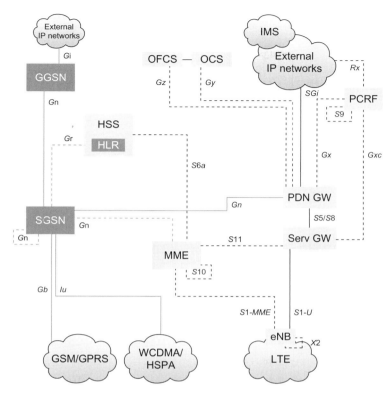

Figure 3.1.3.3 *Interworking between LTE and GSM/GPRS or WCDMA/HSPA.*

This includes both the MME and the PDN GW interfacing the SGSN over the standard packet core Gn interface. It may even be a Gn interface with an older date, that is specified and in operation prior to EPC being designed. This latter case is referred to as a pre-Rel8-SGSN.

In order to get this solution to work, a few key functions are needed.

First of all, the SGSN must be able to distinguish between a terminal that attaches over GSM/GPRS or WCDMA/HSPA but is not capable of moving to LTE, from a terminal that in fact can connect to LTE but is currently attaching to GSM/GPRS or WCDMA/HSPA due to lack of LTE radio coverage. The latter terminal must always be using the PDN GW as the anchor point and never a GGSN since there is no logical connection between the LTE radio network and a GGSN. If such an incorrect choice of IP anchor point would be made, IP sessions would be dropped when changing access network to LTE.

See the example in Figure 3.1.3.4; terminal A has GSM/WCDMA support but is not capable of utilizing LTE access, while terminal B can use all three RANs.

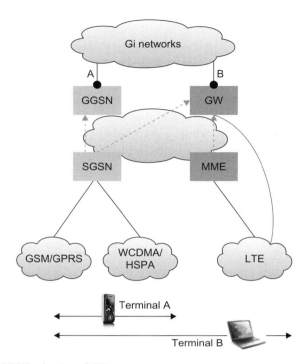

Figure 3.1.3.4 *SGSN selection of GW.*

The simplest case is when terminal B attaches to an LTE radio network. It is then served by the MME which will select a PDN GW and Serving GW (in the figure collectively referred to as 'GW').

When either of the terminals attach over GSM or WCDMA radio it is served by the SGSN. For terminal B this may happen when there is no LTE coverage, while terminal A does not have LTE support so the LTE coverage situation is then irrelevant.

The SGSN may use different ways of choosing either a GGSN or a PDN GW as an IP session anchor point for a terminal, but the most obvious way is to utilize the 'APN' (Access Point Name), which is a part of the configuration data related to a user subscription and is pointing at the preferred external network. Since only terminals which include LTE radio access support (terminal B in the example) may ever move and attach to an LTE RAN, the simplest solution is to make sure that only these subscriptions are configured with an APN that is associated with a PDN GW. This helps the SGSN in taking a correct decision and ensuring the terminal B is using the PDN GW and not the GGSN as the IP anchor point.

The other important part of the solution is to provide a single (or at least consistent) set of user and subscription data. Traditionally, the SGSN has interfaced a logical

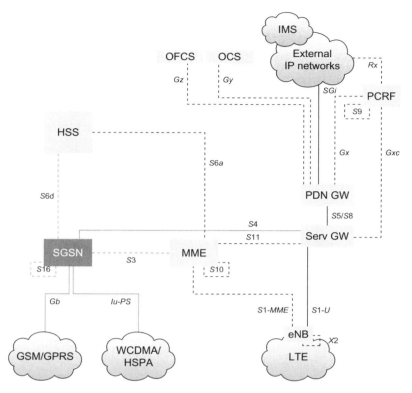

Figure 3.1.3.5 *Interworking using GTPv2 interfaces.*

node called HLR (Home Location Register) which is the main database for user data in GSM and WCDMA. This interface is called Gr. The MME instead interfaces the HSS (Home Subscriber Server) as described above. When moving between GSM/WCDMA and LTE, there must not be inconsistent information in the network about, for example, to what network (A, B or C) a specific terminal is currently attached. This means that the HLR and HSS need either to share a single set of data, or to ensure consistency through other means such as close interaction between the two network functions. 3GPP does not specify any detailed solution to this problem. In fact, the 3GPP specification avoids the problem through defining HLR as a subset of HSS in later versions of the standards. As for the actual solution of ensuring this data consistency, it varies between different vendors of network infrastructure equipment.

However, as you know from above, this is not the only solution for interworking between LTE and GSM/GPRS/WCDMA/HSPA. In fact, 3GPP first defined another solution. This is described in Figure 3.1.3.5.

Just like the first interworking architecture described, this solution naturally also includes an SGSN interfacing the GSM/GPRS and WCDMA/HSPA radio networks using the Gb and Iu-PS interfaces. So far there are no major differences.

Then however, the SGSN implements four new interfaces. Three of these (called S3, S4 and S16) rely on an updated version of the GTP protocol, the protocol that has been used since the old days of GPRS in the late 1990s, and which forms a core part of the 3GPP packet core architecture. All three are used instead of the different variants of the Gn interface present in the 'legacy' packet core architecture. The fourth new interface is S6d which mimics the MME S6a interface towards the HSS for retrieving subscriber data from the HSS, but for the SGSN it is naturally data related to GSM and/or WCDMA, not to LTE. Just as with S6a, the IETF Diameter protocol is used also over S6d, eliminating the need for the SGSN to support SS7/MAP signalling towards the HLR.

S3 is a signalling-only interface. It is used between the SGSN and the MME to support inter-system mobility. S16 is the SGSN-SGSN interface, while S4 is connecting the SGSN and the Serving GW. Note that here is a difference compared to the other solution where the SGSN interfaces the PDN GW and treats this like a GGSN.

Connecting the SGSN with the Serving GW creates a common anchor point for LTE, GSM/GPRS and WCDMA/HSPA in the Serving GW. Since the Serving GW for all roamers is located in the visited network, this means that all user traffic related to one roaming user will pass through this point in the network, regardless of which radio network that is being used. This is new and a difference to how roaming is handled in the old solution where the SGSN itself implements the roaming interface for GSM and WCDMA and the Serving GW only for LTE. With all roaming traffic instead passing through a single point in the network, it allows for the visited network operator to control and monitor the traffic in a consistent way, potentially based on policies. One potential drawback is that user traffic need to pass through one additional network node on its way to the PDN GW, but there is a solution to that, at least for WCDMA/HSPA. This is to utilize a direct connection between the Radio Network Controller (RNC) in the WCDMA radio network and the Serving GW. This interface is called S12, is optional, and if used, it means that the SGSN will only handle the control signalling for WCDMA/HSPA. The primary driver for this is that the network does not have to be scaled in terms of SGSN user capacity as well, important due to the expected large increase of data sent over wireless networks. See Figure 3.1.3.6.

It should be noted that this ability to let the user data bypass the SGSN is in fact also possible with the first solution. This means that the WCDMA RNC directly would interface the PDN GW for the user traffic connections. A difference is

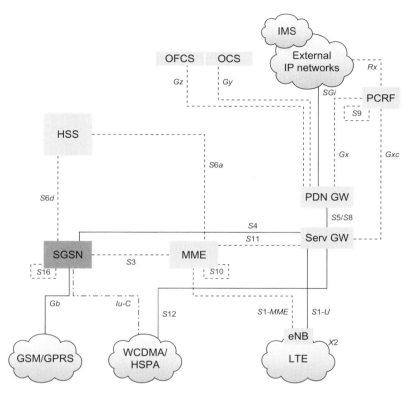

Figure 3.1.3.6 *Direct tunnel support for WCDMA/HSPA.*

that this would not work for roamers since as stated above, roaming traffic here always pass through the SGSN.

A further benefit of utilizing S3/S4 from the SGSN instead of Gn is that the 3GPP specifications then allow for optimization of the signalling load for all terminals in idle mode. This concept is called ISR (short for Idle Mode Signalling Reduction) and will be explained in detail in Section 6.4.

3.1.4 Interworking between LTE and CDMA networks

As stated above, an important part of the objectives behind the creation of EPC has been to allow an efficient interworking with legacy mobile network infrastructure in order to allow for wide area service coverage. As there was significant interest towards using a common EPC that was being developed in 3GPP under the SAE framework, strong efforts were devoted to design a solution linking LTE and the CDMA technologies defined by 3GPP2, allowing for efficient and smooth handovers between the different technologies as the differences in radio coverage require.

Figure 3.1.4.1 shows the EPC solution for LTE-eHRPD interworking.

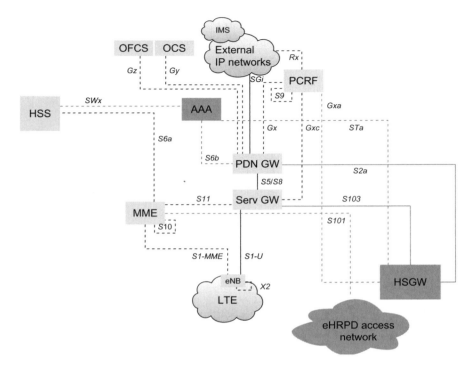

Figure 3.1.4.1 *Interworking between LTE and eHRPD networks.*

It includes the emergence of the eHRPD network, depicted by the darker entities. It shall be noted that this picture excludes all details of the CDMA network itself other than what is relevant for the SAE framework for EPS. In fact, just as with 3GPP networks, there are more than one flavour of CDMA networks – the voice + data network normally referred to as 1xRTT and the data only eHRPD are the two CDMA technology that EPS provides interworking with.

A number of additional interfaces are needed in the EPS architecture to allow LTE-CDMA interworking. Three of these (called S101, S102 and S103) are unique for CDMA networks and used to provide optimal performance during handover, whereas the other three (S2a, Gxa and STa) are generic and may be used for any non-3GPP access interworking, not only with CDMA networks. We will describe these interfaces below.

A fundamental function for allowing efficient interworking between two access technologies is to enable a common set of subscription data to be used – both for authentication purposes as well as for keeping track of which network the user is currently attached to. The core of the solution is to allow for the HSS to act as the common data base for all subscription data. Access authentication for a user attaching over an eHRPD network are handled using 3GPP2 mechanisms which are based on IETF AAA functionality. (AAA is short for Authentication, Authorization and Accounting). For this purpose, the eHRPD network is connected to the EPC architecture over the STa interface. It is terminated in a logical node called the 3GPP AAA Server, which in real-life implementations either may be a software feature inside the HSS or stand-alone AAA equipment interfacing the HSS over the Diameter-based SWx interface. Note that 3GPP already used the AAA mechanism for authentication and authorization before EPS due to interworking solutions for WLAN developed prior to the SAE work.

The PDN GW interfaces the 3GPP AAA Server over the S6b interface. This interface is used by the PDN GW to retrieve certain subscription data. It is also used to store the information regarding the PDN GW the user is connected to, this in order to secure that when the user moves and attaches over LTE, the MME shall be able to select the same PDN GW as was used for eHRPD network, which is a pre-requisite for maintaining the IP session also after the handover.

The PDN GW acts as the common IP anchor point also for users attached over the eHRPD network. The user data between the eHRPD Serving Gateway (HSGW) and the PDN GW are transported over the S2a interface which implements the PMIPv6 protocol, specified by IETF to support intra-network mobility.

The EPC architecture also allows for a common policy controller (the PCRF) to apply policies also in the eHRPD network. This is done over the Gxa interface to the HSGW.

In addition to these core interfaces for the LTE-eHRPD interworking, there are three more interfaces specified. Two of these are used to optimize the handover performance for eHRPD data, while the third is used to support voice services in 1xRTT. This last interface, called S102, will not be further described here, but rather in the section on voice support below.

The S101 interface is used when a packet data handover between LTE and an eHRPD network is to take place. Before the handover itself, the terminal pre-registers in the target network to prepare this network for the handover and thus reduce the perceived interruption time. This pre-registration as well as the actual handover signalling is carried over the S101 interface between the MME and the eHRPD access network. This is described in more detail in Section 6.3.

To further optimize the packet data handover performance between LTE and eHRPD, there is also an S103 interface specified. This is an interface used to forward any IP data packets destined to the terminal that happened to end up in the Serving GW while the user terminal was executing the handover to eHRPD. These packets can then be forwarded to the HSGW in the eHRPD network, thus achieving close to a true loss-less handover performance. The value of this packet forwarding depends on the actual application in use. It can be seen as an optional optimization which may add value in some use cases.

3.1.5   Interworking between 3GPP access technologies and non-3GPP access technologies

The EPC architecture has been designed to allow interconnection with just about any access technology. This creates a common way of treating access to a PDN regardless of the access technology used, meaning that, for example, terminal's IP address assignment, access to general IP services as well as network features like user subscription management, security, charging, policy control and VPN connections can be made independent of the access technology – be it wireless or fixed (Figure 3.1.5.1).

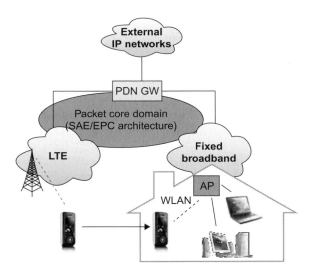

Figure 3.1.5.1   *Interworking between 3GPP access and non-3GPP access technologies.*

Besides the common and access-independent feature set for service treatment, the architecture also allows for interworking between 3GPP technologies (i.e., GSM/GPRS, WCDMA/HSPA and LTE) and non-3GPP technologies (e.g., CDMA (as described above), WLAN or a fixed access technology).

Think of this as a use case: you carry a device which can access among other technologies, LTE and WLAN. You are connected to the LTE/EPC network and move indoors, into your house. There you have a fixed broadband connection connected to a WLAN-capable home router. Depending on preferences, the device may in this situation switch access from LTE to WLAN. The EPS network then includes features to maintain the sessions also during this handover between two quite different access technologies.

The key functionality desired is support for mobility in the PDN GW. Mobile IP was designed in the 1990s by the IETF to provide IP host mobility, which is the ability for a portable computer to connect to a visited IP network, and establish a connection to the home IP network through tunnelling of IP packets. To all corresponding hosts, this computer would appear as still being in the home network. Mobile IP technology has since been used to provide mobility for packet data services in mobile networks based on CDMA technology.

Due to the diversity of the requirements when specifying the SAE framework, this part has (somewhat unfortunately) come out with quite a few options. First of all, either 'Client-based' or 'Network-based' Mobile IP can be used. Client-based means that the Mobile IP client resides in the terminal, and that IP tunnels are established between the terminal and the PDN GW across the access network. Network-based means that there are functions in the access network that acts on behalf of the terminal, and provides mobility support.

The major advantage with a Client-based approach is that it may work over any access network, as long as there is adequate support in the terminal itself. This function may be used totally transparent to the functionality in the access network. The advantage with a Network-based approach is the opposite – it simplifies the terminal client application, but instead requires that there is specific Mobile IP support in the network itself. eHRPD is one access network where the latter approach has been chosen, as described above. One of the key concerns with 'Client-based mobility' was how secure, trusted and efficient such mobility would be. These concerns partly drove the development of the Network-based mobility track in 3GPP and in IETF.

The parts of the EPC architecture that apply to the 'non-3GPP' access support are shown in Figure 3.1.5.2. As can be seen, there are multiple ways to interconnect to a non-3GPP access network.

There are two ways to distinguish between the available options:

1. Is it a connection to 'trusted' or a 'non-trusted network'?
2. Are 'Network-based' or 'Client-based' mobility mechanisms used?

Network-based and Client-based concepts were described above, so what is then a 'trusted' and a 'non-trusted' access network? Simply put, this is really an

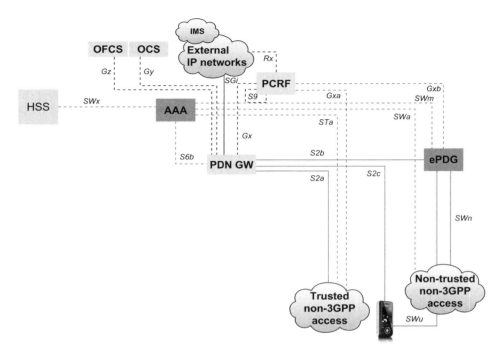

Figure 3.1.5.2   *EPC architecture for non-3GPP accesses.*

indicator on if the 3GPP operator (owning the PDN GW and the HSS) trust the security of the non-3GPP access network. A typical 'trusted' network may be an eHRPD network, while a 'non-trusted' network may be, for example, usage of WLAN in a public café and connecting to the PDN GW over the public Internet.

The S2a, STa and Gxa interfaces were described above for eHRPD and also apply in this context in the same manner. STa and Gxa apply to any trusted non-3GPP access network, and are used for user data and policy control respectively. S2a is used for data connectivity when Network-based mobility schemes are used in combination with trusted networks.

The corresponding interfaces for non-trusted networks are S2b, SWa (not STb) and Gxb. There is however a major difference here. Since the operator may not trust the non-3GPP access network that is used by the device when attaching, the S2b and Gxb interfaces do not interface the access network itself, but instead a new logical node called the ePDG (evolved Packet Data Gateway). This is an evolution of the PDG that is specified in earlier versions of the 3GPP standards to allow interconnection (but not inter-access mobility) of WLAN access to a 3GPP network. Typically, the ePDG belongs to the mobile operator. More information on this can be found in Section 7.3.

Encrypted tunnels are established between the user devices and the ePDG, to ensure that each device can communicate with the network in a secure way. This creates a logical association between the device and the ePDG, referred to as the SWu interface, which both carries signalling needed for management of the tunnel itself, as well as carrying the data. The ePDG then connects to the PDN GW, and data as well as signalling is transferred using the S2b interface between these two nodes. The ePDG also interfaces the PCRF over Gxb to support policy control mechanisms (details of this specific part of the EPC architecture is however not yet defined by 3GPP).

The interface between the non-trusted access network and the ePDG is called SWn. It carries all signalling and data between the two networks – the non-trusted access, and the operator network to which the ePDG belongs. SWu traffic and signalling is hence always routed over the SWn interface.

One final interface to understand in this non-3GPP access solution is SWm. This is a signalling-interface only, and connects the ePDG to the AAA Server. It is used to transport AAA-related parameters between the AAA server (which in itself may get the data from the HSS) and the ePDG, in order to support setup and authentication of the IPsec tunnels between the ePDG and the terminal.

In addition to the Network-based mobility solutions for trusted and non-trusted networks, there is also the Client-based mobility solution which relies on the S2c interface between the mobile device and the PDN GW. This means that an overlay solution is created which does not require any specific support from the underlying non-3GPP access network.

Client-based and Network-based mobility solutions are further discussed in Section 6.3 and Section 11.3.

3.1.6 Support for voice services

In addition to the primary focus on enabling an efficient Mobile Broadband solution, the support for voice services was also given high priority in the 3GPP SAE specification work. From the start, it was however agreed that LTE is a packet-only access network, allowing an optimization for packet services. Since the voice services so far have been realized using circuit-switched technologies, this meant that specific mechanisms were introduced to also allow for voice services for users of the packet data services offered over the LTE access. Section 5.2 explains the subject of voice for LTE in more depth. In this chapter, we want to highlight the applicable parts of the architecture.

Shortly, the two main solutions for voice are to either use the IMS mechanisms and realize voice using the MultiMedia Telephony (MMTel) framework, that is using voice-over-IP, or to stick to the 'old' circuit-switched way of providing voice services. The second option is in 3GPP specifications realized through that users temporarily leave LTE to perform the voice calls over 2G/3G, and then return when the voice call is finished. This may not be the most elegant of solutions, but can be seen primarily as a gap-filler in case IMS infrastructure is not in place.

When a user engaged in an IMS/MMTel voice call moves around, it is not unlikely that the user device may encounter that the LTE radio coverage is being lost. After all, this is a mobile system, and this of course may happen frequently depending on how users move around. For this purpose, 3GPP specified mechanisms to hand over an ongoing voice call in IMS/MMTel. What happens then depends on whether the new (target system) can support IMS/MMTel or not. If this is the case, this will be solved through a packet handover procedure (see Section 12.4) and the IMS/MMTel session will continue after the handover. If this is not the case, the IMS/MMTel session will be handed over to a circuit-switched call in GSM, WCDMA or 1xRTT. To achieve a smooth handover, this procedure involves pre-registration of the terminal in the target system CS domain (i.e., the system that the terminal will be attached to instead of LTE after the handover) and an efficient handover signalling.

If no IMS is present in the network at all or LTE users have to temporarily leave LTE during voice calls as described above, the MME interacts with the circuit-switched infrastructure to achieve this temporary suspension of the LTE services.

Figure 3.1.6.1 highlights the parts of the architecture that applies to voice handovers. Note that this diagram shows the GSM/WCDMA MSC Server and its interfaces to the GSM and WCDMA network. These connections are not included in the overall diagram shown at the beginning of this chapter, purely in order not to make the diagram even more complex.

The case of a packet handover to GSM/WCDMA is supported using handover signalling over the S3 (or Gn) interface between the MME and the SGSN. The corresponding interface for CDMA is called S101 and connects the MME to the RNC used for eHRPD. In order to further optimized the packet handover performance, packet forwarding may be used (it is optional). This means that any packets destined for the user device that may have happened to have been sent 'downwards' from the PDN GW (to either the SGSN, Serving GW or eHRPD HSGW) may be forwarded to the corresponding node in the target system. This is not absolutely required, but may improve the user experience of a handover, since in theory no data need to be lost during the handover. The case of packet forwarding between LTE and GSM/WCDMA is supported over Gn/S4 between SGSN and Serving GW, whereas the case of LTE-eHRPD packet forwarding is supported over the S103 interface between the Serving GW and the HSGW.

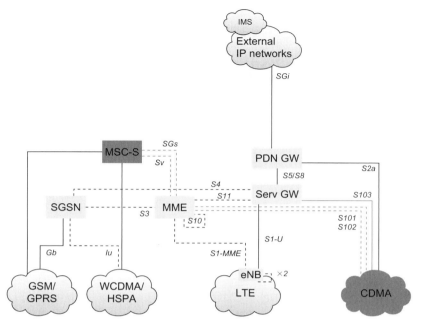

Figure 3.1.6.1 *EPC architecture for voice support.*

In the case MMTel is not supported by the target network, there are also functions for handing over an IMS-based MMTel session to a circuit-switched voice call in GSM, WCDMA or CDMA 1xRTT. This handover is realized through procedures executed over the Sv interface between the MME and the MSC Server (for GSM and WCDMA) or the S102 interface between the MME and the 1xRTT MSC. This procedure is known as 'Single-Radio Voice Call Continuity' (SRVCC).

Finally, support for always falling back to circuit-switched calls even when in LTE coverage (which means that no IMS infrastructure is used) is supported through signalling over the SGs interface between MME and a GSM/WCDMA MSC Server or over the S102 interface connecting the MME to an 1xRTT MSC. This procedure is known as 'Circuit-Switched Fallback' (CSFB).

3.1.7 Miscellaneous features

We are almost through this overview of the 3GPP EPS architecture developed under the SAE work umbrella, including all the logical network nodes and the interfaces. As a final part, let us look at three distinct features in the architecture that may be seen as somewhat outside the core of the architecture (Figure 3.1.7.1).

The first of these functions is called ETWS. This is short for Earthquake and Tsunami Warning System and is considered an important safety feature for countries endangered by nature catastrophes. Simply it means that a warning is

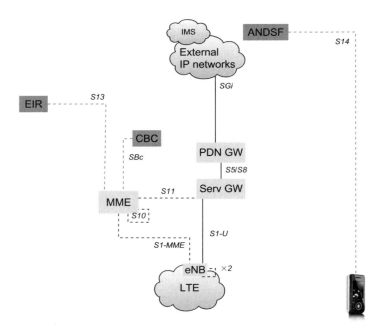

Figure 3.1.7.1   *Miscellaneous features in the EPC architecture.*

received by the Cell Broadcast Centre (CBC) from, say, a government agency monitoring seismic activity and predicting earthquakes. The CBC interfaces the MMEs in the network over the SBc interface. Since all terminals in the network must be reachable for this warning, the MMEs must convey the warning to all terminals that happen to be in idle mode, and whose location is only known with the accuracy of a Tracking Area which may or may not contain lots of base stations and radio cells.

Another feature is the support for the Equipment Identity Register (the EIR) which is optionally used by the MME when a user attaches. The EIR is a database that contains information regarding whether the device used to attach to the network happens to be stolen or not. If that is the case, the MME can reject the attach attempt. The MME interfaces the EIR with the S13 interface.

The final function in the architecture is the ANDSF entity. ANDSF is short for Access Network Discovery and Selection Function, and put simply, it is a function in the network that the operator can use to control how users and their devices prioritize between different access technologies if several non-3GPP access networks are available. It is a means to give the network operator the possibility to control how users attach to the network, based on a number of criteria. The ANDSF logical entity, which interfaces the user device over an interface called S14, will be further described in Section 6.3.

3.1.8   Summing up the architecture overview

The purpose of the description above was to make the overall EPS architecture a bit more comprehensive and understandable. When putting all the pieces back together, we arrive at the complete architecture diagram as shown in the beginning of this chapter (Figure 3.1.8.1).

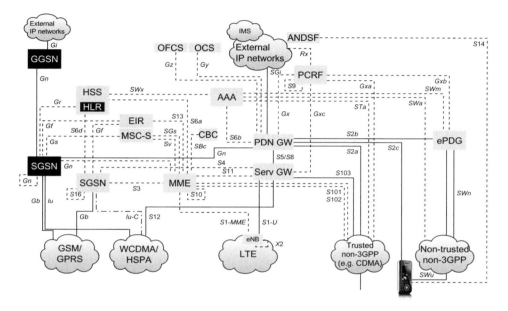

Figure 3.1.8.1   *EPC architecture overview.*

Large parts of the remainder of this book are devoted to describe each of the network elements (or nodes), each of the interfaces and each of the functions in greater detail.

3.2   Mobile devices

This is not a book intending to describe end-user devices in any detail, but in order to understand some of the features and functions included in the EPC, a basic understanding of the capabilities and limitations of the end-user devices is beneficial.

The terms 'Terminal', 'End-User Terminal', and 'User Equipment (UE)' are all used to denote the actual device communicating with the network. Depending on what type of device this is, there may or may not be a human user utilizing the device, and the device may or may not be moving between the cells in the network. Does this sound strange? It will be further elaborated upon below.

3.2.1   Different types of devices

Examples below show devices supporting WCDMA/HSPA and/or GSM/GPRS. The first generation of LTE devices are data modems possible to attach to standard portable computers, sometimes called USB dongles. Over time, devices supporting LTE can be expected to have similar form factors as all of the devices shown below.

Most people would probably associate a 'mobile device' with a mobile phone. This is by far the most common type of end-user device today, and they have been around in at least portable form factors since the 1980s. Today they count in billions and come with different forms and features.

From the start, mobile phones have been designed for voice services, and over time more service capabilities have been added, primarily SMS support as well as e-mail and web surfing.

This is specifically the case for advanced mobile phones, sometimes referred to as Smartphones, which in many cases also include more advanced means for text entry, for example, through a full QWERTY keyboard or through pressure-sensitive screens in combination with hand-writing recognition techniques.

Another type of device that is rapidly increasing in numbers is portable computers equipped with wireless communication support. This can either be standard portable computers to which is added an external modem, connected, for example, over USB, or portable computers which are shipped from factory with built-in support for mobile broadband services. It goes without saying that portable computers of course are very suitable devices for users of data services. It can however be assumed that users of portable computers are on average somewhat less mobile than users of mobile phones. Usage of laptop computers is more common in, for example, a café or in the summer house than in a car, affecting the mobility patterns to some degree.

Also emerging with capabilities and form factors in between Smartphones and portable computers are a group of devices called Mobile Internet Devices (MIDs), which are physically larger than a normal mobile phone but smaller than a laptop. MIDs come with advanced communication capabilities and relatively high processing power and are typically equipped with a large screen for touch-sensitive usage, sometimes in combination with a normal keyboard.

Yet another type of device is the so called 'Fixed Wireless Terminal'. This is a device which is not intended to be carried around, but instead installed permanently in, for example, a private house. A fixed wireless terminal connects to the surrounding world through the mobile network, at the same time as providing standard Ethernet LAN or Wireless LAN connections to the home network. This device can then be used to deliver similar services to the home as is possible with fixed broadband services over copper cables. Fixed Wireless devices and services may be used by mobile operators both to compete with fixed broadband operators as well as to provide broadband services in geographical areas without copper or fibre cabling. Naturally, Fixed Wireless Terminals normally do not move around in the network, reducing the overall signalling load related to handovers.

Yet another type of devices are terminals which are not for usage by or interaction with human users, but instead designed for machine-to-machine (M2M) communications. These devices may come in a wide range of very different form factors, and only the human imagination itself may limit the usage areas. Examples on usage cases for such devices are:

- Built into private cars for communicating service needs, the car's position (retrieved using GPS) as well as receiving up-to-date traffic data for traffic guidance systems
- Built into water or electricity meters for remote control and/or remote meter reading
- Built into street-side vending machines for communicating when goods are out-of-stock or when enough coins are present to justify a visit for emptying
- Built into taxi cars for validating credit cards
- Built into delivery cars for fleet management including optimization of delivery routes and confirming deliveries
- Built into ambulances for sending life-critical medicine data to the hospital prior to arriving in order to increase chances of successful treatments
- Built-into surveillance cameras for home or corporate security purposes.

As said, only the human imagination sets the limits for how wireless communications can be used to enhance business systems or to increase efficiency and security.

Finally, another type of device that is likely to include communication capabilities in the future is digital cameras including video cameras. Photos or films that have just been captured can easily be transferred to the network for backup and storage or for sharing with family and friends. This is not mobile phones with built-in cameras, this is instead cameras with built-in communication capabilities.

3.2.2   Terminals becoming general-purpose devices

A general trend is that more and more 'add-on' functionality is built into mobile phones (including Smartphones) turning these into general-purpose devices. This is functionality that goes beyond the actual communication services, and examples include:

- Media players (MP3 for music, MP4 and other formats for video)
- Camera functionality (photos and video recording)
- FM radio receiver
- GPS receiver for positioning using satellite systems
- Games

To some extent these features may remove the need to carry separate devices such as a digital camera and a music (MP3) player, and some of this is functionality that may be used to generate (photo or video camera function) or consume (music and media player) large amounts of data. Since the resolution and quality of integrated cameras and screens are steadily increasing, as is the memory capabilities of mobile phones, the amount of data managed in the device rapidly increases as well. Some of this data may be downloaded to or uploaded from the device.

There are multiple options for how to get this increasing amount of data into and out from the mobile device. Options include:

- Sending and receiving over the mobile network, for example receiving video streams being part of mobile TV broadcasting, or sending a newly taken photo using a mobile messaging service
- Sending and receiving the data over an alternative access such as WLAN which is supported in more or less 100% of portable computers and is becoming increasingly common in mobile phones
- Downloading and uploading using USB or Bluetooth when connecting to a PC
- Transferring through physically moving a memory card between the mobile phone and a card reader connected to a PC.

Mobile broadband solutions based on HSPA or LTE are optimized to support efficient data transfer to devices regardless of physical location (that is, as long as there is sufficiently good radio coverage).

3.2.3   Some challenges

There are a number of challenges to address for designers of mobile devices. Some of the most important ones are briefly discussed below.

The requirements on end-user terminals, especially mobile phones, are ever-increasing in many areas. More powerful processors are needed to execute more advanced services, support higher data speeds and to feed screens of higher resolution. The ability for users to download new applications themselves for executing for example in a Java Virtual Machine environment puts additional requirements on security, flexibility and processing power.

There are also requirements on decreasing the weight and thickness of the devices, at least the mobile phones. Users naturally tend to prefer light-weight and decent size phones. Before new advanced colour screens with high resolution became available, there was also a push on decreasing the overall physical size, but this of course makes less sense if you want a screen of decent size.

Most devices come with support for multiple radio technologies, supporting for example GSM/GPRS, WCDMA/HSPA, LTE, Bluetooth and WLAN in the same device. To support high speeds of LTE and HSPA, even multiple antennas are used. Some of these radio technologies can be used together, allowing for instance a handover of an IP session between LTE and WLAN. We will discuss the features enabling this in detail in Section 6.3. Some of the radio technologies used in one single device may actually be used simultaneously (for instance WLAN and HSPA or LTE) due to that the power levels and frequency bands are very different and hence no unacceptable interference is generated between the radios. A device that would support for example GSM and LTE simultaneously is however considered more complicated and more expensive to design, requiring more advanced filters to cancel out interference between radio technologies. Instead, solutions have been designed to support efficient handovers between radio technologies. We will cover these in detail in Section 6.3.

The complex situation with respect to different usage of the frequency bands in different countries put an additional burden on terminal vendors. To ensure that a terminal is usable in any country in the world, multiple frequency bands need to be supported. For example, a device may support GSM/GPRS/EDGE in the 850 MHz, 900 MHz, 1800 MHz and 1900 MHz frequency bands. Additionally it may support WCDMA/HSPA in the 900 MHz, 1900 MHz and 2100 MHz bands. This gets even more complex when adding also LTE support in the devices, especially since LTE is specified for a large number of different frequency bands. This wide range of possible LTE frequency bands further complicates the situation for terminal vendors. See the table below that comes from the 3GPP technical specification 36.101 [36.101]

LTE (E-UTRA) band	Uplink (UL) eNode B receive UE transmit	Downlink (DL) eNode B transmit UE receive	Duplex mode
1	1920–1980 MHz	2110–2170 MHz	FDD
2	1850–1910 MHz	1930–1990 MHz	FDD
3	1710–1785 MHz	1805–1880 MHz	FDD
4	1710–1755 MHz	2110–2155 MHz	FDD
5	824–849 MHz	869–894 MHz	FDD
6	830–840 MHz	875–885 MHz	FDD
7	2500–2570 MHz	2620–2690 MHz	FDD
8	880–915 MHz	925–960 MHz	FDD
9	1749.9–1784.9 MHz	1844.9–1879.9 MHz	FDD
10	1710–1770 MHz	2110–2170 MHz	FDD
11	1427.9–1452.9 MHz	1475.9–1500.9 MHz	FDD
12	698–716 MHz	728–746 MHz	FDD
13	777–787 MHz	746–756 MHz	FDD
14	788–798 MHz	758–768 MHz	FDD
...			
17	704–716 MHz	734–746 MHz	FDD
...			
33	1900–1920 MHz	1900–1920 MHz	TDD
34	2010–2025 MHz	2010–2025 MHz	TDD
35	1850–1910 MHz	1850–1910 MHz	TDD
36	1930–1990 MHz	1930–1990 MHz	TDD
37	1910–1930 MHz	1910–1930 MHz	TDD
38	2570–2620 MHz	2570–2620 MHz	TDD
39	1880–1920 MHz	1880–1920 MHz	TDD
40	2300–2400 MHz	2300–2400 MHz	TDD

The fact that there is only partial and no global alignment between different countries and regions on which spectrum to use for mobile communications, leads to a fragmentation in that different regions (and even operators) will need different frequency bands in the devices their customers are using, complicating international roaming scenarios. Organizations like the ITU (International Telecommunication Union) drive a harmonization of spectrum usage between countries and regions, but at the end of the day, it is the decision of the responsible authority of each individual country which frequency bands that are allocated for mobile communications, and how allocation of spectrum to different operators is carried out within these bands. In many countries this happens through spectrum auctions. Let us look at a simple example to illustrate this:

- In country 1 there are the network operators A and B, utilizing bands 4 and 13 respectively

- In country 2 there are the network operators C and D, utilizing bands 8 and 7 respectively.

Depending on how roaming agreements between these operators are set up, this will drive different combinations of frequency bands in the devices, meaning that it is either a multi-band device supporting all required combinations (4 + 13 + 8+7), or different variants of devices, for example supporting only bands 4 + 8 or 13 + 7. Then consider that global roaming requires involvement of operators in hundreds of countries, one can easily understand the potential complexity faced by terminal vendors to design devices for global usage.

On top of all these requirements comes of course the power consumption. User requirements on longer and longer standby and active usage times are steadily increasing, and one important differentiating factor between terminal vendors. To keep the power consumption as low as ever possible is most critical for mobile phones which almost always are powered using a battery. Portable computers are in comparison equipped with high-capacity batteries, and can often be connected to an AC outlet (at least when not moving around). Fixed Wireless terminals and M2M devices are normally always connected to AC power (or powered by a high-capacity car battery in case the M2M device sits in a car).

Battery technology is constantly evolving and increasing battery capacity. This has led to great improvements in terms of battery life as mobile phones have evolved over the years. There are also some features designed in mobile networks to help mobile devices to save power. The ability to enter 'idle mode' is one of these, relieving the device from more frequent signalling with the network on its whereabouts. How this is designed for EPS will be described in Section 6.3.

3.2.4 Concluding words on mobile devices

This has been a short overview of the terminals used with EPC. Even though the rest of this book primarily describes functionality in the network itself, it shall not be forgotten that it is of course a physical terminal that each user (human or not) is utilizing, and that the various requirements put on these devices (as discussed above) naturally impose some constraints and limitations. After all, it is a large part of the functionality of the network that is to be mirrored in a very small device that is also to serve as camera and media player, be light-weight and small, work in any country, and is to run for a long time without requiring charging. A true challenge.

3.3   Relationship of EPC to radio networks

Even though the subject of this book is Core Networks, the functionality of the supported radio access technologies is valuable to understand. A reader with no interest at all in radio technologies may however choose to skip this chapter.

LTE (also known as E-UTRAN) is the latest addition to the radio access technologies specified by 3GPP. LTE relies on EPC for core network functionality, and is naturally closely related to EPC through many inter-dependencies throughout the extensive standardization efforts. The concepts and functionality of LTE is described below. In addition, both GSM and WCDMA are briefly described.

But we will start at a very basic level and describe the basic concepts of mobile radio networks in general.

3.3.1   Overview of radio networks for mobile services

Mobile networks (or cellular networks) consist of a number of base stations, each serving wireless transmission and reception of digital information in one or several 'cells', where a cell refers to a specific portion of the overall geographical area the network serves. In most deployment cases one base station is serving three cells through careful antenna configurations and frequency planning (Figure 3.3.1.1).

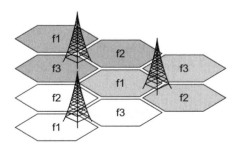

Figure 3.3.1.1   *Cells and base stations.*

The size and the outline of the cell are controlled by a number of factors including base station and terminal power levels, frequency bands (radio signals using lower frequencies propagate over longer distances than radio signals using higher frequencies if the same power level is used) and antenna configurations.

The radio wave propagation environment also has a significant effect on the cell size, there is a large difference depending on if there are lots of buildings, mountains, hills or forests in the area, compared to if the surrounding area is fairly flat and mostly uninhabited.

A fundamental ability of a cellular network is to allow the usage of the same frequency in multiple cells. See Figure 3.3.1.1 where f1 denotes one specific frequency. As can be seen, this frequency can be reused in multiple cells. This means that the total capacity of the network is greatly increased compared to the case if different frequencies would be needed for every site. The most intuitive way of allowing this frequency reuse is to make sure that base stations supporting cells using exactly the same subset of the available frequencies are geographically located sufficiently far apart, in order to avoid radio signals from interfering with each other. However, GSM, WCDMA and LTE have functionality that also allows adjacent cells to use the same frequency sets.

Base stations are located at sites which are carefully selected in order to optimize the overall capacity and coverage of the mobile services. This means that in areas where many users are present, for example in a city centre, the capacity needs are met through locating the base station sites more closely to each other and hence allowing more (but smaller) cells, while in the countryside where not so many users are present, the cells are normally made larger to cover a large area with as few base stations as possible.

Base stations are connected to other network nodes through transmission links, referred to as the RAN backhaul network. GSM and WCDMA radio networks include a centralized node (Base Station Controller (BSC) and Radio Network Controller (RNC) respectively) implementing some of the radio network functionality, while LTE radio networks rely solely on the base stations to provide the complete set of radio functions. While the exact functional division between base stations and BSC/RNC is out of the scope of this book. A brief overview of the most important functions of any digital cellular radio network is described in the following section.

3.3.2 Functionality of radio networks

Although the three radio technologies specified by 3GPP so far (GSM, WCDMA and LTE) are different in several ways, the fundamental functionality of radio networks is common.

The most important functions of a cellular radio network include:

- Transmission and reception of data over radio carriers. This goes without saying – wireless transmission is naturally a key feature of a radio network. The characteristics of radio transmission are dependant on many parameters such as distance from transmitter, used frequency, if any party is moving, transmission power that is used, height of antenna, and so on. Detailed fundamentals of electromagnetic wave propagation is however out of scope of this book.
- Modulation and demodulation of the radio carriers. This is a fundamental feature of both analogue and digital wireless transmission. For a digital system, this means that the flow of bits related to a specific service flow (e.g. a video stream) which may arrive at speed of, let us say 2 Mbit/s, are influencing a high-frequency radio carrier through different means. This means that one or more fundamental characteristics of this radio carrier is changed (modulated) at constant time intervals depending on the next set of bit or bits – 0's or 1's. This could be, for example, the phase, amplitude or the frequency of the carrier. Every change corresponds to a 'symbol' which may consist of one, two or more bits. Today's most advanced mobile systems (HSPA and LTE) allow for the usage of up to six bit symbols, which means that there are 64 different symbols used. As an example, '001010' may mean a specific phase and a specific amplitude of the carrier, while '111011' may mean the same phase but a different amplitude. See Figure 3.3.2.1 where the two dark points represents these two symbols, the arrow represents the carrier amplitude and the circles represent the phase of the carrier.
  Such more advanced symbol-modulation schemes put high requirements on the quality of the radio channel in order to avoid too frequent misinterpretations by

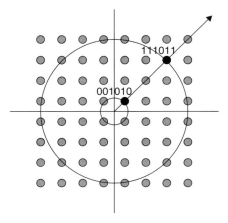

Figure 3.3.2.1   *Modulation.*

the receiver which is converting the radio carrier changes into a flow of bits on the receiver side, the process naturally known as demodulation.

- Scheduling of transmission of data from multiple users. This includes buffering of data from different users or applications while waiting for free radio capacity, and may include different priorities between queues of data, allowing different QoS to be applied to different flows of data. There are a number of different scheduling algorithms proposed for optimum sharing of the available transmission capacity based on the needs of services and users.

- Error correction schemes in order to ensure that the number of bit errors, inevitably occurring over any transmission link, are minimized. There are two main approaches to error correction – FEC (Forward Error Correction) and ARQ (Automatic Repeat reQuest). FEC means that additional data bits (called redundancy bits) are added to the user data through which one or multiple bit errors may be detected and/or corrected. ARQ means that errors detected in a received block of data, for example, through looking at a checksum, are triggering a request to the other party to resend the data. FEC and ARQ are normally combined to increase the performance of the radio channel in that some errors can be corrected through FEC while larger errors need a retransmission through ARQ. How to best balance FEC and ARQ mechanisms in order to maximize the performance is dependant on the requirements of the service and the characteristics of the actual radio channel, and normally also varies over time. Advanced radio communication systems here deploy adaptive coding, whereby the coding protection varies based on knowledge of the radio channel characteristics.

- Paging of idle terminals. This allows terminals to save battery power through entering what is called 'idle mode'. This means that the network no longer require the terminals to tell the network every time they move from one cell to another cell. Instead terminals can move within a larger geographical area (defined by the network operator) without contacting the network so frequently and thus saving battery power. This is of course only possible when no services are active. If services are triggered from the network (e.g., through an incoming voice call), the terminal is told to re-attach through the usage of broadcasting paging messages. If services are triggered from the end-user terminal (e.g., the user wants to do a voice call), the terminal is first triggered to leave idle mode including notifying the network of exactly which cell it will be using for the upcoming voice call.

- Mobility (handover) support to allow for continuous service coverage even if user devices have to change base station or cell due to physical movements in the area. This is of course a very common scenario and very valuable to the end-users. No need to stand still when using a mobile network service.

- Interference management in order to minimize the disturbance between multiple user devices or cells that share the same or neighbouring frequency bands.

- Encryption of user data and signalling to protect the integrity and contents of the user transmissions as well as to protect the network from hostile attacks.
- Power control to efficiently utilize the available power as well as to minimize interference between terminals.

3.3.3 GSM

GSM is the first generation of digital radio technologies specified by ETSI and later on inherited by 3GPP. Since the first generation of cellular networks was based on analogue transmission, GSM is normally referred to as a 2G (second generation) technology. The development of GSM was driven by a wish to specify a standard that could be supported globally, and was supported by many operator and telecom equipment vendors. The work started during the 1980s and the first GSM networks came on air in 1991. Since then, the success and global adoption of GSM has been tremendous. In April 2009, the number of GSM users around the world passed 2.3 billion, making GSM by far the most successful mobile technology to date.

Multiple users are sharing the capacity of one GSM radio carrier which occupy 200 kHz spectrum in any of the supported frequency bands. The most common GSM bands are 900 MHz and 1800 MHz, but also 1900 MHz and 850 MHz are supported and used in some countries.

GSM is a TDMA (Time-Division Multiple Access) technology, meaning that the radio channel is divided into radio 'frames'. A radio frame is, simply put, an exact number of bits being transmitted on the channel. Each user is then allocated one or more time slots of each radio frame. When GSM is used for telephony, one time slot is needed for every voice call. One GSM radio channel consists of eight time slots per radio frame, meaning that in this case, up to eight users can share one channel simultaneously. It is actually possible to squeeze in up to 16 users into one GSM channel through utilizing what is called half-rate coding, at the expense of fewer bits being available per user and hence potentially a somewhat degraded voice quality (Figure 3.3.3.1).

Figure 3.3.3.1   *GSM timeframes.*

GSM also contains support for packet data services. These services are referred to as GPRS (General Packet Radio Service), and were specified as an add-on to GSM during the mid-1990s. The first GPRS services came on air in the late 1990s. Since the number of bits possible to transmit during one time slot is quite limited due to the small bandwidth of the GSM radio carrier, GPRS allows for one user to temporarily use more than one time slot in order to support higher data speeds. With the addition of EDGE technology (Enhanced Data rates for GSM Evolution) which adds more advanced signal processing technologies to the GSM channel, the peak rates of packet data services over GSM may reach above 400 kbits/s under favourable radio conditions and given that all eight time slots are allocated to one single user.

3.3.4 WCDMA

WCDMA is a third generation (3G) radio technology, and the first version was specified by 3GPP during the late 1990s (it is normally referred to as Rel-99 WCDMA).

WCDMA is specified for 5 MHz wide channels, that is significantly more than GSM. Besides the more advanced signalling processing deployed with WCDMA, this also means that higher data rates can be supported. Rel-99 WCDMA could in theory support user downlink data rates of 2 Mbits/s, but in practice the limit in these networks is 384 kbits/s.

WCDMA is fundamentally different to GSM in that TDMA is not used as a means to separate the traffic from multiple users. Instead the concept of CDMA is deployed, meaning that a specific code is allocated to each terminal. This code is combined with the data to be transmitted, and used to modulate the radio carrier. All terminals are transmitting on the same 5 MHz channel and are separated due to the nature of the code instead of being allocated different frequencies or time slots. In order to allow communication with terminals further away from the base station site, the WCDMA concept includes advanced power control mechanisms which control the power levels of all terminals in the cell 1500 times a second.

Another feature of WCDMA is the ability to support soft handover and macro diversity, mechanisms allowing the terminal to communicate with more than one base station or cell simultaneously, enhancing the performance for the end-user, especially towards the outer parts of the cell.

As a later add-on to WCDMA, 3GPP specified HSPA technologies, allowing a more efficient sharing of the available radio capacity for packet services, and allowing much higher data rates for the end-users.

The WCDMA/HSPA technology is evolved further in 3GPP Rel-8 and beyond, allowing for even higher end-user data rates. Peak data rates exceeding 40 Mbits/s in the downlink direction (from network to mobile terminal) is possible given good radio conditions and moderate system load. This is made possible through introduction of more advanced modulation schemes and 'MIMO' techniques (which is further described in the LTE section below). In the uplink direction, HSPA Rel-8 can support up to 11.6 Mbits/s (MIMO (Multiple Input, Multiple Output) is not used). As an additional means of supporting higher data rates, the concept of Multi-Carrier WCDMA can also be used, meaning that more than one 5 MHz WCDMA carrier is used for downlink transmission and that data sent over these carriers is received and combined in the mobile device. This greatly improves the data rates for WCDMA/HSPA.

3.3.5 LTE

Hand-in-hand with the 3GPP System Architecture Evolution work, the work on the next generation RAN was carried out in the LTE study.

Just like the outcome of the SAE work was specifications of EPC, the outcome of the LTE work was specifications of E-UTRAN, short for Evolved UTRAN.

However, names and terms that have been used for some time tend to stick in peoples' minds. LTE is now the official term for the radio access technology used for E-UTRAN.

The work on LTE started off in late 2004 and early 2005 with defining a set of targets for the upcoming technical study and subsequent specification work. These targets can be found in the 3GPP technical report 25.913 [25.913]. A summary of the most important targets include:

- Downlink and uplink peak data rates of at least 100 and 50 Mbits/s respectively, assuming that a 20 MHz wide spectrum is used.
- The time it takes to change a user device from idle to active state shall not be more than 100 ms.
- The latency (delay) of user data shall not be more than 5 ms in the radio access network.
- Spectrum efficiency of 2–4x compared to a Rel-6 3G network (spectrum efficiency is measured as the cell throughput in bits/s/MHz).
- Interruption time during a handover from LTE to GSM or WCDMA of maximum 300 or 500 ms for non-real-time and real-time services respectively.
- Support for both FDD and TDD multiplexing schemes with the same radio access technology (FDD means transmission and reception on different

frequencies while TDD utilizes the same frequency but with transmission and reception separated in the time domain).

- Support for a wide range of channel bandwidths, ranging from 1.4 MHz to 20 MHz.

Implementations of LTE as defined in 3GPP Rel-8 meets and in many respects even exceeds these requirements. This is made possible through careful selection of technologies including utilization of advanced signalling processing mechanisms.

The LTE radio network is connected to the EPC through the S1 interface, a key interface in the EPS architecture (Figure 3.3.5.1).

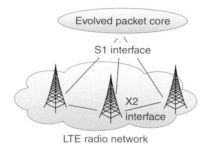

Figure 3.3.5.1 *LTE and simplified connection to EPC.*

LTE base stations are optionally also interconnected to each other through the X2 interface. This interface is used to optimize performance, for example, for handover between base stations or cells.

The S1 interface is divided into two parts:

1. S1-MME carries signalling messages between the base station and the MME. In addition it also carries signalling messages between the terminal and the MME which are relayed via the base station and piggybacked on radio interface signalling messages over the air interface.
2. S1-U carries user data between the base station and the Serving GW.

The S1 interface will be further described in Chapter 10 (Figure 3.3.5.2).

A key technology for LTE is the OFDM (Orthogonal Frequency Division Multiplexing) transmission scheme which is used for downlink transmission, that is, from the base stations to the end-user devices. This is key to meeting the spectrum flexibility requirements. The basic concept of OFDM is that the total available channel spectrum (e.g., 10 MHz) is sub-divided into a number

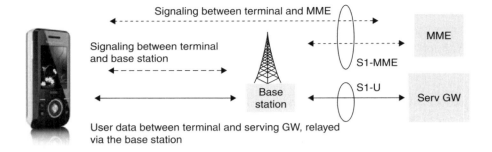

Figure 3.3.5.2   *S1-MME and S1-U interfaces.*

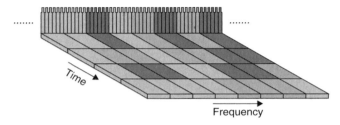

Figure 3.3.5.3   *Downlink channel-dependent scheduling in time and frequency domains.*

of 15 kHz channels, each carrying one sub-carrier. The available capacity (the usage of these sub-carriers) can be controlled in both the time and frequency domains at the same time (Figure 3.3.5.3).

OFDM also has the benefit of being very robust against *multipath fading*, that is, the variations in signal strength which are typical for mobile communications and are caused by the signal between transmitter and receiver is propagating over multiple paths at the same time. Reflections of the radio waves in various objects mean that multiple copies of the signal arrive at the receiving antenna, since these are not synchronized in time due to slightly different propagation distances (Figure 3.3.5.4).

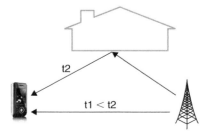

Figure 3.3.5.4   *Multipath propagation.*

In the uplink direction, that is, from the end-user device to the base station, a slightly different multiplexing scheme is deployed for LTE. As opposed to the LTE downlink, uplink transmission relies on only one single carrier. The key benefit of this is a lower peak-to-average ratio, that is, the amplitude of the power used for transmission is not varying as much as in the case of OFDM. This means a more efficient terminal power-amplifier operation in the end-user terminal, allowing for lower overall power consumption and hence a longer battery life. In order to efficiently multiplex many users with the need to transmit at the same time, LTE allows for allocation of only a subset of the available uplink channel to each user. Still, it is a single carrier per user, as opposed to the downlink transmission scheme.

High peak rates in general rely on radio channels with good characteristics, that is, with low noise and interference, as well as limited load on the cell that is being used. To take advantage of such channels, LTE allows for advanced modulation schemes, collectively referred to as HOM (Higher Order Modulation) schemes. 64QAM is one such scheme, allowing for six bits ($2^6 = 64$) to be transmitted for every symbol change on the radio carrier. Another technique is to simultaneously utilize multiple antennas in either the downlink or both the uplink and downlink directions. This technique is called MIMO, and further boosts capacity. With a combination of HOM and MIMO, really high peak rates are made possible with LTE – above 300 Mbits/s in the downlink direction. In the uplink direction, LTE in 3GPP Rel-8 does not include MIMO and the peak rates may reach 75 Mbits/s.

A detailed description of LTE is outside the scope of this book, but an excellent source of information of advanced 3GPP radio technologies including LTE can be found in [Dahlman 2008].

4

EPS deployment scenarios and operator cases

An attentive reader may question the need for the evolution of the core networks – it is, after all, possible to provide a relatively diverse set of applications and services using the existing 3G networks. Why, then, might an operator decide that they wish to evolve their network towards LTE and EPC? There are two main factors that may impact such a decision. First, the national regulatory authority may decide to make spectrum available for LTE – an existing operator may therefore act to protect their business interests and make a bid for that spectrum in order to ensure that their competitors or potential new entrants do not get access to that spectrum and offer broadband services.

The other factor that may drive the decision is market-driven; an existing or new operator may wish to provide their end-users with the high-bandwidth services that mobile broadband will offer in order to differentiate their service offering from other players in the industry.

Evolved Packet System (EPS) provides operators the ability to deliver broadband services via a 'common core network', which then combined with 'common IP Multimedia Subsystem' enables 3GPP developed common packet and IP Multimedia Subsystem (IMS) core networks not only for 3GPP accesses but towards any non-3GPP access technologies as well. As will be discussed in Part III, EPS has the ability to provide a migration path for various types of operators' deployment scenarios, including those who have deployed 3GPP-based systems as well as operators who have not (e.g. operators with High Rate Packet Data (HRPD)). In addition EPS provides opportunities for greenfield (=new) operators to start with a new mobile broadband network.

Naturally, there are several different existing radio and core network technologies already rolled out in operator networks. When upgrading to LTE and EPC, therefore, there are several different migration paths that an operator can follow depending on their existing deployed network. Specifically, there are five distinct deployment scenarios we have considered, as outlined below:

1. LTE and EPC for operators with existing 3GPP installations
2. LTE and EPC for Greenfield operators
3. LTE and EPC for operators with existing 3GPP2 installations
4. Operators providing interworking to non-3GPP accesses, for example WiMAX and WLAN
5. EPC-only (without deploying LTE) for operators who use existing packet core networks and 3GPP access technologies like WCDMA and High Speed Packet Access (HSPA).

4.1 Scenario 1: EPS with LTE deployment with existing 3GPP installations

This scenario covers the situation where an operator has an existing GSM or WCDMA radio network and core network installation. In many cases, the operator will also have commenced the deployment of mobile broadband through the use of HSPA technology. The services in this instance can therefore be pure IP services, or based on the IMS, or some other proprietary mechanism. The key focus for such an operator is therefore to enable end-users to have access to higher bandwidth but also to ensure that the existing services available to GSM or WCDMA end-users are still accessible to the LTE subscribers where applicable.

A common deployment scenario is when an operator has an existing HSPA mobile broadband installation up and running. Spectrum is made available for LTE and the operator is then faced with the choice of using spectrum for building an increasingly dense 3G network, or to start deploying LTE in parallel in order to provide support for high-bandwidth data services. In this case, LTE can be rolled out in order to provide high-speed data services in 'hotspot' areas, for example city centres, airports or any other places that might require an operator to provide extra high bandwidth. Running LTE in parallel in this fashion allows an end-user to always fall back to HSPA coverage, which may offer up to, say, 40 Mbit/s per user. In this manner, the end-user can enjoy relatively high-speed coverage through a wide-area HSPA network combined with, say, up to 100 Mbits/peak rate in LTE hotspots.

Another possible deployment is where the operators of existing 3GPP accesses have not decided on HSPA and instead decided to roll out LTE-based mobile broadband network. The typical operators in these cases have GSM/GPRS and/or WCDMA networks deployed currently and thus deploying EPS with LTE. The same reasoning can be the deterministic factor for such deployment scenario, to restate: LTE can be rolled out initially in order to provide high-speed data services in 'hotspot' areas, for example city centres, airports or any other places that might require an operator to provide extra high bandwidth. And then the operator has a choice of either continuing with LTE over a wider coverage, that is for the rest of the operator's network, or to rely on the existing deployed network (for example HSPA), deploy HSPA with EPC.

In terms of devices, most initial LTE devices are data centric in the form of modems for laptops (e.g. wireless data cards and dongles), but over time there will also be a gradual increase in the number of handheld terminals that support the LTE radio access.

In the initial phase of the roll out of LTE, it is natural for the operator to keep the LTE and the 2G/3G network infrastructures separate and introduce EPS in a step-wise manner. The main reason for this is to ensure that the existing revenue-generating infrastructure is not affected in the first phase of LTE deployment.

So, how would an operator decide on the mechanism to roll out LTE and EPC in parallel with their existing 2G or 3G network infrastructure? Initially the operator would need to assess where in their network they would want to improve the broadband capacity or where they wish to offer even high-peak bandwidth to their end-users. The operators would need adequate number of eNodeBs and Mobility Management Entities (MMEs) to handle the control plane signalling and mobility management for Intra and Inter Radio Access Technology (RAT) and also an adequate number of Serving Gateways and PDN Gateways. Since EPS is an all-IP network, an appropriate IP network infrastructure needs to be in place, for example Routers, DNS servers, FireWalls, etc. Additional decisions that would be needed is the protocol choice for S5/S8 reference point and whether dynamic policy control and charging would be deployed, since these decisions also increase the number of entities that need to be deployed. There are now several additional decisions that they need to take in order to deploy the network, but nothing out of the ordinary than what needs to done for any new infrastructure deployment.

Firstly, do operators need to separate the PDN gateway for the new subscribers or not? An operator may have the option of replacing existing GGSNs with PDN GWs because PDN GWs will support the GGSN functionality through the Gn–Gp interfaces from the SGSN. These interfaces use the GPRS Tunnelling Protocol (GTPv1) which allows interworking with legacy SGSNs. The PDN connectivity in the PDN GW in this case should be similar or equal to the PDN connectivity offered by the

GGSN in order to ensure that the resident services on the existing 2G/3G network are available to the LTE subscribers during mobility to and from 2G/3G access. Of course, operators may choose not to supply some of the existing 2G/3G services to the LTE subscribers; for example, dedicated PDNs for a particular company might not be prioritized in the first phase of the LTE roll out. Another option is to deploy new PDN GWs in parallel with existing GGSNs, allowing existing terminals to continue using GGSNs as anchor points, while LTE-capable terminals would rely on the PDN GWs. Selection between these anchor nodes would be handled by the SGSNs, as previously discussed in Chapter 3, and which will be further described in Section 9.2.

The Serving GW may or may not be integrated with the PDN GW. Even in the cases where they are integrated, however, there will be situations where a subscriber will use the Serving GW on one node and the PDN GW on another node. For example, when a user is roaming, they will use the Serving GW in the roaming network and the PDN GW of their home network in case of Home Routed Traffic (see 'Architecture Overview' Chapter 3 for more on the various roaming models supported in EPS). Another example where a Serving GW and PDN GW may be located in separate nodes is when a subscriber connects to multiple service networks (e.g. Corporate network and IMS) via separate APNs leading to 2 different PDN GWs.

GTP may be used on the interfaces between the PDN GW and the Serving GW; S5 is the interface used internally, while S8 is the interface used while roaming. GTP is selected in this example to illustrate further as it is already well established for roaming in 2G and 3G networks.

Secondly, an operator will need to decide whether to deploy stand-alone MMEs, or combined MMEs and SGSNs. Note that such deployment of combined node is more natural for operators with existing 2G/3G networks and there are some benefits to such deployment option.

Essentially, therefore, an operator is most likely to add LTE as a stand-alone network, get some LTE terminals into their network and then start offering LTE access to subscribers. Subscribers can then request the operator to upgrade their subscription, which an operator will do through updating the end-users subscription in the Home Subscriber Server (HSS) and/or with a new UICC card. This means that when an end-user accesses the network via GSM or WCDMA, the instruction can be given to the SGSN that it needs to select a PDN GW, rather than a GGSN. The simplest way of doing this is through using different Access Point Names (APNs) depending on if the terminal is LTE capable or not. There are also features which allow the SGSN to select a Serving GW/PDN GW instead of GGSN if the terminal is LTE capable or select a GGSN if the User Equipment (UE) is only 2G/3G capable, based on information from the UE. The use of DNS and other terminal capabilities for such selection is further described in Chapter 9. Use of DNS and other terminal capabilities for such selection can be found in Part III – Selection Functions. When an LTE subscriber starts in GSM coverage but roams into LTE coverage, they will need to have their PDN connection established via a PDN GW, rather than the legacy GGSN.

The operator now has a functional but fairly stand-alone LTE network with EPC in parallel with a 2G/3G packet core. Note that if the operators have not upgraded to a Rel-8 EPS network for 2G/3G networks, then Inter RAT handover will only work in the direction from LTE towards the 2G/3G network. When more and more of the subscribers are migrated by the operator to allow LTE access as well, this leads to a situation that is administratively inefficient for the operator to maintain two separate core network architectures (i.e., one with Gn/Gp based core network with SGSN/GGSN and its associated infrastructure and the other being the upgraded EPC architecture) for the different radio network accesses.

As long as the nodes are installed within the deployed network, there should be little problem in maintaining the parallel networks unless an operator needs to update them regularly; two parallel networks should function perfectly well. There may, therefore, be a number of GGSNs that will remain installed within an operator's network for some time. At some point, however, it may become more economically viable for the operator to migrate to a full EPS architecture and decommission the remaining core network infrastructure originally installed for the 2G/3G networks. However, it may be possible to upgrade the existing GGSNs to PDN GWs, in which case, the migration path will be a relatively easy process.

The S4-SGSN, first defined in Rel-8, has been specified to handle the situation where handover from and to LTE is required as well as provide functions developed for the EPC in order to connect via a Serving GW and a PDN GW. Essentially, the S4-SGSN provides a control plane interface between the S4-SGSN and the MME and a user plane interface between SGSN and Serving GW. The S4-SGSN also enables Idle Mode Signalling Reduction (ISR) function (as described in detail in Part III – Session Management and Mobility, Section 6.4), which reduces signalling overhead over the air between the UE and the network as a user moves between 2G/3G and LTE radio networks. The capability to support ISR and support of handover both to and from 2G/3G access are the essential differences between an S4-SGSN and a Gn-SGSN.

With the introduction of mobile broadband, it is expected that there will be an exponential increase in data traffic, this has a significant effect on the cost of a network in terms of scalability – in essence, an operator will need to ensure that their networks are also configured in the most cost-effective manner possible. Take the SGSN, for example, which is responsible for handling control plane signalling and maintaining session data whilst also keeping track of the UE; this implies that the SGSN should be scalable for processing, handling security associations and the states of millions of users. The User Plane nodes, meanwhile, for example Serving GW, need to be scaled to match the expected data volumes. The PDN GW is another example of a user plane node; this may also need to be scalable for Deep Packet Inspection (DPI) in addition to data volumes. The PDN GW also needs to interface the Policy and Charging Control (PCC) infrastructure in order to download filters for Quality of Service (QoS) control and therefore also needs some additional control plane capacity for this.

It is likely, therefore, that an operator will want to avoid having the SGSN heavily involved in the user plane as it is responsible for signalling and does not have any responsibility for the handling of subscriber traffic. An operator wanting to avoid the cost of upgrade to scale the SGSN in order to handle the vast amounts of data traffic that a mobile broadband network brings, such operator will naturally want to remove this 'hop' in their network as it will reduce their overall cost.

For operators with an existing WCDMA/HSPA installations, it is possible for them to bypass the SGSN for the user plane and take a subscriber's traffic directly from the Radio Network Controller (RNC) into the GGSN in case of non-roaming scenarios for existing packet core before EPC. With S4-SGSN it is possible to do a similar bypass, but in this case the subscriber traffic is taken directly to the Serving GW/PDN-GW and as Serving GW provides the mobility anchor for 3GPP subscribers, the bypass of user plane traffic from SGSN is supported for roaming and non-roaming scenarios via Serving GW in EPC (in case of GPRS, only non-roaming bypass is supported).

This allows for an operator's upgrades to the SGSN to be directly related to the upgrades required for the control plane. Through bypassing the SGSN for user plane traffic, the operator will no longer need to match the control plane and user plane capacity to one another. Through implementing them in separate nodes, it is possible to scale them independently.

In this deployment scenario, though upgrading of the HSS is definitely desirable for LTE subscribers, it is not mandatory due to the fact that both 3G UICC

(SIM card) and the existing HSS may be reused for LTE subscribers but with the use of Interworking Function (IWF) where mapping of functions as well as protocols (Diameter to Mobile Application Part (MAP) and vice versa) may be required. The use of IWF may have its limitations in addition to protocol mapping during critical time-consuming signalling like the Attach/Registration procedures, but it can be useful for operators in this scenario where the HSS upgrades may not be possible in a timely fashion. Different vendors may have different solutions for the support of both MAP and Diameter towards the User Management infrastructure.

4.2   Scenario 2: LTE and EPS for Greenfield operators

If we consider a pure Greenfield scenario where the operator starts LTE and EPC without having to consider any existing deployed networks and interworking, then issues arise which are purely related to roaming considerations and interworking aspects as well as the use of GRX (GPRS Roaming Exchange, this is the Roaming network deployed by operators who have roaming agreements for GPRS, more information available at: www.gsmworld.com) networks that an operator's environment requires them to use. Also the need for national roaming is likely due to that LTE coverage may be assumed to be spotty at the beginning of the deployment. The Greenfield operator may hence need to rely on partners within the same country for offering wide area service coverage for its subscribers.

An operator in this scenario will start with the LTE access network connected to an EPC. As the operator does not have any pre-existing requirements, they are in a position to go with all the 'bells and whistles' an EPS can provide. A 'fully fledged' IP-based network would need to include the properly configured DNS infrastructure for 3GPP networks and MMEs, Serving GWs, PDN GWs (combined or stand alone), HSS and other entities depending on the operator's choice such as Equipment Identity Registers (EIRs), Policy Control and Charging Rules Functions (PCRFs). Of course here we are only talking about the core EPS functions, while it is assumed that operators will also deploy the necessary entities that are required to support charging, security and legal intercepts as well as routers, firewalls and the IP backbone as demanded for an IP-based network. The users will have the UICC (SIM cards) and appropriate handsets enabling the higher level of security that has been designed for EPS. This is the simple part of the operator's deployment scenario, but it gets interesting when the operator starts to consider LTE roaming as well as supporting, for example, 3G/LTE-capable subscribers as inbound roamers to their network. In such scenarios, operators need to consider supporting the necessary GRX network requirements, which may require support of SS7-based MAP protocols to/from HSS (via IWFs) as well as necessary GTP variants running on the GRX networks. When considering Greenfield operators, it can be assumed that they will not deploy 2G/3G radio access networks and thus they would not need to support any SGSN or GGSN functionality except what is required for roaming. But their own subscribers shall be able to roam into operators' networks where LTE/2G/3G are supported or only 2G/3G are supported. Assuming the roaming support requires that it is mobile broadband only, since LTE is not capable of supporting circuit-switched services, the roaming considerations are only for Packet Core data traffic, the Greenfield operator must be able to handle queries to and from these networks over roaming interfaces that may require support of SS7-based protocols as well as pre-Rel-8 protocols. There should not be any other significant requirements on the Greenfield operators, and while it may not be as simple as it may sound to enable this type of roaming, it is very much feasible and one of the key interworking supported in 3GPP systems as can be seen from the analysis of Scenario 1.

EPS also provides a Greenfield operator the possibility to consolidate any other access networks via a common Packet Core and Policy architecture.

4.3   Scenario 3: LTE and EPS deployment for 3GPP2 operators

These operators would fall into the category of 2G/3G operators who have well-established CDMA networks with a substantial customer base and investments. Thus, depending on the strategy for the long-term commitment towards LTE and the strategy for existing CDMA networks, the migration timeline and deployment intensity may be quite wide-ranging and different across the 3GPP2 operator community.

Even though in a sense this kind of operator can be seen as Greenfield operator for LTE and EPC (e.g. 3GPP developed system), their requirements are not the same as the 'pure' Greenfield operators. One of the key aspect to consider is the strategy and plans for their existing CDMA networks and how 'tightly coupled' the LTE deployment will be with the evolution of the HRPD networks. In addition, the operator's existing customer base and roaming partners need to be supported and a well thought out 'full migration' to LTE may well be the most convenient but not likely scenario at the beginning of LTE/EPC deployment. When it comes to LTE/EPC deployment, the 3GPP2 operators have exactly the same requirements as a Greenfield operators, that is the need for deployment of E-UTRAN, HSS, MME, Serving GW and PDN GW, as well as the security and charging entities and DNS and IP infrastructure. But in addition to these basic considerations, an incumbent

3GPP2 operator must also consider the interworking support for their existing HRPD networks as well as securing dual-mode terminals (CDMA and LTE) as well as deployment of UICC. Depending on the type of interworking, there are different requirements on the network. In the case of 'tight interworking' (also known as optimized handover support), the 3GPP2 operator needs to upgrade their HRPD access networks as well as deploy HSGW (replacing or upgrading existing Packet Data Serving Node (PDSN) components of their network). Also upgrades of the Authentication, Authorization and Accounting (AAA) infrastructure and support functions in its MME and Serving GW are needed including S101 and S103 interfaces that allow for optimized handover between these two networks. In the case of 'loose interworking' (also known as non-optimized handover), the HRPD Access network needs to be able to support broadcast of certain parameters as well as configuration of LTE and HRPD cell information in order to facilitate the handover efficiently. In this case there are no specific functions needed in the MME, but the PDSN (e.g. the Access GW for 3GPP2 HRPD system) needs to be upgraded to an HSGW including EPS-related functions. The AAA infrastructure used in 3GPP2 systems also has to be upgraded to EPS AAA infrastructure and the UICC would be required for the handset.

Depending on what kind of inbound and outbound roaming agreements will be supported, the operator may choose to deploy additional functions (like in the case of Greenfield operators) to enable roaming with 2G/3G networks.

It is expected that the operator's 3GPP2- and 3GPP-based networks (i.e. LTE/EPC) would be running in parallel for an extended period of time and this time-line would vary from one operator to another operator depending on the operator strategies including whether they plan to use LTE as a hotspot mobile broadband access in strategic locations like city centres and major urban centres or if they plan a ubiquitous LTE network in the near term.

In the long run, it can be expected to be more efficient to run a common network providing full functionality for the operator instead of running two networks with different technologies in parallel for a long time. In addition, as EPS is being developed to provide a common packet core network for the future and for any access technology able to connect using one of the mechanism EPS provides, EPS also then opens up opportunities for further consolidating additional access networks via a common packet core.

Terminal availability supporting multiple access technologies for one common core network would also be more desirable than multiple ones. Thus a consolidation of the networks also allow for simplification of the handsets and to focus more on attractive services and applications towards end-users. Availability of terminals supporting LTE as well as CDMA 1xRTT and HRPD technologies may also contribute to a prolonged maintenance period of the CDMA networks even after considered undesirable by a specific operator's long-term goal.

But for this scenario, the operator also has to consider the migration path for their existing customer base and that may vary from one operator to another operator depending on the strategy of what to do with the circuit-switched service offerings and corresponding infrastructure. Some operators may choose LTE/EPC for a much wider deployment and also utilize the IMS and Multimedia Telephony (MMTel) for voice multimedia services, while an operator intending to keep the existing circuit switched services may rather use the 1xRTT infrastructure and use the CS Fallback mechanism to continue to provide the voice services this way. See Chapter 5 for more information on the realization of voice services for different operator scenarios. One may consider reading Part II – Services in EPS, in order to better understand the opportunities in a glance.

Note that depending on the protocol choice possible for the roaming interfaces, the complexity of the deployment may also increase for an operator.

4.4   Scenario 4: WiMAX and WLAN operators

Even though it might be questioned whether operators operating exclusively in these access technologies would see any motivation to migrate towards EPC (without LTE), the benefits can be seen from a common core network as well as handover/session continuity perspective with existing and well-established access technologies with huge number of subscribers (over 3 billion and counting…). The EPS could also open up the possibility to get into roaming relationship that exists, for example, via GRX with international community. One may consider the scenario that an LTE operator establishes partnership with, for example, a WLAN access provider, connected via the EPC networks and providing indoor coverage in certain local network environment. Since handover/session continuity with IP and session preservation is supported, users can easily move in and out of such coverage while maintaining their sessions via EPC. Depending on the scenario, such non-3GPP access network providers may desire to establish a full EPC network using AAA infrastructure, Access Gateways and PDN Gateways (and optionally Serving GWs in a very specific situations). Alternatively, an operator may choose to deploy AAA infrastructure and Access Gateways to connect via an existing operator's EPS networks, where special business relationship may be established with such access providers. This type of symbiotic relationship may benefit both operators business, depending on the market environment they are operating on (e.g. wide usage of WLAN in a certain city centre).

An additional key question is of course the availability of dual/multi mode handsets, providing support for both LTE (and/or other 3GPP accesses) and the non-3GPP access that the operator is using. In the case of WLAN this combination can be expected to be very common in terminals.

4.5 Scenario 5: Consideration for EPC-only deployment with existing 2G/3G accesses

Existing 2G/3G operators may choose to deploy/upgrade to EPC without the necessity of deploying LTE. Some aspects of this scenario has been already touched upon in Scenario 1, and here we highlight a few of these aspects. Some of the main benefits have already been mentioned such as support for handover to/from LTE, support for handover with non-3GPP access networks, the ability to provide local breakout more efficiently, and the built-in support of optimized user plane traffic (also known as Direct Tunnel for 3G packet core) for both roaming and non-roaming scenarios. Other benefits include allowing IP-only networks thus avoid maintaining SS7 networks for the packet core, being prepared for support/inclusion of other non-3GPP access network connection (such as explained in Scenario 4), efficient network operations and maintenance and enhanced QoS and Policy Control and Charging support. Since the terminals that support existing 2G/3G procedures will be supported without any problems, operators can continue to serve their existing subscribers. In addition, the selection of the GW (GGSN vs. Serving GW/PDN GW) based on the terminal's network capability (which indicates if the terminal can support LTE or not) can be used to divert the subscriber towards a Serving GW/PDN GW (when LTE capable) or towards a GGSN (when not LTE capable). As it becomes increasingly costly for vendors as well as operators to continue to maintain multiple tracks of architecture which leads to multiple tracks of products, it would serve the overall community to converge over time towards a minimal set of product variants.

There are obviously other possible scenarios that we have not discussed here. The purpose of this section has been to explore some possible key scenarios. During 3G deployment, some interesting aspects came about such as incumbent national operators not getting the license for 3G operations. Such obstacles also led to the creation of new solutions in 3GPP, for example the Network sharing feature allowing an operator with a 3G license to share their radio network with an operator without a 3G license and thus able to provide services using 3G to their subscribers. LTE and EPC have been developed when keeping such scenarios in mind, meaning that radio network sharing as well as sharing of MMEs by multiple operators are supported in the standard specifications.

5
Services in EPS

While GSM was revolutionizing communications and taking mobile telephony to the masses, the World Wide Web was having a similar effect on the Internet. The tremendous popularity of the Internet and its value to end-users in providing relevant applications and information is driving a demand for mobile Internet. While a detailed discussion of the types of services that Evolved Packet Core (EPC) will be used for is beyond the scope of this book, this chapter attempts to provide a brief overview of the services that the mobile broadband revolution will enable for end-users. In particular, we cover the realization of voice services in an all-IP environment; for while the Internet services will provide a large number of innovative services, support for voice services and applications will remain a necessary and important part of an operator's network.

The functionality of the EPC enables the mobile network operator to offer a new set of services to the users through its flat architecture and enables products and network deployments to be built for bandwidth-intensive services from the very start. In addition, EPC provides a number of features to the operator in order to support provisioning, monitoring, control and charging of these services.

The following sections provides a brief overview of the services that EPC enables: data services, voice services and messaging services.

5.1 Data services

The LTE and EPC are designed for IP services; this means that, in theory, almost any application relying on IP communication can utilize the IP access service offered by EPC. IP networks higher layer functionality is implemented in client and server applications residing on a terminal and a network server respectively. The role of the radio network and packet core network is to provide IP communication between the two end-points, the two IP hosts.

The IP-based applications being made accessible for mobile users through the EPC IP access service may either be:

- provided by the mobile operator
- residing in a corporate IP network or
- accessible over the Internet.

In practice, these cases can and will of course co-exist, for example some applications may be provided by the operator while others applications will reside on the Internet. An end-user will in theory be able to access any application as long as it is based on IP.

The service offering to users normally is a combination of services provided by the operator, for example access to specific web content, downloading of music or similar, and Internet access. The operator normally refers to these offerings as services. In order to make the functionality of EPC clear, we instead differentiate between applications and services.

An application refers to functions realized through software on top of the IP host stack in the mobile device. Regardless of what functionality this application offers to the user, the service offered by the network is IP access, enabling routing of IP packets between the mobile device and external IP networks. The functionality of the application (e.g. a web session) is transparent to the mobile network (Figure 5.1.1).

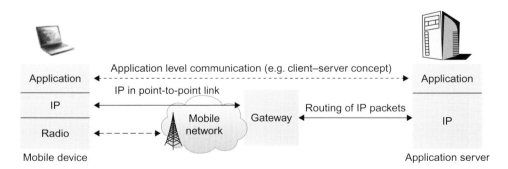

Figure 5.1.1 *Applications and services on mobile broadband.*

The IP access service includes authenticating the user (or to be formally correct – the SIM card), authorizing the usage of the IP access service, allocating an IP address to the device, and enabling IP packets to be sent to and from this device.

Compare to attaching a computer to an Ethernet network, the gateway node in the EPC network becomes the first hop IP router for the IP host residing on the terminal. There is a difference in the wireless case in that the terminal is connected to the gateway over a point-to-point link. This means that all communication between mobile users will always pass the gateway.One advantage of the EPC solution is that this stable IP anchor point survives mobility within and between different radio networks, meaning that the same IP address is maintained regardless of which access network that is used.

5.1.1   A note on application development

As mentioned previously, mobile broadband is about more than providing just IP connectivity for end-users. With the ability of mobile devices to connect directly to the Internet, the nature of application development becomes much more dynamic; in essence the mobile networks are opened up to the same potential for innovative application development as has been experienced on the Internet; a plethora of different applications will become available, created by companies, developer communities and individuals [Mulligan, 2009]. A detailed discussion of application development is beyond the scope of this book – a brief overview of application development trends is provided below in order to frame the role of EPC in delivering the promise of mobile broadband applications and services that end-users will pay for.

Perhaps the most fundamental paradigm shift that will occur in the mobile industry due to the advent of mobile broadband and an all-IP core is therefore the location of application development. For nearly 30 years, applications have been developed specifically for mobile phone networks, for example voice mail or voice conferencing services. These applications were developed either in-house or by application developers skilled in telecommunications protocols and network architecture. The operator networks were closed entities. Thanks to EPC and advances in processing capacities of the terminals, this limitation on who will be creating the applications will be lifted; now any developer who has an understanding of IP technology will be able to create an application. As software development becomes simpler and new tools are developed, even end-users will be able to create their own applications. The tremendous success of products such as the Apple iPhone and the emerging open source initiatives is the key indicator to the potential for growth in these markets. As the industry progresses more and more such services and devices will be released.

The need to provide appropriate tools for the different emerging technologies has not gone unnoticed within the mobile industry and several initiatives have been developed in this area; some of these are from established players in the mobile telecommunications arena, while others are more Internet based. Those readers who wish to have more information about cutting-edge application development environments for mobile broadband and its services are referred to Appendix C.

It should also be noted that voice and data services will, over the years, no longer come to be viewed as separate services. They will be combined together in many different ways, from multimedia communications that link in, or mash-up, content from the Internet to more complex applications such as combining voice services with the deep web or semantic web technologies that are emerging today on the Internet. The limit to the type of applications that will emerge is only the imagination. Mobile broadband allows the tremendous innovative potential of the Internet for application development to be harnessed in a mobile terminal. The delivery of voice services is covered in the next section; it should be remembered that it is fully possible to combine the voice services described below with other applications and services running on the all-IP infrastructure that EPC provides.

5.2 Voice services

Voice services have been the primary source of revenue for mobile operators since the dawn of basic mobile service offerings back in the 1950s. The emergence of GSM technology in the early 1990s was the starting point for the unprecedented global adoption of mobile communication services. By mid-2009, the number of GSM users in the world was close to 3.5 billion and was still growing by many millions each month. Hundreds of new models of handsets and other devices supporting GSM are released each year.

Given the importance voice services have to billions of users, it is not unexpected that while the majority of efforts behind designing the EPS architecture and procedures primarily targeted an efficient IP access service, the importance of efficient voice support was acknowledged right from the start of the work item.

This section introduces and describes the different services as a basis for understanding the detailed descriptions and signalling flows found in later chapters, particularly Sections 12.6 and 12.7.

There are two fundamentally different ways that voice services can be realized for LTE users; using circuit-switched or IP Multimedia Subsystem (IMS) technologies. We will describe the two different technologies and the differences between them in the following sections.

Readers familiar with these concepts can skip to Section 5.2.3, where we explain how these two technologies are linked to the EPS architecture and how they can be used to deliver voice services.

5.2.1   Voice services based on circuit-switched technology

Circuit switching is the traditional technology used in a telephony network. In this technology, a continuous link is established between two end-users in a phone call.

A central part of the circuit-switched network architecture is the Mobile services Switching Centre (MSC). This is the core network function supporting voice calls, handling both the signalling related to the calls and switching the actual voice calls. Modern deployments of circuit-switched core networks are normally designed with a separation of the signalling functions (handled by the MSC-Server) from functions handling the media plane (handled by the Media Gateway). Figure 5.2.1.1 shows a simplified architecture.

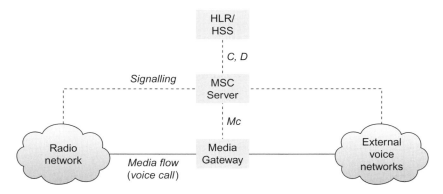

Figure 5.2.1.1   *Circuit-switched architecture.*

Here, the MSC-Server includes call control and mobility control functions, while the media, that is the actual data frames making up the voice calls, flows through a Media Gateway that can convert between different media and transport formats, as well as invoke specific functions into the voice calls, such as echo cancellation or conferencing functions. The MSC-Server controls the actions taken by the Media Gateway on a specific call and interacts with the Home Location Register/Home Subscriber Server (HLR/HSS) which handles subscription data for users of circuit-switched services.

While voice calls in mobile networks have been converted into streams of digital data since the early 1990s, the data frames themselves are not sent between mobile devices and the networks using shared channels or IP technology.

This means that unique resources in the network need to be dedicated to each voice call throughout the duration of the call. The connection is established at call setup, and is maintained until call termination when the network resources are released. Circuit-switched connections therefore consume network resources with a fixed bandwidth and a fixed delay for the duration of the call. This is also valid if no actual communication takes place, that is neither part has anything to say. As long as the call is ongoing, the allocated network resources are not available to other users. There is no obvious way to optimize these resources across multiple users.

It should be noted that this, however, is somewhat of a simplification. In order to improve the resource usage for circuit-switched services, some mechanisms have been designed to enable a somewhat more efficient usage of the available bandwidth, for example through taking advantage of silent periods in the voice calls and enabling multiplexing of several users onto a common channel. Also, in a wireless system, the available bandwidth varies to some extent, due to that the characteristics of the radio channel as such changes during the call. This may result in variations of the voice quality as the voice coder adapts to the changing radio environment.

Since the voice data for circuit-switched services are not transported using IP packets between the devices and the network, there is also no way to multiplex several services onto the same service stream, nor to provide a standard Application Programming Interface (API) towards other services or applications in the device.

The packet data services in GSM, WCDMA and LTE, however, offer IP connectivity between the mobile device and a gateway node. This IP connectivity can be used for any IP-based application and may be utilized by multiple applications simultaneously. One of these applications is naturally voice. Furthermore, the call as such can be more than a voice call and consist of several media components in addition to the voice media itself. In the EPC, carrier-grade multimedia services are provided with a technology called IMS, which is covered in the next section.

5.2.2   Voice services with IMS technology

IMS (IP Multimedia Subsystem) was originally designed by 3GPP in order to ena-ble IP-based multimedia services over GSM and WCDMA systems, but was later on expanded to support also other access networks. The IMS concept is based around the Session Initiation Protocol (SIP), defined in IETF RFC3261. This pro-tocol was designed by the IETF as a signalling protocol for establishing and man-aging media sessions, for example voice and multimedia calls, over IP networks. As mentioned in Chapter 2, the 3GPP re-uses appropriate protocols from other standards bodies within their specifications and SIP is one such protocol. It should also be noted that after the merger of the work between TISPAN, 3GPP2 and 3GPP into the 'Common IMS', all IMS specifications are handled by 3GPP. Both fixed and mobile networks use the same IMS specifications. It is also important to note that the voice services described within this chapter are able to be used with non-3GPP accesses, as well as the ones specified by 3GPP. As stated several times, one of the key design points for the EPC was ensuring that both 3GPP and non-3GPP accesses would be able to connect to and utilize such services.

IMS is defined as a subsystem within the mobile network architecture. It con-sists of a number of logical entities interconnected via standardized interfaces. Note that these are logical entities, and that vendors of IMS infrastructure equip-ment may combine some of these entities on the same physical product or prod-ucts (Figure 5.2.2.1).

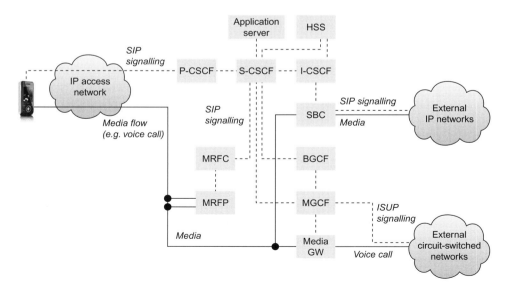

Figure 5.2.2.1   *IMS architecture.*

At the core of the IMS subsystem is the Call Session Control Function (CSCF). This is the node handling the SIP signalling, invoking applications and controlling the media path. The CSCF is logically separated in three different entities:

- The Proxy-CSCF (P-CSCF)
- The Serving-CSCF (S-CSCF)
- The Interrogating-CSCF (I-CSCF).

These three entities may very well reside as different software features on the same physical product.

The primary role of the P-CSCF is as a SIP proxy function. It is in the signalling path between the terminal and the S-CSCF, and can inspect every SIP message that is flowing between the two end points. The P-CSCF manages quality-of-service and authorizes the usage of specific bearer services in relation to IMS-based services. The P-CSCF also maintains a security association with the terminal as well as may optionally support SIP message compression/decompression.

The S-CSCF is the central node of the IMS architecture. It manages the SIP sessions and interacts with the HSS (Home Subscriber Server) for subscriber data management. The S-CSCF also interacts with the application servers (AS).

The primary role of the I-CSCF is to be the contact point for SIP requests from external networks. It interacts with the HSS to assign an S-CSCF that handles the SIP sessions.

The HSS manages IMS-related subscriber data and contains the master database with all subscriber profiles. It includes functionality to support access and service authorization, mobility management and user authentication. It also assists the I-CSCF in finding the appropriate S-CSCF.

The Media Resource Function Processor (MRFP) is a media plane node that can (but does not need to) be invoked to process media streams. Examples of use cases where the media data is routed via the MRFP are conference calls (where mixing of multiple media streams are needed) and transcoding between different IP media formats.

The Media Resource Function Controller (MRFC) interacts with the CSCF and controls the actions taken by the MRFP.

The Breakout Gateway Control Function (BGCF) handles routing decisions for outgoing calls towards circuit-switched networks. It normally routes the sessions to a Media Gateway Control Function (MGCF).

The Media Gateway (MGW) provides interworking including conversion and transcoding between different media formats used for IMS/IP and circuit-switched networks.

The MGCF provides the logic for IMS interworking with external circuit-switched networks. It controls the media sessions through ISUP signalling towards the external network, SIP signalling towards the S-CSCF, and through controlling the actions of the MGW.

The Session Border Controller (SBC) is an IP gateway between the IMS domain and an external IP network. It manages IMS sessions and provides support for controlling security and quality of the session. It also supports functions for firewall and NAT traversal, that is when the remote IMS terminal resides behind a device (for instance, a home or corporate router) which provides IP address conversion.

The AS (Application Server) implements a specific service and interacts with the CSCF in order to deliver it to end-users. Services may be defined within 3GPP, but thanks to the use of IP technology, it is not necessary for all services to be standardized. One example of a service that is defined by 3GPP is Multimedia Telephony (MMTel). MMTel has been designed to support voice calls using IMS, but MMTel can provide more than telephony in that other media can be added to the voice call, turning this into a complete multimedia session.

It is beyond the scope of this book to provide a more detailed description of IMS. For a more in-depth description, see, for example [Camarillo, 2008].

5.2.3 *Realization of voice over LTE*

The LTE radio access has been designed to be optimized for IP-based services. This means that LTE has no support for dedicated channels optimized for voice calls. It is a packet-only access with no connection to the circuit-switched mobile core network. This is different to GSM, WCDMA and CDMA, which support both circuit- and packet-switched services and it naturally impacts the technical solution for how to deliver voice to LTE users.

Depending on the network operator build-out plans and the frequency bands used for LTE, the radio coverage can be assumed to be non-continuous or even spotty, at least in the initial stages of LTE deployment. Voice as a service, however, relies on a continuous service coverage. In a mobile network, the support for continuous service coverage is realized through handovers between radio cells and between base stations.

For EPC, two basic approaches have been guiding the work in defining voice service support. Simply put, either voice services for LTE users are produced using the circuit-switched infrastructure that is used for voice calls in GSM, WCDMA and CDMA, or alternatively, IMS technology and the MMTel application are used.

5.2.4 Voice services using IMS technology

MMTel is the standardized IMS-based service offering for voice calls. It offers more possibilities than a traditional circuit-switched voice call, in that additional media components, for example video or text, may be added to the voice component, thus enhancing the communication experience and value for the end-user. Since EPS is designed to efficiently carry IP flows between two IP hosts, MMTel is a natural choice for offering voice services when in LTE coverage.

In order to realize full-service coverage as well as service continuity, however, one cannot rely on the fact that LTE coverage is present everywhere the user may want to make a voice call. Full voice service coverage therefore relies on the following facts:

- other access networks are complementing the LTE access network in terms of coverage
- the device used to make the voice call (a traditional mobile phone or another device) also supports these access technologies and the technology used for voice calls in that technology (like circuit-switched procedures in, e.g., GSM)
- inter-system handovers are possible.

Figure 5.2.4.1 shows the dark small areas that illustrate LTE coverage while the large lighter area that illustrates a technology with much better radio coverage.

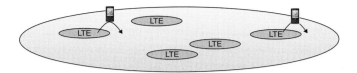

Figure 5.2.4.1   *Voice services and the need for mobility support.*

There are three different use cases that need to be considered:

1. A voice call is established when in LTE coverage (dark area), and the user is not moving outside LTE coverage during the duration of the call. For this use case, MMTel would be used to provide the voice service over LTE.
2. A voice call is established when outside LTE coverage (light area). The call would then instead be established using circuit-switched access over, for example, GSM. Depending on the solution, the call could be converted into a SIP-based call and handled by the IMS system, or it can be handled as a traditional circuit-switched call by the MSC.
3. A voice call is established when in LTE coverage (dark area) and during the voice call, the user moves outside LTE coverage. If the system depicted as a light area can support IMS/MMTel voice services, this would be handled through a 'Packet Handover' between LTE and the other system (e.g. WCDMA/HSPA or eHRPD) and the voice service would continuously be served as an IP-based service and handled by the IMS infrastructure. If this is not the case, specific measures are needed to secure service continuity when LTE coverage is lost. The 3GPP solution for this is called Single-Radio Voice Call Continuity (SRVCC).

5.2.5   Single-radio voice call continuity

SRVCC is designed to allow for the handover of a voice call between a system that supports the IMS/MMTel voice service and a second system where there is insufficient radio access support for carrying the MMTel service. This could be, for instance, due to insufficient bandwidth for IP services, or insufficient QoS support in the network.

SRVCC hence defines a solution for how an IP-based voice call in 'system A' (dark grey area) can be handed over to 'system B' (light grey area) which serves the voice call using circuit-switched procedures.

So why is this called a 'Single Radio' procedure? Additional complexity of this handover procedure comes from the fact that a 'normal' mass market terminal (the end-user device) cannot be connected to both system A and system B at the same time. It instead has to execute a very quick handover in order not to cause a serious service degradation such as an annoyingly long interruption during the voice call. This is because the end-user device would require more complex and expensive radio filters, antennas and signal processing if simultaneous connections to two systems would need to be maintained. This is where the 'Single Radio' comes in. It extends the 3GPP Rel-7 VCC solution which allows for handovers between IMS-based service over WLAN and circuit-switched services over, for example, GSM. In Rel-7 VCC, the assumption is, however, that dual-radio is used, that is the terminal is simultaneously connected to both WLAN and GSM at the same time. This is possible due to the difference between a system with local coverage and low transmitting power (WLAN) and a system with wide coverage and relatively high transmitting power (GSM).

3GPP has specified the following combinations for SRVCC (system A to system B):

- LTE to GSM
- LTE to WCDMA
- WCDMA (HSPA) to GSM
- LTE to 1xRTT.

The solution is based on that IMS is kept as the system serving the user for the complete duration of the call (it is the 'service engine' for the voice call), also when the user is served by system B. SRVCC includes interaction between the MME of the EPC core network and the MSC-Server of the circuit-switched core network, as well as an IMS VCC Domain Transfer Function (DTF).

The details of SRVCC are described in Section 12.6.

It should be noted that in some countries, some features are needed from a regulatory perspective in order to offer a telephony service. These are scheduled for completion in Rel-9 of the 3GPP specifications. This is specifically the support of location services as well as the support for prioritization of IMS-based emergency calls.

5.2.6 Circuit-switched fallback

Circuit-switched fallback (CSFB) is an alternative solution to using IMS and SRVCC to provide voice services to users of LTE. The fundamental differences are that IMS is not part of the solution, and in fact, voice calls are never served over LTE at all. Instead, CSFB relies on a temporary inter-system that switches between LTE and a system where circuit-switched voice calls can be served.

The solution is based on the fact that LTE terminals 'register' in the circuit-switched domain when powered and attaching to LTE. This is handled through an interaction between the MME and the MSC-Server in the circuit-switched network domain.

There are then two use cases to consider – voice calls initiated by the mobile user or voice calls received by the mobile user:

1. If the user is to make a voice call, the terminal switches from LTE (system A) to a system with circuit-switched voice support (system B). Any packet-based services that happened to be active on the end-user device at this time are either handed over and continue to run in system B but on lower data speeds or suspended until the voice call is terminated and the terminal switches back to LTE again and the packet services are resumed. Which of these cases that apply will depend on the capabilities of system B.
2. If there is an incoming voice call to an end-user that is currently attached to LTE, the MSC-Server will request a paging in LTE for the specific user. This is done via the interface between the MSC-Server and the MME. The terminal receives the page, and temporarily switches from LTE to system B where the voice call is received. Once the voice call is terminated, the terminal switches back to LTE.

The details of CSFB are described in Section 12.7.

5.2.7 Comparing SRVCC and CSFB

The two approaches on how to offer voice services to LTE users are fundamentally different.

The main strengths of IMS/MMTel and SRVCC include:

- Allowing for simultaneous usage of high-speed packet services over LTE and voice calls
- MMTel offers an enhanced experience to the end-user, enabling the addition of extra media components within the voice call itself.

While the main strengths of CSFB include:

- No need to rely on a deployment of IMS infrastructure and services before offering voice as a service to LTE users

- Same feature and service set offered for voice services when in LTE access as when in a system supporting circuit-switched voice calls. The circuit-switched core network infrastructure can be utilized also for LTE users.

As discussed above, both approaches rely on the end-user device used for the voice call being capable of supporting access to not only LTE but also systems with presumed wider radio coverage (e.g. GSM) as well as the capabilities to execute circuit-switched voice calls.

It should also be noted that both solutions can be simultaneously supported in the same network, and it can be assumed that operators initially deploying CSFB may over time migrate towards the MMTel + SRVCC solution.

5.3 Messaging services

The ability to send messages to users of mobile devices has grown immensely popular since the introduction of Short Messaging Service (SMS) in GSM. The introduction of more advanced messaging services such as Multimedia Messaging Service (MMS) has offered the ability to also include photos, graphics and sound in addition to text in the messages. Instant messaging and chat-like services have also been introduced as a means to further enhance the messaging experience for the users.

Just as for voice, there are two fundamentally different ways of realizing messaging support with EPC – either using an IP-based solution (like IMS-based messaging or SMS-over-IP) or using the circuit-switched infrastructure that is normally used to deliver SMS messages over GSM and WCDMA. The fact that LTE is a packet-only radio access calls for some specific mechanisms to be included in the latter case.

For the case where messages are sent based on IP, there are no specific features needed in the EPC. Messages are sent transparently through the network from a messaging server to the client, and are treated just like any IP packet by the EPC. How the messaging application as such is realized is independent of EPC (as long as IP as used as the transport technology) and beyond the scope of this book. Any sort of media (text, video, sound, graphics, etc.) could be included in messages sent using IP.

For the case of using the circuit-switched infrastructure for delivering messages, the MME interacts with the MSC Server. The MSC Server is normally connected to a messaging centre for delivery of SMS messages over control channels in, for example, GSM and WCDMA, and through the interaction with MME, this solution can be used also for LTE. Messages are then included in NAS signalling messages between MME and the mobile device. This solution supports only SMS text messaging, meaning that other types of messages (e.g. MMS) need to be based on IP, just as for GSM and WCDMA.

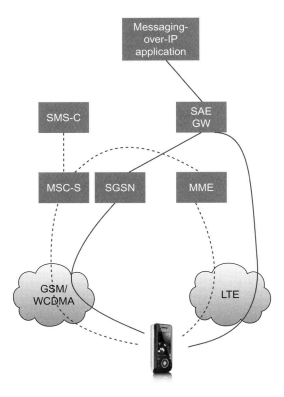

Figure 5.3.1   *Options for messaging services.*

The two variants of messaging transmission are illustrated in Figure 5.3.1 where the dotted lines denote SMS transmission using signalling interfaces and the solid lines denote Messaging-over-IP transmission.

Part III
Key Concepts

6
Session management and mobility

6.1 IP connectivity and session management

LTE and the Enhanced UTRAN introduces significant improvements in the amount of bandwidth provided for wireless transmission using cellular technology, truly paving the way for mobile broadband. In order to provide support for the type of applications and traffic that these access networks would enable, it was necessary to prepare the core network. In conjunction with the evolution of the access networks provided with LTE, the convergence of technologies at the IMS level meant that it was now also possible to provide a common packet core with appropriate policy, security, charging and mobility. This provides end-users with ubiquitous access to network services across different access networks and also provides them with session continuity across the access technologies. This section covers the IP connectivity and the management of the sessions within EPC. A high-level view of the steps involved is shown in Figure 6.1.1.

6.1.1 The IP connection

6.1.1.1 General

The most fundamental task of the EPC is to provide IP connectivity to the terminal for both data and voice services. This is maybe an even more significant task than in GSM/WCDMA systems where the circuit-switched domain is also available. With EPS, only the IP-based packet-switched domain is available. Even though there are different possibilities to interwork towards the circuit-switched domain as described in Chapter 5, only the IP-based packet-switched domain is natively supported when using E-UTRAN.

The IP connectivity will have certain properties and characteristics depending on the scenario and the type of services that the user wants to access. First, the IP connectivity will be provided towards a certain IP network. This would in many cases be the Internet but it could also be a specific IP network where the

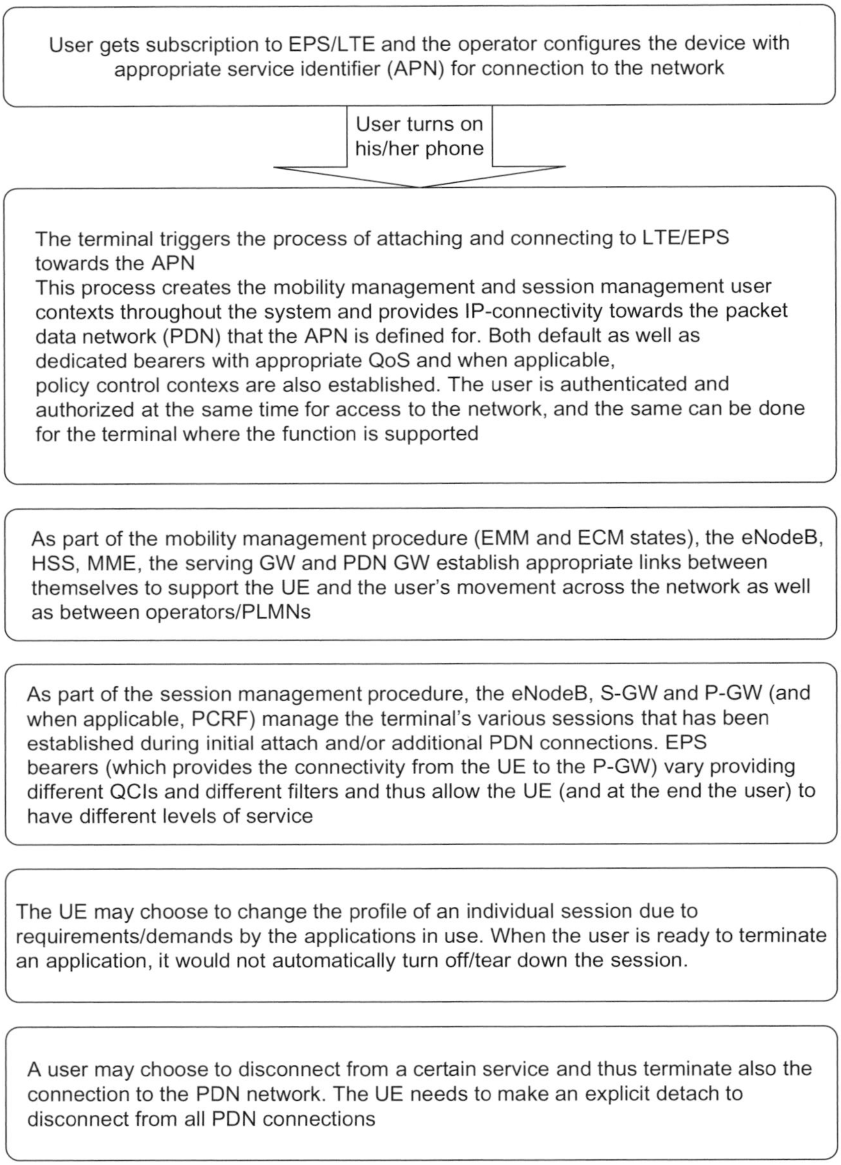

User gets subscription to EPS/LTE and the operator configures the device with appropriate service identifier (APN) for connection to the network

User turns on his/her phone

The terminal triggers the process of attaching and connecting to LTE/EPS towards the APN
This process creates the mobility management and session management user contexts throughout the system and provides IP-connectivity towards the packet data network (PDN) that the APN is defined for. Both default as well as dedicated bearers with appropriate QoS and when applicable, policy control contexs are also established. The user is authenticated and authorized at the same time for access to the network, and the same can be done for the terminal where the function is supported

As part of the mobility management procedure (EMM and ECM states), the eNodeB, HSS, MME, the serving GW and PDN GW establish appropriate links between themselves to support the UE and the user's movement across the network as well as between operators/PLMNs

As part of the session management procedure, the eNodeB, S-GW and P-GW (and when applicable, PCRF) manage the terminal's various sessions that has been established during initial attach and/or additional PDN connections. EPS bearers (which provides the connectivity from the UE to the P-GW) vary providing different QCIs and different filters and thus allow the UE (and at the end the user) to have different levels of service

The UE may choose to change the profile of an individual session due to requirements/demands by the applications in use. When the user is ready to terminate an application, it would not automatically turn off/tear down the session.

A user may choose to disconnect from a certain service and thus terminate also the connection to the PDN network. The UE needs to make an explicit detach to disconnect from all PDN connections

Figure 6.1.1   *IP Connectivity flow diagram.*

telecom operator provides certain services, for example based on IMS. Then, the IP connectivity would be provided using one or both of the available IP versions, that is IPv4 and/or IPv6. Additionally, the IP connectivity should fulfil certain QoS requirements depending on the service being accessed. For example, the connection may need to provide a certain guaranteed bit rate or allow a prioritized treatment over other connections. The following sections discuss how the above concepts are solved in EPS, both for the 3GPP family of accesses and for other accesses connected to EPC.

6.1.1.2   The PDN connectivity service

As mentioned above, the EPS provides the user with connectivity to an IP network. In EPS, as well as in 2G/3G Packet Core, this IP network is called a 'Packet Data Network' (PDN). In EPS, the IP connection to this PDN is called a 'PDN connection'. Why do we use fancy names such as 'Packet Data Network' and 'PDN connection'? Why not just simply say 'IP network' and 'IP connection'? After all, isn't this what we actually mean? There are two reasons for this terminology, the first technical and the second historical. First of all, the PDN connection comprises more than just the basic IP access; QoS, charging and mobility aspects are all parts of a PDN connection. Also, when GPRS was originally specified, there was a desire to support different packet data protocol (PDP) types besides IP, such as Point-to-Point Protocol (PPP) (with support for different network-layer protocols) and X.25. Since GPRS could provide access to other PDN types than just IP networks, it made sense at that time to refer to them as PDNs rather than just IP networks. In reality, however, IP-based PDNs became the most commonly used variant in the vast majority of deployments on the market. It has therefore been decided that EPS only supports access to IP-based PDNs using IPv4 and/or IPv6. The name 'Packet Data Network', however, remains.

The operator may provide access to different PDNs with different services. One PDN could, for example, be the public Internet. If the user establishes a PDN connection to this 'Internet PDN', the user can browse websites on the Internet or access other services available on the Internet. Another PDN could be a specific IP network set up by the telecom operator to provide operator-specific services, for example based on IMS. In summary, if the user establishes a PDN connection to a specific PDN, he or she would only get access to the services provided on that PDN. It is of course possible to provide multiple services in a single PDN. The operator chooses how to configure its networks and services.

A terminal may access a single PDN at a time or it may have multiple PDN connections open simultaneously, for example, to access the Internet and the IMS services simultaneously if those services happen to be deployed on different PDNs. In the latter case, the terminal would have multiple IP addresses, one (or two if both IPv4 and IPv6 are used) for each PDN connection. Each PDN connection represents a unique IP connection, with its own IP address (or pair of IPv4 address and IPv6 prefix) (Figure 6.1.1.2.1).

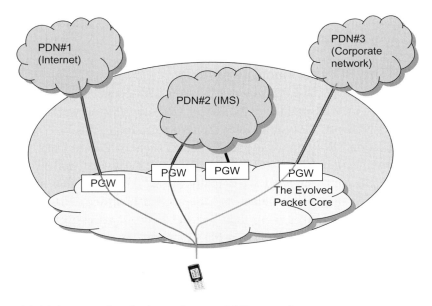

Figure 6.1.1.2.1   *UE with multiple simultaneous PDN connections.*

One PDN connection is always established when the terminal attaches to the EPS (refer to Chapter 12 for a high-level description of the attach procedure for the different accesses). During the attach procedure, the terminal may provide information about the PDN that the user wants to access. The information is carried in a parameter called 'Access Point Name' (APN). The APN is a character string that contains a reference to the PDN where the desired services are available. The network uses the APN when selecting the PDN for which to set up the PDN connection; the operator defines what APNs (and corresponding PDNs) are available to a user as part of their subscription. In case the terminal does not provide any APN during the attach procedure, the network will use a default APN defined as part of the user's subscription profile in the HSS in order to establish the PDN connection. It can be noted that the APN is used to select not only PDN but also the PDN GW providing access to that PDN. These selection functions are described in more detail in Chapter 9.

Additional PDN connections may be established when the terminal is attached to EPS. In this case the terminal sends a request to the network to open a new PDN connection. The request must always contain an APN to inform the network about what PDN that the user wants to access.

The terminal may at any time close a PDN connection.

6.1.1.3 Relation between EPC, application and transport layers

The PDN connection provides the user with an IP connection to a PDN. When a PDN connection is established, context data representing the connection is created in the UE, the PDN GW and, depending on access technology used, also in other core networks nodes in between, for example the MME and Serving GW for E-UTRAN access. The EPS is concerned with this 'PDN connection layer' and the associated functions such as IP address management, QoS, mobility, charging, security, policy control, etc.

The PDN connection is a logical connection between a specific IPv4 address and/or IPv6 prefix allocated to a UE and a particular PDN. The user data belonging to the PDN connection is transported between the terminal and the base station over the underlying radio connection. The user data is also carried by an underlying transport network between the network entities in the EPC. At the same time, the application (or service) that the user may be running is transported on top of the PDN connection. Here, and in the following few sections, we use the term 'application' in a generic manner, including protocol layers on top of IP.

The transport network in the EPC provides an IP transport that can be deployed using different technologies such as MPLS, Ethernet, wireless point-to-point links, etc. Over the radio interface, the PDN connection is transported on top of a radio connection between the UE and the base station. Figure 6.1.1.3.1 provides an illustration of the relation between application layer, PDN connection 'layer' and transport layer. The IP transport layer entities in the backbone network, such as IP routers and layer 2 switches, are not aware of the PDN connections as such. In fact, these entities are typically not aware of per-user aspects at all. Instead they operate on traffic aggregates and if any traffic differentiation is needed, it is typically based on Differentiated Services (DiffServ) and techniques operating on traffic aggregates.

Figure 6.1.1.3.1 provides a schematic illustration of the application layer, PDN connection 'layer' and transport layer for EPS when the UE is connected over a GERAN, UTRAN or E-UTRAN access. It can be noted that the user IP connection (the PDN connection) is separate from the IP connection between the EPC nodes (the transport layer). This is a common feature in mobile networks where the user plane is tunnelled over a transport network in order to provide per-user security, mobility, charging, QoS, etc. This distinction between PDN connection and transport layer is also important when we come into aspects related to IP address allocation, QoS, etc., below.

Figure 6.1.1.3.1 also illustrates the different layers in the case where the UE is connected to EPC over other accesses. In general, an IP-based transport layer

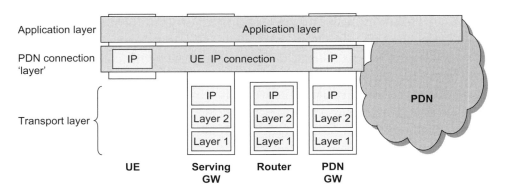

Figure 6.1.1.3.1   *Schematic view of application, PDN connection and transport layers for 3GPP family of accesses.*

below the PDN connection layer is present for all cases. It should, however, be noted that the details regarding the protocol layers below the PDN connection layer depends on what mobility protocol is used and what other protocols are used when accessing over a non-3GPP access.

6.1.1.4   IP addresses

A key task of the EPS is to provide IP connectivity between a UE and a PDN. The IP address allocated to the UE comes from the PDN where the UE is accessing. It should be noted that this IP address, and the IP address domain of the PDN, is different from the IP network (or backbone) that provides the IP transport between nodes within the EPC. The backbone providing the IP transport in the EPC can be a purely private IP network used solely for the transport of PDN connections and other IP-based signalling in the EPC, either for a single operator in non-roaming cases or between operators in roaming scenarios. The PDN is, however, an IP network where a user gets access and is provided services, for example the Internet. This section is only concerned with the IP addresses allocated to the UE.

Each PDN may provide services using IPv4 and/or IPv6. A PDN connection must thus provide connectivity using the appropriate IP version. Currently, the majority of the IP networks where end-users get access, for example using GPRS or fixed broadband accesses, is based on IPv4. That is, the user will be assigned an IPv4 address and access IPv4-based services. Also, most services available on the Internet are IPv4 based. It is likely, however, that IPv6 deployments will become more common in the EPS/E-UTRAN time frame and it is thus important that EPS provides efficient support for both IPv4 and IPv6, for example, to allow easy migration and co-existence.

Usage of IPv6 instead of IPv4 is primarily motivated by the vast number of IPv6 addresses available for allocation to devices and terminals. Some operators are already experiencing or will soon experience shortages of IPv4 addresses to varying degrees. The allocation of IPv4 addresses across the world and between organizations differs greatly. IPv6 does not have this problem since the addresses are 128 bits long, in theory providing 2^{128} addresses. In comparison to the 32 bits used for IPv4, IPv6 therefore provides significantly more addresses.

However, since the IP infrastructure and applications on both private networks and the Internet are still mostly based on IPv4, the introduction of IPv6 is a great challenge in terms of migration and smooth introduction. This is because IPv4 and IPv6 are not interoperable protocols; IPv6 implements a new packet header format, designed to reduce the amount of processing an IP header requires. Thanks to this fundamental difference in headers, workarounds are needed to get them functioning on the same network. Multiple mechanisms exist allowing, for example, devices using IPv6 to communicate with applications based on IPv4 (as they are on the Internet) as well as transporting IPv6 packets over IPv4 infrastructure. These solutions all have their pros and cons but details about these are beyond the scope of this book. Interested readers are referred to the many excellent books on IPv6 available; some examples are [Li, 2006] and [Blanchet, 2006] but many others exist.

In the 2G/3G core network, each PDN connection (i.e. PDP context) supports one IP address only. In order for a terminal to request both an IPv4 and an IPv6 address/prefix, it has to activate two PDN connections (two 'primary' PDP contexts), one for each IP version. With EPS, this has changed. A terminal activating a PDN connection in EPS may request an IPv4 address, an IPv6 prefix or both for that PDN connection. That is, EPS supports three types of PDN connections; IPv4 only, IPv6 only, as well as dual stack IPv4/IPv6. It can be noted that in 3GPP Rel-9, support for dual stack PDP context is also introduced in the 2G/3G GPRS core network.

EPS supports different ways to allocate an IP address. The IP address may be assigned using different protocols depending on what access is used. The detailed procedure for allocating an IP address also depends on deployment aspects as well on the IP version (v4 or v6). This is explained in more detail in the following sections.

IP address allocation in 3GPP accesses

The method to allocate IPv4 addresses and IPv6 prefixes are different. Below we will describe how IPv4 addresses and IPv6 prefixes are allocated in EPS. There are two main options for allocating in IPv4 address to the UE in 3GPP access:

1. One alternative is to assign the IPv4 address to the UE during the attach procedure (E-UTRAN) or PDP context activation procedure (GERAN/UTRAN).

In this case the IPv4 address is sent to the UE as part of the attach accept message (E-UTRAN) or Activate PDP Context Accept message (GERAN/UTRAN). This is a 3GPP-specific method of assigning an IP address and this is the way it works in most 2G/3G networks. The terminal will also receive other parameters needed for the IP stack to function correctly (e.g. DNS address) during the attach (E-UTRAN) or during PDP context activation (GERAN/UTRAN). These parameters are transferred in the so-called Protocol Configurations Options (PCO) field.

2. The other alternative is to use DHCPv4 (often referred to as just DHCP). In this case the UE does not receive an IPv4 address during attach or PDP context activation. Instead the UE uses DHCPv4 to request an IP address after attach (E-UTRAN) and PDP context activation (for GERAN/UTRAN) is completed. This method to allocate IP addresses is similar to how it works, for example, in Ethernet and WLAN networks where terminals use DHCP after the basic layer 2 connectivity has been set up. When DHCP is used, the additional parameters (e.g. DNS address) are also sent to the UE as part of the DHCP procedure.

Whether alternative 1 or 2 is used in a network depends on what is requested by the UE as well as what is supported and allowed by the network. It should be noted that both these alternatives are supported already in 2G/3G core network standards even though alternative 1 is used in the vast majority of existing 2G/3G networks. One difference is, however, that in 2G/3G core networks, the selected IPv4 allocation method (alt 1 or 2) is configured per APN. This means that only one method is supported for each APN. In EPS, however, this has been made more flexible to allow both methods to co-exist for the same APN. It is still possible to deploy an EPS network so that only one method per APN is used, if desired.

We now proceed to the IP address allocation procedure for IPv6. The only method currently supported in EPS is stateless IPv6 address auto configuration. For more details on basic IPv6 features, please refer to [Hagen, 2006]. When IPv6 is used, a /64 IPv6 prefix is allocated for each PDN connection and UE. The UE can utilize the full prefix and can construct the IPv6 address by adding an Interface Identifier to the IPv6 prefix. Since the full /64 prefix is allocated to the UE and the prefix is not shared with any other node, the UE does not need to perform Duplicate Address Detection (DAD) to verify that no one else is using the same IPv6 address.

With stateless IPv6 address auto configuration, attach and (for GERAN/UTRAN) PDP context activation are completed first. The GW then sends an IPv6 Router Advertisement (RA) to the UE after attach and PDP context activation are completed. The RA contains the IPv6 prefix that is allocated to this PDN connection.

The RA is sent over the already established PDN connection and is therefore sent only to a specific terminal. This is different compared to some non-3GPP access networks where many terminals share the same layer 2 link (e.g. Ethernet). In these networks the RA is sent as a broadcast to all connected terminals. After completing the IPv6 stateless address auto configuration, the terminal can use stateless DHCPv6 to request other necessary parameters, for example, DNS address. The option to allocate IPv6 prefix(es) using stateful DHCPv6 is currently not supported by EPS.

IP address allocation in other accesses

The way by which IPv4 addresses and/or IPv6 prefixes are assigned in other accesses differs depending on what access is used and what mobility protocol (PMIPv6, MIPv4 or DSMIPv6) is used.

When the terminal attaches from a trusted non-3GPP access and PMIPv6 is used on the S2a interface, the address allocation is quite similar to how it works in 3GPP accesses. An access may, for example, have access specific means to deliver IPv4 address to the UE, or DHCPv4 may be used. For IPv6, stateless IPv6 address auto configuration is typically supported. The IP layers are illustrated in Figure 6.1.1.4.1.

When the terminal attaches in an untrusted access and PMIPv6 is used on S2b interface, the terminal receives the IP address from the PDN during the IKEv2-based authentication with the ePDG. It can be noted that before IKEv2 is performed and the IPSec tunnel is setup, an additional IP address is involved. The reason is that the terminal needs local IP connectivity from the untrusted non-3GPP access in order to communicate with the ePDG. This local IP address does,

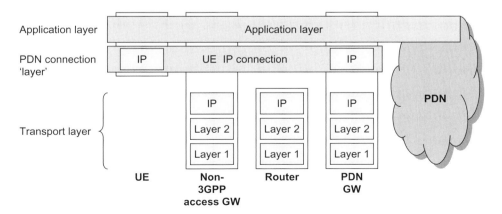

Figure 6.1.1.4.1 *Schematic view of application, PDN connection and transport layers for trusted non-3GPP accesses when PMIP is used on S2a.*

however, not come from a PDN but is only used to setup the IPSec tunnel. This is illustrated in Figure 6.1.1.4.2. For more details, see [Ch 12.1].

From an IP address allocation point of view, the situation is rather similar when DSMIPv6 is used. The terminal receives its IP address (also referred to as Home Address in Mobile IP terminology) during the DSMIPv6 bootstrapping with the PDN GW. However, the terminal first needs to acquire a local IP address to be used as Care-of Address. Therefore the UE has two IP addresses, one for the local connection (Care-of Address) and one for the PDN connection (Home Address). This is illustrated in Figure 6.1.1.4.3.

The most involved case, which is not shown in the figure, is when DSMIPv6 is used over untrusted non-3GPP accesses. In this case the terminal uses three IP

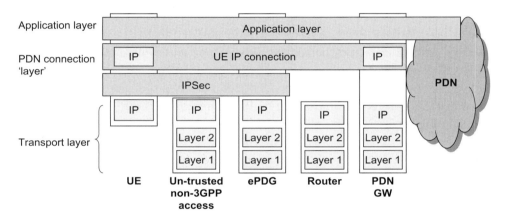

Figure 6.1.1.4.2 *Schematic view of application, PDN connection and transport layers for untrusted non-3GPP accesses when PMIP is used on S2b.*

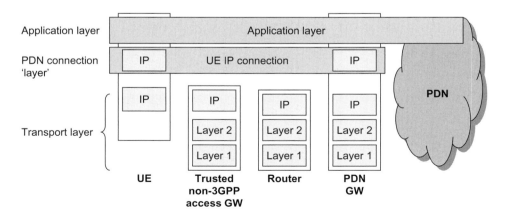

Figure 6.1.1.4.3 *Schematic view of application, PDN connection and transport layers for trusted non-3GPP accesses when DSMIPv6 (S2c) is used.*

addresses, a local IP address to establish the IPSec tunnel towards an ePDG, the IP address received from ePDG which is used as a Care-of Address and then finally the IP address for the PDN connection received during DSMIPv6 bootstrapping with the PDN GW.

For more details on DSMIPv6, see [Chapter 11.3] and [Chapter 12.1].

6.2 Session management, bearers and QoS aspects

6.2.1 General

Providing PDN connectivity is not just about getting an IP address; it is also about transporting the IP packets between the UE and the PDN in such a way that the user is provided a good experience of the service being accessed. Depending on whether the service is a voice call using Voice over IP, a video streaming service, a file download, a chat application, etc., the QoS requirements for the IP packet transport are different. The services have different requirements on the bit rates, delay, jitter, etc. Furthermore, since radio and transport network resources are limited and many users may share the same available bandwidth, efficient mechanisms must be available to partition the available (radio) resources between the applications and the users. The EPS needs to ensure that all these different service requirements are supported and that the different services receive the appropriate QoS treatment in order to enable a positive user experience.

This section describes the basic functions in EPS to manage the user plane path between the UE and the PDN GW. One key task of the session management features is to provide a transmission path of a well-defined QoS. The basic principles around session management are covered in this chapter, but the more detailed QoS aspects can be found in the subsequent chapters on QoS.

In the subsections below we introduce the 'EPS bearer' which is a central concept in E-UTRAN and EPS both for providing the IP connection as such and for enabling QoS. We also look at several aspects related to how GERAN and UTRAN accesses are connected to EPS, however, without going into the same level of detail regarding session management for GERAN/UTRAN. Finally we look at similar aspects for other accesses connected to EPS.

6.2.2 The EPS bearer for E-UTRAN access

For E-UTRAN access in EPS, one basic tool to handle QoS is the 'EPS bearer'. In fact, the PDN connectivity service described above is always provided by one or more EPS bearers (also denoted just 'bearer' for simplicity). The EPS bearer provides a logical transport channel between the UE and the PDN for transporting IP traffic. Each EPS bearer is associated with a set of QoS parameters that describe the properties of the transport channel, for example bit rates, delay and bit error rate, scheduling policy in the radio base station, etc. All conformant traffic sent over the same EPS bearer will receive the same QoS treatment. In order to provide different QoS treatment to two IP packet flows, they need to be sent over different EPS bearers. All EPS bearers belonging to one PDN connection share the same UE IP address. The QoS aspects and its relation to the EPS bearers will be discussed in more detail in Section 8.1.

6.2.2.1   Default and dedicated bearers

A PDN connection has at least one EPS bearer but it may also have multiple EPS bearers in order to provide QoS differentiation to the transported IP traffic. The first EPS bearer that is activated when a PDN connection is established is called the 'default bearer'. This bearer remains established during the lifetime of the PDN connection. Even though it is possible to have an enhanced QoS for this bearer, in most cases the default bearer will be associated with a default type of QoS and will be used for IP traffic that does not require any specific QoS treatment. Additional EPS bearers that may be activated for a PDN connection are called 'dedicated bearers'. This type of bearer may be activated on demand; for example, in case an application is started that requires a specific guaranteed bit rate or a prioritized scheduling. Since dedicated bearers are only set up when they are needed, they may also be deactivated when the need for them no longer exist, for example, in case the application that needs the specific QoS treatment is no longer running.

6.2.2.2   User plane aspects

The UE and the PDN GW (for GTP-based S5/S8) or Serving GW (for PMIP-based S5/S8) use packet filters to map IP traffic onto the different bearers. Each EPS bearer is associated with a so-called Traffic Flow Template (TFT) that includes the packet filters for the bearer. These TFTs may contain packet filters for uplink traffic (UL TFT) and/or downlink traffic (DL TFT). The TFTs are typically created when a new EPS bearer is established, and they are then modified during the lifetime of the EPS bearer. For example, when a user starts a new service, the filters corresponding to that service can be added to the TFT of the EPS bearer that will carry the user plane for the service session. The filter content may come either from the UE or from the PCRF (see Section 8.2 on PCC for more details).

The TFTs contain packet filter information that allows the UE and PGW to identify the packets belonging to a certain IP packet flow aggregate. This packet filter information is typically a 5-tuple defining the source and destination IP addresses, source and destination port as well as protocol identifier (e.g. UDP or TCP). It is also possible to define other types of packet filters based on other parameters related to an IP flow. The filter information may contain the following attributes:

- Remote IP Address and Subnet Mask
- Protocol Number (IPv4)/Next Header (IPv6)
- Local Port Range
- Remote Port Range
- IPSec Security Parameter Index (SPI)
- Type of Service (TOS) (IPv4)/Traffic class (IPv6)
- Flow Label (IPv6).

The word 'remote' refers to the entity on the external PDN with which the UE is communicating while 'local' refers to the UE itself. The UE IP address is not contained in the TFT since it is understood that the UE is only assigned a single IP address, or possibly a single IP address of each IP version, per PDN connection.

Some of the above-listed attributes may co-exist in a packet filter while others mutually exclude each other. Table 6.2.2.2.1 lists the different packet filter attributes and possible combinations. Each packet filter in a TFT is associated with a precedence value that determines in which order the filters shall be tested for a match.

An example for how a TFT is used could be that the UE starts an application that connects to a media server in the PDN. For this service session, a new EPS bearer may be set up with the appropriate QoS parameters and bit rates. At the same time, packet filters are installed in the UE and the PDN GW that directs all

Table 6.2.2.2.1 *Valid packet filter attribute combinations.*

Packet filter attribute	Valid combination types		
	I	II	III
Remote address and subnet mask	X	X	X
Protocol number (IPv4)/next header (IPv6)	X	X	
Local port range	X		
Remote port range	X		
IPSec SPI		X	
TOS (IPv4)/traffic class (IPv6) and mask	X	X	X
Flow label (IPv6)			X

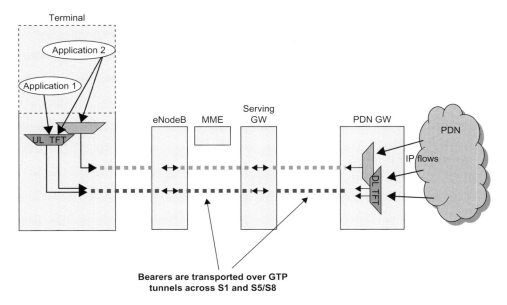

Figure 6.2.2.2.1 *EPS bearer for GTP-based system.*

traffic for the corresponding media onto that newly established EPS bearer. The Policy and Charging Control (PCC) system may be used at service establishment to ensure that the right QoS and TFT is provided.

When an EPS bearer is established, a bearer context is created in all EPS nodes that need to handle the user plane and identify each bearer. For E-UTRAN and a GTP-based S5/S8 interface between Serving GW and PDN GW, the UE, eNodeB, MME, Serving GW and PDN GW will all have bearer context. The exact details of the bearer context will differ somewhat between the nodes since the same bearer parameters are not relevant in all nodes. Furthermore, as will be seen further below, when PMIP-based S5/S8 is used, the PDN GW will not be aware of the EPS bearers.

Between the core network nodes in EPC, the user plane traffic belonging to a bearer is transported using an encapsulation header (tunnel header) that identifies the bearer. The encapsulation protocol is GTP-U. When E-UTRAN is used, GTP-U is used on S1-U and can also be used on S5/S8. The other alternative, to use PMIP on S5/S8 is described further below. The GTP-U header contains a field that allows the receiving node to identify the bearer the packet belongs to Figure 6.2.2.2.1 illustrates two EPS bearers in a GTP-based system. A user plane packet encapsulated using GTP-U is illustrated in Figure 6.2.2.2.2. For more information on GTP, see Section 11.2.

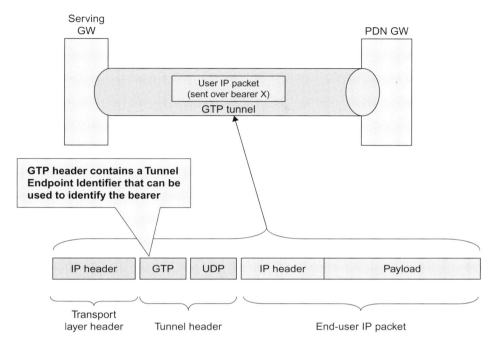

Figure 6.2.2.2.2 *EPS bearer transport for GTP-based system.*

6.2.2.3 PDN connections, EPS bearers, TFTs and packet filters – bringing it all together

In this chapter we have so far described several different concepts used in EPS and E-UTRAN to provide IP connectivity to a PDN and provide an appropriate packet transport; PDN connections, EPS bearers, TFTs and packet filters. Before going into more details on bearer procedures in EPS, it may be useful to see how they all relate to each other. Figure 6.2.2.3.1 provides an illustration of how the UE, PDN connection, EPS bearer, TFT and packet filters within the TFT relate to each other.

6.2.2.4 Control plane aspects

There are several procedures available in EPS to control the bearers. These procedures are used to activate, modify and deactivate bearers as well as to assign QoS parameters; packet filters, etc., to the bearer. Note, however, that if the default bearer is deactivated the whole PDN connection will be closed. EPS has adopted a network-centric QoS control paradigm, meaning that it is basically only the PDN GW that can activate, modify and deactivate an EPS bearer and decide which packet flows are transported over which bearer. It may be noted that this network-centric approach is different from pre-EPS GPRS. In GPRS it was originally only the UE that would take the initiative to establish a new bearer

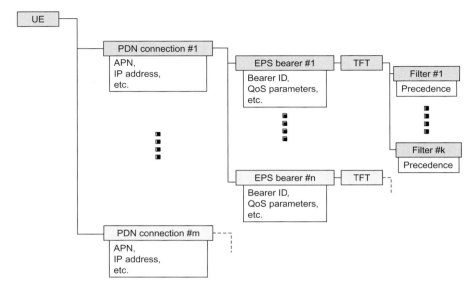

Figure 6.2.2.3.1 *Schematic relation between UE, PDN connection, EPS bearer, TFT and packet filters.*

(or PDP context as the bearers are called in GPRS) and decide on what packet flows to transport over that PDP context. In 3GPP Rel-7, the NW-initiated bearer procedures were introduced by specifying a new procedure with the long name 'network-requested secondary PDP context activation procedure'. In this procedure it is the GGSN that takes the initiative to create a 'dedicated bearer', known as secondary PDP context in 2G/3G packet core, as well as to assign packet filters. The move towards a network-centric approach has been taken one step further with EPS, since it is now only the PDN GW that can activate a new bearer and decide which packet flows are transported over which bearer. For more information on the network-centric QoS control paradigm, see Section 8.1.

It should be noted that when an EPS bearer is established or modified, the state in the radio access may also be modified to provide an appropriate radio layer transport for each active EPS bearer. More information about this can be found in [Dahlman, 2008] and Section 8.1.

6.2.2.5 Bearers in PMIP- and GTP-based deployments

In the illustration for bearers in GTP-based systems above, the EPS bearers extend between the UE and the PDN GW. This is how it works when GTP is used between Serving GW and PDN GW. As explained in Chapter 3, it is also possible to deploy EPS with PMIP between Serving GW and PDN GW. While GTP is designed to support all functionality required to handle the bearer signalling as well

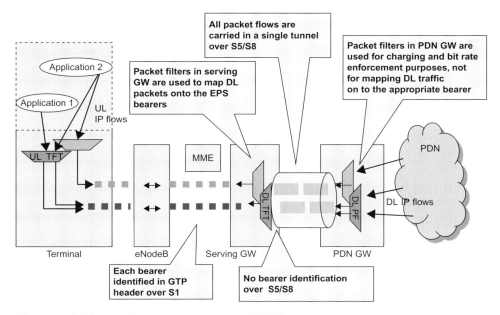

Figure 6.2.2.5.1 *EPS bearer when PMIP-based S5/S8 is used.*

as the user plane transport, PMIP was designed by IETF to only handle functions for mobility and forwarding of the user plane. PMIP thus had no built-in features to bearers or QoS-related signalling. During 2007 there were long discussions in 3GPP whether PMIP should be extended to support bearer-related signalling as well as to allow user plane marking to identify the EPS bearers, similar to GTP. It was eventually decided, however, that the PMIP-based reference points would not be aware of the EPS bearers. This means that it is not possible for the bearers to extend all the way between UE and PDN GW. Instead the bearers are only defined between UE and Serving GW when PMIP-based S5/S8 is used. Consequently, it is not sufficient to have the packet filters only in UE and PDN GW. Without bearer markings between S-GW and PDN GW, also the Serving GW would need to know the packet filters in order to map the downlink traffic onto the appropriate bearer towards the UE. This is illustrated in Figure 6.2.2.5.1.

The observant reader has noticed from Figure 6.2.2.5.1 that the PDN GW still uses the packet filters, just as with the GTP-based S5/S8. Why is this so? Isn't it enough that Serving GW has the packet filters in this case? It is true that the PDN GW does not need the packet filters to do the bearer mapping of downlink traffic, since this is instead done by the Serving GW when PMIP-based S5/S8 is used. The PDN GW, however, still performs important functionality such as bit rate enforcement and charging for the different IP flows. This is not directly related to the EPS bearers but it is the reason why PDN GW has packet filter knowledge also for PMIP-based S5/S8. These functions of the PDN GW, common to both PMIP-based S5/S8 and GTP-based S5/S8, are further described in Section 8.2 on PCC.

6.2.3 Session management for EPS and GERAN/UTRAN accesses

The 2G/3G Core Network uses the concept of PDP contexts to provide PDN connectivity and QoS management in the core network. The PDP context is defined between the UE and the GGSN and defines all the information used for a certain connection, including PDP address, QoS class, etc. The PDP context corresponds to EPS bearers for E-UTRAN.

The PDP context concept is partially maintained also when 2G/3G accesses are connected to EPC. In principle it would have been possible to replace the PDP context procedures for 2G/3G with the EPS bearer procedures but it was considered preferable to maintain the PDP context procedures, at least between UE and the SGSN. Within the EPC, however, the EPS bearer procedures are used also when the UE is in 2G/3G access. The SGSN provides the mapping between PDP context and EPS bearer procedures, and maintains a one-to-one mapping between PDP contexts and EPS bearers. There are several reasons why this architecture is preferable:

- By using PDP context procedures between UE and SGSN, the UE can use similar ways to connect when the 2G/3G access connects to EPC as when it connects to the GPRS architecture.
- By using the EPS bearer in the EPC also for 2G/3G access, it is easier for the PDN GW to handle mobility between E-UTRAN and 2G/3G. Since there is a one-to-one mapping between PDP context and EPS bearer, the handover between 2G/3G access and E-UTRAN access is simplified.

The interfaces where PDP context procedures are used and where EPS bearer procedures are used respectively are illustrated in Figure 6.2.3.1.

An SGSN using S4 maps the UE-initiated PDP context procedures (activation/ modification/deactivation) over GERAN/UTRAN into corresponding EPS bearer procedures towards the Serving GW. One aspect that complicates this mapping is the fact that PDP contexts are controlled by the UE while EPS bearers are controlled by the NW. One consequence is that the PDN GW needs to be aware that the UE is using 2G/3G and adjust its behaviour appropriately. For example, in case the UE is using GERAN/UTRAN and has requested activation of a

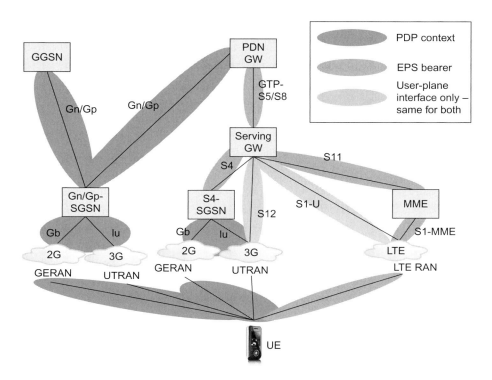

Figure 6.2.3.1 *Usage of PDP context procedures and EPS bearer procedures for EPS as well as for EPS interworking with Gn/Gp-based SGSNs.*

secondary PDP context, the PDN GW must activate a new EPS bearer corresponding to the PDP context. This is not the case if the UE is using E-UTRAN. In this case the UE cannot directly request activation of an EPS bearer, the UE can only make a request for certain resources and the PDN GW can decide whether or not to activate a new EPS bearer or modify an existing EPS bearer.

The SGSN also needs to map parameters provided from the UE (e.g. QoS parameters defined according to pre-rel-8 GPRS) into corresponding EPS parameters towards the Serving GW. See Section 8.1 for further details on QoS.

Another aspect that is worth mentioning in this context is that a Gn/Gp-SGSN may select either a PDN GW (supporting Gn/Gp) or a GGSN. See Chapter 9 for more details on the selection functions.

6.2.4 Session management for other accesses

In the previous subsections we have described session management for the 3GPP family of accesses and the use of bearers (i.e. EPS bearers and PDP contexts) to handle the user plane path between the UE and the network. It was found that the bearer is the basic enabler for traffic separation, that is, providing differential treatment for traffic with differing QoS requirements. The bearer procedures are specific to the 3GPP family of accesses but other accesses may have similar features and procedures to manage the user plane path and to provide traffic separation between different types of traffic. The details of the QoS mechanisms and the terminology used may differ between the different access technologies, but the key function to provide differentiated treatment for traffic with differing QoS requirements is common to all accesses that are able to provide QoS support. There are also accesses that basically only support best effort delivery of packets, without any differentiated treatment. Most vanilla IEEE 802.11b WLAN networks fall into that category.

We will not describe the access specific QoS 'bearer' capabilities and procedures that may be supported by the different accesses outside the 3GPP family of accesses. We will, however, look at how QoS is managed using the PCC architecture when these accesses interwork with EPS as described in Section 8.2.

6.3   Mobility principles

6.3.1   General

In the early days of GSM, the system supported a single radio access technology (GSM) and there was no mobility to/from other technologies. Since then 3GPP has developed WCDMA and LTE. 3GPP2 has developed CDMA (1xRTT and HRPD) and in addition other foras have developed a multitude of access technologies, such as WLAN or WiMAX, and fixed access, such as xDSL, Passive optical Networks (PON) or Cable. With such a range of access technologies available to the users to pick from and also for operators to select as their preferred system of choice, mobility has become quite complex. There is definitely a need and desire to find a 'common' set of tools allowing the end-user's devices to converge towards supporting a core set of mobility mechanisms.

With EPS, 3GPP aimed to provide not only a common core network for all access technologies but also *mobility* between heterogeneous access technologies. EPS is the first complete realization of multi-access convergence; a packet core network that supports full mobility management, access network discovery and selection for any type of access network.

In this section we will give an overview of the mobility functionality in EPS, starting from the mobility functionality for LTE, WCDMA and GSM, then adding on HRPD, WLAN and other radio technologies.

Mobility is the core feature of mobile systems, and many of the major system design decisions for EPC are derived directly from the necessity of ensuring mobility. The functionality of mobility management is required to ensure the following:

- That the network can 'reach' the user, for example, in order to notify the terminal about incoming calls
- That a user can initiate communication towards other users or services such as Internet access, and
- That ongoing sessions can be maintained as the user moves.

Associated functionality also ensures the authenticity and validity of the user's access to the system. It authenticates and authorizes the subscription and prepares the network and the user's device (i.e. the UE) with subscription information and security credentials.

6.3.2 Mobility within 3GPP family of accesses

6.3.2.1 Cellular idle-mode mobility management

Idle-mode mobility management in cellular systems like LTE, GSM/WCDMA and CDMA is built on similar concepts. To be able to reach the UE, the network tracks the UE, or rather the UE updates the network about its location, on a regular basis. The radio networks are built by cells that range in size from tens and hundreds of metres to tens of kilometres. It would not be practical and would cause a lot of signalling to keep track of a UE in idle mode every time it moves in between different cells. It is also not practical to search for the UE in the whole network for every terminating event (e.g. an incoming call). Hence the cells are grouped together into 'registration areas' (Figure 6.3.2.1).

The base stations broadcast registration area information and the UE compares the broadcasted registration area information with the information it has previously stored. If the registration area information does not match the information stored in the UE, it starts an update procedure towards the network to inform that it is now in a different registration area.

For example, when a UE that was previously in registration area 1 moves into a cell in registration area 2, it will notice that the broadcast information includes a different registration area identity. This difference in the stored and broadcasted information triggers the UE to perform a Registration update procedure towards the NW. In the Registration update procedure, the UE informs the network about the new registration area it has entered. Once the network has accepted the registration update, the UE will store the new registration area.

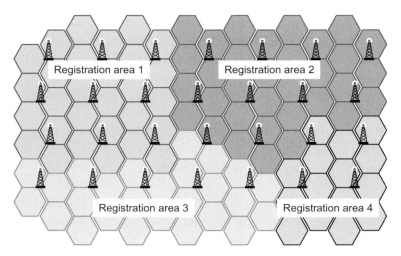

Figure 6.3.2.1   *Registration areas.*

In EPS the registration areas are called Tracking Areas (TA). In order to distribute the registration update signalling the concept of tracking area lists was introduced in EPS. The concept allows a UE to belong to a list of different TAs. Different UEs can be allocated to different lists of tracking areas. As long as the UE moves within its list of allocated TA list, it does not have to perform tracking area update. By allocating different lists of tracking areas to different UEs, the operator can give UE's different registration area borders and by that reduce peaks in registration update signalling, for example, when a train passes a TA border.

In addition to the registration updates performed when passing a border to a TA where the UE is not registered, there is also a concept of periodic updates. Periodic updates are used to ensure that the NW does not continue to page for UEs that are out of coverage or have been turned off.

The size of the tracking areas/tracking area lists is a compromise between registration update load and the paging load in the system. The smaller the areas you have, the fewer the cells you need to page the UEs in. On the other hand you will have frequent TA updates. The larger the area you have, the higher the paging load in the cells, but less tracking area updates signalling. In LTE, the concept of tracking area lists can also be used to reduce the frequency of tracking area updates. If you, for example, can predict the movement of UEs, the lists can be adapted for an individual UE to ensure that they pass fewer borders and UEs that receive lots of paging messages can be allocated smaller TA lists, while UEs that are paged infrequently can be paged less frequently.

In GSM/WCDMA there are two registration area concepts: one for PS domain (Routing Areas, RA) and the other for CS domain (Location Areas, LA). GSM and WCDMA cells may be included in the same Routing and Location Areas, allowing the UE to move between technologies without performing Routing Area Updates (RAU). The Routing Areas are a subset of the Location Areas and can only contain cells from the same LA. There is no support for lists of routing or location areas in GSM/WCDMA and hence all UEs share the same RA/LA borders. There is, however, another optimization that has been introduced in GSM/ WCDMA, since the RA is a subset of the LA; the UE can perform combined RAU/LAU where the UE is tracked on an RA basis. Since the RA is a subset of the LA, the network also knows in which LA the UE is. The UE can hence perform the combined updates when crossing the RA borders and no extra LAU procedures are needed. This optimization does, however, require support in both the UE and the NW. The combined update procedure is gaining momentum in GSM deployments while so far it has not been deployed much in WCDMA networks.

Generic concept	EPS	GSM/WCDMA GPRS	GSMWCDMA CS
Registration area	List of tracking areas (TA list)	Routing Area (RA)	Location Area (LA)
Registration area update procedure	TA Update procedure	RA Update procedure	LA Update procedure

Summary of Idle mobility procedure in EPS:

- A TA consists of a set of cells
- The registration area in EPS is a list of one or more TAs
- The UE performs TA Update when moving outside its TA list
- The UE also performs TA Update when the periodic TA Update timer expires. An outline of the Tracking Area Update procedure is shown in Figure 6.3.2.2 and it contains the following steps:

1. When the UE reselects a new cell and realizes that the broadcasted TA ID is not in their list of TAs, the UE initiates a TAU procedure to the network. The first action is to send a TA update message to the MME.
2. Upon reception of the TA message from the UE, the MME check if a context for that particular UE is available, if not it checks the UE's temporary identity to determine which node that keeps the UE context. Once this is determined the MME asks the old MME for the UE context.
3. The old MME transfers the UE context to the new MME.
4. Once the MME has received the old context, it informs the HSS that the UE context is now moved to a new MME.
5-6. The HSS cancels the UE context in the old MME.

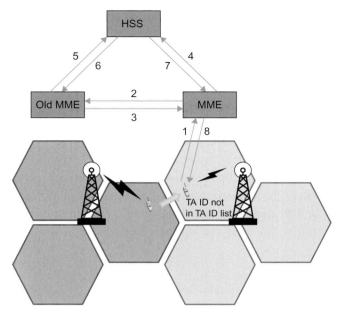

Figure 6.3.2.2   *TAU procedure.*

7. The HSS acknowledge the new MME and inserts new subscriber data in the MME.
8. The MME informs the UE that the TAU was successful and as the MME was changed it supplies a new GUTI (where the MME code points back to the new MME).

6.3.2.2   Paging

Paging is used to search for idle UEs and establish a signalling connection. Paging is, for example, triggered by downlink packets arriving to the Serving GW. When the Serving GW receives a downlink packet destined for an Idle UE, it does not have an eNodeB address to which it can send the packet. The Serving GW instead informs the MME that a downlink packet has arrived. The MME knows in which TA the UE is roaming and it sends a paging request to all eNodeBs within the TA lists. Upon reception of the paging message, the UE responds to the MME and the bearers are activated so that the downlink packet may be forwarded to the UE.

6.3.2.3   Cellular active-mode mobility

Great effort has been put into optimized active-mode mobility for cellular systems. The basic concept is rather similar across the different technologies with some variations in the functional distribution between UE and networks. While in active mode, the UE has an active signalling connection and one or more active bearers and data transmission may be ongoing. In order to limit interference and provide the UE with a good bearer, the UE changes cell through handover when there is a cell that is considered to be better than the cell that the UE is currently using. To save on complexity in the UE design and power, the systems are designed to ensure that the UE only needs to listen to a single base station at a time. Also, for inter RAT handover (e.g. E-UTRAN to UTRAN HO) the UE shall only need to have a single radio technology active at a time. It may need to rapidly switch back and forth between the different technologies, but at any instance of time only one of the radio technologies are active.

To determine when to perform handover the UE measures the signal strength on neighbour cells regularly or when instructed by the network. As the UE cannot send or receive data at the same time as it measures on neighbour cells, it receives instruction from the network on suitable neighbour cells that are available and which the UE should measure on. The network (eNodeB) creates measurement time gaps where no data is sent or received to/from the UE. The measurement gaps are used by the UE to tune the receiver to other cells and measure the signal strength. If the signal strength is significantly stronger on another cell, the handover procedure may be initiated.

In E-UTRAN the eNodeBs can perform direct handover via the direct interface (X2) between eNodeBs. In the X2-based HO procedure, the source eNodeB and the target eNodeB prepare and execute the HO procedure. At the end of the HO execution, the target eNodeB requests the MME to switch the downlink data path from the source eNodeB to the target eNodeB. The MME in turn requests the S-GW to switch the data path towards the new eNodeB.

In case there is downlink packets sent before the S-GW has switched the path towards the new eNodeB, the source eNodeB will forward the packet over the X2 interface.

In case the X2 interface is not available between eNodeBs, the eNodeB can initiate a handover involving signalling via the core network. This is called S1-based handover. The S1-based HO procedure sends the signalling via the MME and may include change of MME and or SGW.

6.3.3 Mobility between E-UTRAN and HRPD

As already mentioned, one important goal of EPS is to support efficient interworking and mobility with CDMA/HRPD networks. For HRPD networks, there is a significant subscriber base already out there with major North American and

Asian operators operating their networks. Even though the two technologies (one developed in 3GPP and the other in 3GPP2) have been competing over the last 20 years, the two bodies have also cooperated in many areas in order to develop common standards that are strategically important to operators, examples of these are the IMS and PCC development. Driven by a major CDMA operator, a number of vendors cooperated extensively in order to develop special optimized HO procedures between E-UTRAN and HRPD access which would have efficient performance and reduced service interruption during handover. This work was then brought into mainstream 3GPP standards under the SAE work item umbrella and further enhanced and aligned with the mainstream 3GPP work ongoing for SAE and thus produced the so-called Optimized Handover between E-UTRAN and HRPD. HRPD networks then became known as evolved HRPD (eHRPD) to highlight the changes required for interoperability and connectivity with EPC and E-UTRAN.

Mobility between E-UTRAN and HRPD has been specified to allow an efficient handover with minimal interruption time also for those terminals that can only operate a single radio at a time. There is thus no need for the terminal to operate both HRPD and E-UTRAN interfaces simultaneously. Despite the fact that these terminals support multiple radio technologies, this property of being able to operate only one radio at a time is sometimes referred to as 'single radio capability'.

Note that in the early deployment of E-UTRAN in an existing HRPD network, it is considered more prevalent, and thus more important, to support E-UTRAN to HRPD handover then the reverse direction, since it is assumed that the HRPD networks would have sufficient coverage to keep a user within HRPD system. In order to support a network-controlled handover from E-UTRAN to HRPD, the eNodeB can be configured with HRPD system information that is sent over E-UTRAN in order to assist the terminal in preparation for cell reselection or handover from the E-UTRAN to HRPD system. The terminal also makes appropriate measurements on HRPD cells while being connected to E-UTRAN. Similarly to the active-mode mobility for 3GPP accesses described in the previous section, measurement time gaps have to be provisioned to the terminal in order to allow the terminal to only use a single radio at a time, that is either E-UTRAN or HRPD. The measurements are reported to the eNodeB to allow the E-UTRAN to make the appropriate handover decisions.

The purpose of the optimized procedures is to minimize the total service interruption time experienced at the UE, by having the UE prepare the target access system before actually leaving the source access system. The preparation in the target access system is done by enabling the UE to exchange access-specific signalling with the target access over the source access. The S101 interface between the MME and HRPD Access Network is used to tunnel the signalling between UE and target access system. A benefit with letting the UE prepare the target access via a tunnel over the source access is that the direct exchange of UE context between the different access networks can be minimized. The impact on the access networks can be minimized since neither access needs to adapt its signalling towards another access technology.

The handover between E-UTRAN and HRPD is performed in two phases:

1. Pre-registration (or preparation) phase where the target access and specific core network entity for the specific access (MME for E-UTRAN and HRPD S-GW or HSGW for eHRPD access) is prepared ahead of the actual handover
2. Handover (or execution) phase where the actual access network change occurs.

In the pre-registration/preparation phase for E-UTRAN to HRPD handovers, the UE communicates with the HRPD access network via the E-UTRAN access and the MME. The HRPD signalling is forwarded transparently by the E-UTRAN and MME between the UE and the HRPD RNC. This is illustrated in Figure 6.3.3.1. In the E-UTRAN to HRPD direction, there is no time limit of how long a UE may be pre-registered in the HRPD system before the handover takes place. Therefore pre-registration may take place well in advance of the actual handover.

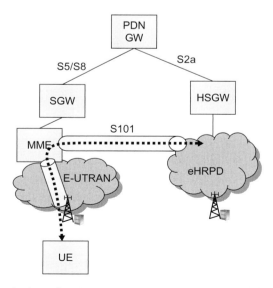

Figure 6.3.3.1 *Terminal performing pre-registration in HRPD while being connected to E-UTRAN. The dashed line illustrates HRPD specific signalling transparently forwarded by E-UTRAN and MME.*

When the decision to handover from E-UTRAN to HRPD is taken, the handover phase is executed. During this phase, some additional preparation of HRPD target access is performed before the actual user plane path switch takes place.

The handover from HRPD to E-UTRAN is also performed with a preparation phase where E-UTRAN-specific signalling is exchanged between UE and MME via the S101 interface. This time the signalling is transported transparently via the HRPD access network. A difference with handover from HRPD to E-UTRAN is that the pre-registration/preparation takes place immediately prior to the terminal switches from HRPD to E-UTRAN radio, that is just before the actual handover is performed. There is thus no long-lived pre-registration state in target access in this case. For more details on the HO procedures between E-UTRAN and HRPD, please see Section 12.4.

6.3.4 Generic mobility between 3GPP and non-3GPP accesses

Generic mobility between heterogeneous access types, such as between a 3GPP access and WLAN or WiMAX, are also supported by EPS. Generic mobility is, as the name suggests, generic in the sense that the mobility procedures are not specifically adapted to any particular access technology. Instead the procedures are generic enough to be applicable to any non-3GPP access technology, such as WLAN and WiMAX, as long as the accesses support some basic requirements. The generic mobility procedures are, for example, also applicable for HRPD in case the optimized mechanisms described in the previous section are not deployed.

Since the generic mobility procedures are supposed to work with any access, they are also not optimized towards any specific access technology. In 3GPP and other standards fora, generic mobility is also referred to as non-optimized handover. However, a goal has still been to provide efficient HO procedures also in this generic case. In particular, for a terminal that is able to operate in multiple access technologies simultaneously – often referred to as 'dual-radio capable' terminals, it is possible to prepare the target access before performing the actual handover. The terminal can, for example, perform authentication procedures in target access while still using the source access to transfer user plane data. This is in a sense similar to the pre-registration described for HRPD with the difference that the 'pre-registration' is done in the actual target access instead of via the source access.

A key difference compared to the mobility mechanisms within and between 3GPP accesses and the optimized interworking between LTE and HRPD is that generic mobility does not assume any interaction between the two access networks. Instead the source and target access networks are fully decoupled.

The generic handover is always triggered by the terminal, that is, there are no measurements of cells in the target access being reported to the radio access network where the UE is attached. There are also no handover commands from the source access to trigger the handover. Instead it is up to the terminal to decide when to initiate the handover. The terminal may, for example, base its decision on measurement of signalling strength of available access networks. The operator may also use the Access Network Discovery and Selection Function (ANDSF) to provide the terminal with information about access networks as well as policies for access network selection. The ANDSF is described further in a following section.

The EPS supports two different mobility concepts for generic mobility between 3GPP and non-3GPP accesses; host- and network-based mobility. Host-based mobility is a term often used to denote a mobility scheme where the terminal (or host) is directly involved in the movement detection and mobility signalling. Mobile IP, defined by IETF, is one example of such a mobility protocol. In this case the terminal has IP mobility client software.

Another type of mobility protocol and mobility scheme is the network-based mobility management scheme. In this case the network can provide mobility services for a terminal that is not explicitly exchanging mobility signalling with the network. It is a task of the network to keep track of the terminal's movements and ensure that the appropriate mobility signalling is executed in the core network in order for the terminal to maintain its session while moving. Proxy Mobile IPv6 (PMIPv6) is an example of a network-based mobility protocol. GPRS Tunnelling protocol (GTP) is another example of a network-based protocol that is used to support mobility.

EPS supports multiple mobility protocol options using host- and network-based mobility protocols. EPS supports two host-based mobility schemes; Dual-stack Mobile IPv6 (DSMIPv6) and Mobile IPv4 (MIPv4) over non-3GPP accesses. Over the 3GPP accesses host-based mobility is not used. Instead it is always assumed that the 3GPP access is the 'home link', in the host-based Mobile IP sense. Please see Section 11.3 for further descriptions of this and other Mobile IP concepts. Over non-3GPP accesses, it is also possible to use network-based PMIPv6 as was shown in Chapter 3. For 3GPP accesses, only network-based mobility protocols are used; either GTP or PMIPv6.

For a more detailed description of the mobility protocols and mobility mechanisms, see Chapter 11 for protocol descriptions and Chapter 12 for the details of the HO procedures.

Host- and network-based mobility schemes have different properties. A host-based scheme requires support for the mobility protocol in the terminal. Host-based mobility is often generic in the sense that it assumes nothing or very little about any particular access network. It can therefore be used also over those access networks not supporting mobility at all. A consequence is that basic host-based mobility is not optimized for any particular access network.

Network-based mobility on the other hand requires that the access network supports the mobility protocol. GTP, for example, is developed by 3GPP to support mobility between 3GPP accesses and includes support for context transfer and other features needed to provide a seamless handover within and between 3GPP accesses.

Even though the division between host- and network-based mobility schemes often provides a useful classification, it can be noted that in a complete system solution both host- and network-based mobility components are included. It is possible to handover from an access where host-based mobility protocol is used to an access where a network-based mobility protocol is used. Also with network-based mobility protocols, the terminals often need to be mobility aware and handle mobility even if they do not explicitly participate in the mobility signalling. Therefore, even though the division into host- and network-based mobility is a useful categorization on a high level and when discussing one specific protocol, it is also important to remember that in reality the full solution typically contains pieces of both. EPS supports different mobility protocols in different accesses and there may be change in mobility protocol when moving between access networks. The terminal may, for example, move from a non-3GPP access where DSMIPv6 is used to a 3GPP access where GTP or PMIPv6 is used. It is the task of the PDN GW to ensure that IP session continuity as provided also when different mobility protocols are used in the different accesses.

6.3.4.1   IP mobility mode selection

The EPS specifications allow both host- and network-based mobility; different operators may make different choices about which of the two to deploy in their networks. An operator may also choose to deploy both. In 3GPP accesses, there is only a choice between two network-based mobility protocols (PMIPv6 or GTP) and the selection of one protocol over the other has no impact on the terminal since the network choice of protocol is transparent to the UE. It should be noted that even if multiple mobility protocols are supported in a network deployment, only a single protocol is used at a time for a given UE and access type.

Terminals may support different mobility mechanisms. Some terminals may support host-based mobility and thus have a Mobile IP client installed (Dual

Stack Mobile IPv6 and/or Mobile IPv4 client). Other terminals may support IP level session continuity where network-based mobility protocols are used. There may also be terminals that support both mechanisms. In addition, some terminals may neither have a Mobile IP client nor support IP session continuity using network-based mechanisms. These terminals will not support IP level session continuity but could still attach to EPC using different access technologies.

There is thus a need for selecting the right mobility mechanism when a terminal attaches to the network or is making a handover. If network-based mobility protocols are selected, there also needs to be a decision if session continuity (i.e. IP address preservation between the accesses) shall take place.

EPS has defined different means for how this selection can be done. The rules and mechanism for selecting the appropriate mobility protocol is referred to as IP Mobility Mode Selection (IPMS).

One option with IPMS is to statically configure the mobility mechanism to use in the network and the terminal. This is, for example, possible if the operator is only supporting a single mobility mechanism and if the operator can assume that the terminals used in its network supports the deployed mobility mechanism. If a user switches to another terminal not supporting the mobility protocol deployed by the operator, IP level session continuity may not be possible.

The other option is to have a more dynamic selection where the decision to use either network-or host-based mobility is made as part of attach or HO procedures. It should be noted that over 3GPP accesses only network-based mobility is supported using either PMIPv6 or GTP. Therefore, mobility mode selection is only needed when the terminal is using a non-3GPP access.

IPMS is performed when the user attaches in a non-3GPP access or sets up an IPSec tunnel towards an ePDG, before an IP address is provided to the terminal. The terminal may provide an indication about its supported mobility schemes during network access authentication by using an attribute in the EAP-AKA and EAP-AKA' protocols [24.302]. The indication from the UE informs the network if the terminal supports host-based mobility (DSMIPv6 or MIPv4) and/or if it supports IP session continuity using network-based mobility. The network may also learn about the terminal capabilities using other mechanisms. For example, if a terminal has attached in 3GPP access and performed bootstrapping for DSMIPv6, the network implicitly understands based on the already performed bootstrapping procedure that the terminal is capable of using DSMIPv6. Based on its knowledge about the terminal and the capabilities of the network, the network decides which

Figure 6.3.4.2.1 *ANDSF architecture.*

mobility mechanism to use for the particular terminal. In case the network has
no knowledge about a terminal's capabilities, the default is to use network-based
mobility protocols. Host-based mobility can only be selected by the network if
it knows that the terminal supports the appropriate mobility protocol (DSMIPv6
or MIPv4).

6.3.4.2   Access network discovery and selection

The Access Network Discovery and Selection Function (ANDSF) is a server that
has been defined in 3GPP TS 23.402 [23.402]. The ANDSF is a server in the network that provides the UE with policies and network selection information. The
UE can query the server for information about other access networks; for example
non-3GPP access networks, specifically it contains the data management and control functionality necessary to provision the discovery of different networks and
selection procedures. These procedures are defined by the operator. Either the UE
can request the information or the ANDSF can initiate the data transfer to the UE.

As depicted in Figure 6.3.4.2.1, the ANDSF function contains a reference point
S14 to the UE to provide the UE with access network selection policies and/or
access discovery hints.

The information provided on the S14 interface can be used by the UE to understand which network to scan for and also the operator policies with regard to
handover – for example whether to stay on 3GPP access or whether to perform
the handover to another access network. Please see Section 10.11 for further
details on ANDSF.

6.4   Idle mode signalling reduction (ISR)

Idle mode signalling reduction is a feature that allows the UE to move between
LTE and 2G/3G without performing Tracking Area (TA) or Routing Area (RA)
update once ISR has been activated. As idle mode mobility between RATs may
be rather common, especially in deployments with spotty coverage, ISR can be
used to limit the signalling between the UE and the network as well as signalling within the network.

In 2G/3G there is also similar functionality which allows the UE to move between GERAN and UTRAN cells in idle mode without performing any signalling. This is implemented through using a common SGSN and common RAs; that is the GERAN and UTRAN cells belong to the same RA and can be paged for terminating events in both GERAN and UTRAN cells that belong to the RA.

It would have been possible to keep a similar concept for SAE but that would have required a combined SGSN/MME node. Due to this difference in architecture (SGSN vs. MME) and the area concepts (RA vs. list of TA), it was decided to implement the ISR functionality in a different way between E-UTRAN and GERAN/UTRAN than between GERAN and UTRAN.

The ISR feature enables signalling reduction with separate SGSN and MME and also with independent TAs and RAs. The dependency between EPC and 2G/3G is minimized at the cost of ISR-specific node and interface functionality.

The idea behind the ISR feature is that the UE can be registered in a GERAN/UTRAN RA at the same time as it is registered in an E-UTRAN TA (or list of TAs). The UE keeps the two registrations in parallel and run periodic timers for both registrations independently. Similarly the network keeps both registrations in parallel and it also ensures that the UE can be paged in both the RA and the TA(s) it is registered in.

6.4.1 ISR activation

A prerequisite for ISR activation is that the UE, SGSN, MME, Serving GW and HSS support ISR. The ISR support is specified to be mandatory for the UE but optional for network entities. ISR also requires an S4-SGSN, that is it is not supported with a Gn/Gp-SGSN.

At the first attach to the network, ISR is not activated. ISR can only be activated when the UE has first been registered in an RA on 2G/3G and then registers in a TA or vice versa.

If the UE first registers on GERAN/UTRAN and then moves into an LTE cell, the UE will initiate a TA update procedure. In the TA update procedure, the SGSN, MME and Serving GW will communicate their capabilities to support ISR, and if all nodes support ISR, the MME will indicate to the UE that ISR is activated in the TAU accept message.

A simplified example of a TA update procedure with ISR activation is shown in Figure 6.4.1.1. Before the TAU procedure is performed, the UE has attached in

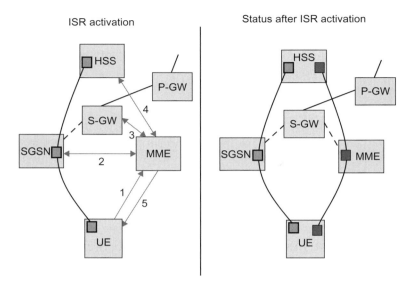

Figure 6.4.1.1   *Outline of ISR an ISR activation procedure.*

GERAN/UTRAN and have an active registration with an SGSN. There is hence an active MM context in UE, SGSN and HSS, and the SGSN has a control connection with the Serving GW.

In Step 1 the UE initiates a TAU procedure

In Step 2 the MME requests the UE context and indicates to the SGSN that it is ISR capable. The SGSN responds with the UE context and indicates that it supports ISR

In Step 3 the Serving GW is informed that the UE is registered with the MME and that ISR is activated

In Step 4 the HSS may be updated with the MME address. The update type indicates that HSS shall not cancel the SGSN location

In Step 5 the MME informs the UE that the TAU procedure was successful and that ISR is activated.

As shown in Figure 6.4.1.1, when ISR is activated, the UE is registered with both MME and SGSN. Both the SGSN and the MME have a control connection with the Serving GW and the MME and SGSN are both registered in HSS. The UE stores MM parameters from SGSN (e.g. P-TMSI and RA) and from MME (e.g. GUTI and TA(s)) and the UE stores session management (bearer) contexts that are common for E-UTRAN and GERAN/UTRAN accesses. SGSN and MME store each other's address when ISR is activated.

When ISR is activated, a UE in idle state can reselect between E-UTRAN and GERAN/UTRAN (within the registered RAs and TAs) without any need to signal with network.

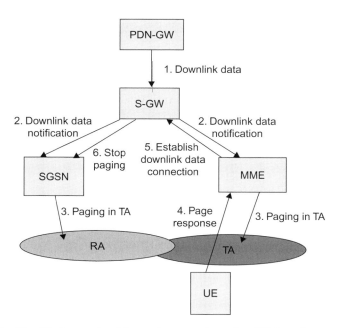

Figure 6.4.2.1 *Simplified example of the paging procedure when ISR is active.*

6.4.2 Paging

When the UE is in idle mode and ISR is active and downlink data arrives in Serving GW, it will send a downlink data notification to both the MME and the SGSN. The MME will then initiate paging in the TA, where the UE is registered, and the SGSN will initiate paging in the RA, where the UE is registered. When the UE receives the paging message, it will perform a service request procedure on the RAT it is currently camping on. As part of the service request procedure, the Serving GW will be requested to establish a downlink data connection towards an eNodeB in case the UE was camping on E-UTRAN and towards the SGSN/RNC in case the UE was camping on GERAN/UTRAN. When the Serving GW receives this request to establish the downlink data connection, it will also inform the SGSN or MME to stop paging on the other RAT.

A simplified example of the paging procedure, when ISR is active and the UE camps on E-UTRAN, is outlined in Figure 6.4.2.1.

6.4.3 ISR deactivation

ISR activation has to be refreshed at every RAU and TAU procedure and the UE will deactivate ISR if it does not receive ISR active indication in the RAU accept and TAU accept messages.

The UE and the network run independent periodic update timers for GERAN/
UTRAN and for E-UTRAN. The UE will perform RA update if it camps on
GERAN/UTRAN when the periodic RAU timer expires and it will perform
TA update if it camps on E-UTRAN when the periodic TAU timer expires. If
the UE camps on a different RAT when the periodic timers expires, it will not
perform the update procedure; for example, if it camps on E-UTRAN when
the periodic RAU timer expires, it will remain camping on E-UTRAN and not
perform a RAU update.

When the MME or SGSN do not receive periodic updates, the MME and
SGSN may decide to implicitly detach the UE. The implicit detach removes
session management (bearer) contexts from the node performing the implicit
detach and it removes also the related control connection from the Serving
GW. Implicit detach by one CN node (SGSN or MME) deactivates ISR in the
network.

When the UE cannot perform periodic updates in time, it starts a Deactivate ISR
timer. When this timer expires and the UE was not able to perform the required
update procedure, the UE locally deactivates ISR.

ISR shall be deactivated by MME or SGSN, that is by omitting the signalling of
'ISR activation', when:

– CN node change resulting in context transfer between the same type of CN
  nodes (SGSN to SGSN or MME to MME)
– Serving GW change.

There are also situations where the UE needs to deactivate ISR locally, for
example:

– modification or activation of additional bearers
– after updating either MME or SGSN about the change of the DRX parameters or UE capabilities
– E-UTRAN selection by a UTRAN-connected UE (e.g. when in URA_PCH
  to release Iu on UTRAN side).

6.5   Identifiers and corresponding legacy IDs

Permanent and temporary subscriber identities are constructed to identify not
only a particular subscriber but also the network entities where the permanent
and temporary subscriber records are stored.

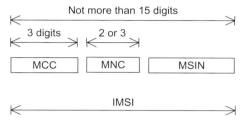

Figure 6.5.1.1 *Structure of IMSI.*

6.5.1 Permanent subscriber identifiers

Subscriptions (the USIM cards, the little piece of plastic you get from the operator) are identified with an IMSI (International Mobile Subscriber Identity). Each USIM card is assigned a unique IMSI. The IMSI is an E.164 number (basically a string of digits like a phone number) with a maximum length of 15 digits. The IMSI is constructed by an MCC (Mobile Country Code) and MNC (Mobile Network Code) and an MSIN (mobile subscriber identity) (Figure 6.5.1.1).

The MCC identifies the country and the MNC identifies the network within the country. The MSIN in turn is a unique number for each subscriber within a particular network.

The IMSI is the permanent subscription identifier and it is used as a master key in the subscriber database (HSS). By its construction it allows any network in the world to find the home operator of the subscriber, specifically it provides a mechanism to find the HSS in the home operator network.

The IMSI is also used in 2G/3G networks and there is no change of the purpose or format of the IMSI for SAE/LTE.

6.5.2 Temporary subscriber identifiers

Temporary subscriber identifiers are used for several purposes. They provide a level of privacy since the permanent identity does not need to be sent over the radio interface. But more importantly they provide a mechanism to find the resources where the subscriber's temporary context is stored. The temporary context for the subscriber is stored in an MME (or SGSN in the 2G/3G case) and, for example, the eNodeB needs to be able to send signalling from a UE to the correct MME where the subscriber's context resides.

Pooling of MMEs was an integral part of the SAE/LTE design (as opposed to 2G/3G where pooling was a feature added a few years after the original design). Hence the

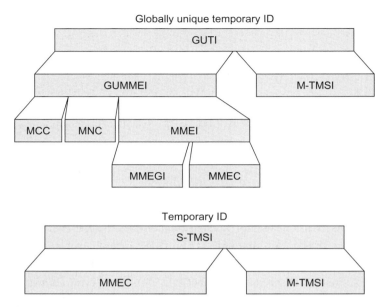

Figure 6.5.2.1   *Structure of GUTI and S-TMSI.*

temporary identities in SAE/LTE could be designed with pooling in mind. This has resulted in a cleaner design of the temporary identities for SAE/LTE.

Figure 6.5.2.1 illustrates the temporary identifiers. The GUTI (Globally Unique Temporary ID) is a worldwide unique identity that points to a specific subscriber context in a specific MME. The S-TMSI is unique within a particular area of a single network. The UE can use the S-TMSI when communicating with the network as long as it stays within a TA that is part of the TA list it has received.

The GUTI consists of two main components: (1) the GUMMEI (Globally Unique MME Identifier) that uniquely identifies the MME which allocated the GUTI and (2) the M-TMSI (MME Temporary Subscriber Identity) that identifies the subscriber within the MME.

The GUMMEI is in turn constructed from MCC (country), MNC (network) and MMEI (MME identifier, the MME within the network). The MMEI is constructed from an MMEGI (MME group ID) and an MMEC (MME Code).

The GUTI is a long identifier and to save radio resources, a shorter version of the GUTI is used whenever possible. The shorter version is called S-TMSI and it is unique, only within a group of MMEs. It is constructed from the MMEC and the M-TMSI. The S-TMSI is used for, for example, paging of UEs and service request.

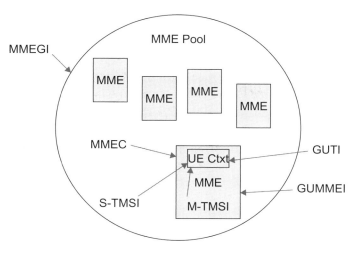

Figure 6.5.2.2 *Identifiers as pointers.*

One can imagine the different temporary identities, a set of pointers, to network resources in EPS (Figure 6.5.2.2).

6.5.3 Relation to subscription identifiers in 2G/3G

Why is there a need for a relation between the EPS and the 2G/3G identifiers? The main reason is mobility between GSM/WCDMA and LTE. When the UE moves from one access to the other, it should be possible to locate the node where the UE context is stored, for example, when moving to LTE the MME needs to find the UE context in the SGSN and vice versa. Another reason is that the implementation of combined SGSN and MME nodes will be available and it is preferable to map the temporary identifier used on LTE and the temporary identifier used on GSM/WCDMA so that they point to the same combined node.

The IMSI – the common permanent subscription identifier across GSM, UMTS and EPS – and the same subscription can be used to access all technologies.

The temporary identities in 2G/3G looks a little bit different, primarily since the original design of GSM/UMTS assumed a strictly hierarchical system where an RA was controlled by a single SGSN node. Hence the GSM/UMTS original design did not explicitly include any SGSN pool and SGSN node identifiers. When pooling was added in the GSM/UMTS system, these node identifiers were encoded inside the temporary identifier.

In GPRS the P-TMSI is the identifier that identifies the subscriber context inside an SGSN. The globally unique identifier in GPRS has no explicit name but by combining the RA ID (RAI) and the P-TMSI you get a globally unique identifier. The RAI is constructed by MCC, MNC, LAC and RAC.

Pooling is discussed in another chapter but the basic principle for pooling in 2G/3G is that the identities P-TMSI range is divided among the SGSNs in a pool. There are a set of bits inside the P-TMSI called NRI (Network Resource identifier) that points to specific SGSNs, and each SGSN in a pool is assigned one (or more) unique NRIs. The NRI in 2G/3G hence corresponds to the MME code and the NRI identifies as an SGSN in a pool just as the MME code identifies an MME in an MME pool.

The operator needs to ensure that the MMEC is unique within the MME pool area, and if overlapping pool areas are in use, they should be unique within the area of overlapping MME pools.

The GUTI is used to support subscriber identity confidentiality, and in the shortened S-TMSI form, to enable more efficient radio signalling procedures (e.g. paging and Service Request).

6.6 Pooling and overload protection

EPC was designed from the start with pooling of network elements as a foundation of the system, in contrast to 2G and 3G systems, where pooling was added as an afterthought. The pooling mechanisms in EPC are efficient for an operator's network through allowing them to centralize and pool a group of their signalling nodes. It should be noted that the network elements are stateful; as a result the UE context is stored in every node involved in handling the UE.

In 2G and 3G networks, the core network was designed as a hierarchical system; when a terminal or UE was located in a particular cell, it was only able to connect to one Base Station (BS). While there was one-to-many relationships between the SGSN and Base Station Controller (BSC) and also between the BSC and BS, in practice, the UE was only able to be connected to one BS, BSC and therefore only one SGSN (Figure 6.6.1).

The same hierarchical structure applied in WCDMA networks for the Radio Network Controller (RNC) and node B. Naturally, this does not utilize the capacity of the nodes to their full potential; in the strictly hierarchical systems of the 2G and 3G networks, the capacity of the nodes can never be perfectly balanced; for example when users move in and out of cities. The network must be dimensioned according to the peak load in a particular area, which may be much higher than the

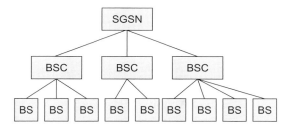

Figure 6.6.1 *Original hierarchical system design for GSM/GPRS.*

average load in that area. Through applying a pooling mechanism, it is possible to dimension the whole pooled capacity to the peak rates of a much larger region; for example it is now possible to create a pool of network nodes for the entire London region, rather than the need to split the network up into many smaller regions.

So, how is this implemented in an actual network configuration?

As described in Section 6.5, an MME has several different identifiers associated with it, which help to manage the pooling mechanism. As a brief review, the *MME Group* identity refers to the name of the pool which the MME belongs to, while the *MMEC* identifies the actual node within a group; the *MMEI* is therefore formed by combining the MMEGI and MMEC together.

In LTE, the eNodeB knows which MMEs it can communicate with. When a terminal that is already registered enters a cell, it sends a service request to the network, which contains a GUTI, within which the MMEI is encapsulated. The eNodeB checks the GUTI code and if the MMEI is within the pool of MMEs for which it has a connection, it simply uses the MMEI to send it to the correct one. When a GUTI is available, therefore, it makes everything reasonably simple – the eNodeB can always route you to the correct MME. If the GUTI is absent, however, or the GUTI points to an MME that is not within the pool associated with that particular eNodeB, the eNodeB selects a new MME and forwards the service request to the new MME. The MME then returns a GUTI to the eNodeB which naturally includes its own MMEI.

The eNodeB cannot just randomly choose any old MME in the pool to send the service request to, however, because all of the MME nodes may have different capacities. Randomly allocating UEs to them, therefore may overload nodes with lower capacity. When configured, therefore, the MME is assigned a 'weight factor', which indicates its capacity. The MME then informs all of the eNodeBs in its pool what this weight factor is. The eNodeBs can therefore distribute UEs across the pool of MMEs accordingly.

There are certain situations where it may be useful to manipulate the weighting factor of an MME. Take the example of an already established pool of eNodeBs and MMEs that you wish to insert a new MME into. Initially that particular MME can inform the eNodeBs that is has a higher weight factor than it actual will during 'normal' operations. This means that the particular MME will then be allocated some UEs by eNodeBs. Once the new MME reaches a suitable load, it will update its weight factor and inform the eNodeBs of this change.

The implementation of pooling in the core network architecture also implies some changes with regard to network dimensioning. In the hierarchical systems, when you dimension your system, you will generally apply some general formula, for example peak load plus an additional 20% capacity in order to handle extra load. This is dimensioned across an area, however, not one particular region as with a pooled network architecture. This means that the numbers of subscribers that you are dealing with when calculating excess capacity are significantly different.

Take the example of a 3G system for which you are dimensioning a particular region for say a peak capacity of 100,000 users. Accordingly, you will dimension the nodes to handle peak capacity plus 20%, that is 120,000 subscribers for that particular area. When calculating across a region, however, the figures start to get a lot larger; for example 1,000,000 subscribers for say a big city; this implies a significantly larger 'excess' capacity needs to be factored in. However, this excess capacity is across many nodes, rather than across just a few. While the most natural thing when planning for redundancy in a pooled system would be to implement N + 1 nodes, where N is the number of nodes required for peak capacity of that region, since nodes can handle millions of users, the issue of the necessity of such a large cost may be raised. It is important, however, that there is capacity available in case a network node is to be taken out of service.

7
Security

7.1 Introduction

Providing security is one of the key aspects of mobile networks. One of the more obvious reasons is that the wireless communication can easily be intercepted by anyone within a certain range of the transmitter. There is thus a risk that the data that is transferred is eavesdropped, or even manipulated, by third parties. There are also other threats, for example, an attacker may trace a user's movement between radio cells in the network or discover the whereabouts of a specific user. This may constitute a significant threat to users' privacy. Apart from security aspects directly related to the end-users, there are also security aspects related to the network operator and the service providers as well as security between network operators in roaming scenarios. For instance, there should be no doubt regarding which user and roaming partner were involved in generating certain traffic in order to assure a correct and fair charging of the subscribers.

There are also regulatory requirements related to security and these may differ between countries and regions. The regulations can, for example, be related to exceptional situations where law enforcement agencies can request information about the activities of a terminal as well as intercept the telecommunications traffic. The framework in a telecommunications system for supporting this is called 'lawful intercept' and is described in a separate chapter. There may also be regulations to ensure that end-users' privacy is protected when using mobile networks. Requirements like these are in general captured in the national and/ or regional laws and regulations by the responsible authorities for that specific nation/region.

Below we discuss different aspects of security in mobile networks, starting with a brief discussion on key security concepts and security domains. Then security aspects related to both end-users as well as within and between network entities are discussed.

7.2 Security services

7.2.1 Introduction

Before we go into the actual security mechanisms of EPS, it may be useful to briefly go through some basic security concepts which are important in cellular networks.

Before a user is granted access to a network, *authentication* in general has to be performed. During authentication the user proves that he or she is the one he/she claims to be. Typically, *mutual authentication* is desired, where the network authenticates the UE and the UE authenticates the network. Authentication is in general done with a procedure where each party proves that it has access to a secret known only to the participating parties, for example, a password or a secret key.

The network also verifies that the subscriber is *authorized* to access the requested service, for example, to get access to EPS using a particular access network. This means that the user must have the right privileges (i.e. a subscription) for the type of services that is requested. Authorization for an access is often done at the same time as authentication. It can be noted that different kinds of authorization may be done in different parts of the network and at different instances during an IP session. The network may, for example, authorize the use of a certain access technology, a certain QoS profile, a certain bit rate, access to certain services, etc.

Once the user has been granted access, there is a desire to protect the signalling traffic and user plane traffic between the UE and the network and between different entities within the network. *Ciphering* and/or *integrity protection* may be applied for this purpose. With ciphering (i.e. encryption and decryption) we ensure that the information transmitted is only readable to the intended recipients. To accomplish this, the traffic is scrambled so that it becomes unreadable for anyone who manages to intercept it, except for the entities that have access to the correct cryptographic keys. Integrity protection on the other hand is a means to detect whether traffic that reaches the intended recipient has not been modified, for example, by an attacker between the sender and the receiver. If the traffic has been modified, the integrity protection ensures that the receiver is able to detect it. Ciphering and integrity protection serves different purposes and the need for ciphering and/or integrity protection differs depending on what traffic it is. Furthermore, the data protection may be done on different layers in the protocol stack and as we will see, EPS supports data protection features on both layers 2 and 3 depending on scenario.

In order to encrypt/decrypt as well as to perform integrity protection, the sending and receiving entities need *cryptographic keys*. It may seem tempting to use the same key for all purposes, including authentication, ciphering, integrity protection, etc. However, using the same key for several purposes should in general be avoided. One reason is that in case an attacker manages to recover the key by breaking, for example, the encryption algorithm, the attacker would at the same time learn the key used also for authentication and integrity protection. Furthermore, the keys used in one access should not be the same as the keys used in another access. If they would be the same, the keys recovered by an attacker in one access with weak security features, could be reused also to break accesses with stronger security features. The weakness of one algorithm or access thus spreads to other procedures or accesses. To avoid this, keys used for different purposes and in different accesses should be distinct, and an attacker who manages to recover one of the keys, shall not be able to learn anything useful about the other keys. This property is called *key separation* and, as we will see, this is an important aspect of EPS security design. In order to achieve key separation, the UE and the EPC derives distinct keys during the authentication process that are used for different purposes.

By *privacy protection* we here mean the features that are available to ensure that information about a subscriber does not become available to others. For example, it may include mechanisms to ensure that the permanent user ID is not sent in clear text over the air link. If done, it would mean that an eavesdropper could detect the movements and travel patterns of a certain user.

Laws and directives of individual nations and regional institutions (e.g. European Union), typically define a need to intercept telecommunications traffic and related information. This is referred to as *lawful intercept* and may be used by law enforcement agencies in accordance with the laws and regulations.

7.2.2   Security domains

In order to describe the different security features of EPS it is useful to divide the complete security architecture into different security domains. Each domain may have its own set of security threats and security solutions. The 3GPP TS 33.401 divides the security architecture into different groups or domains:

1. Network access security
2. Network domain security
3. User domain security
4. Application domain security
5. Visibility and configurability of security.

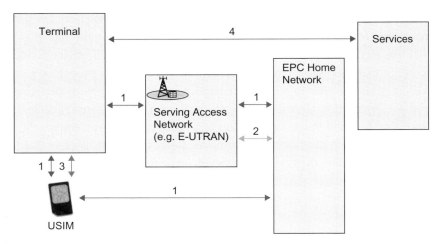

Figure 7.2.2.1   *Schematic figure of different security domains.*

The first group is specific to each access technology (e.g. E-UTRAN, GERAN, UTRAN, etc) , whereas the others are common for all accesses. Figure 7.2.2.1 provides a schematic illustration of different security domains.

1. Network access security

By network access security we mean the security features that provide a user with a secure access to the EPS. This includes mutual authentication as well as privacy features. In addition, protection of signalling traffic and user plane traffic in the particular access is also included. This protection may provide confidentiality and/ or integrity protection of the traffic. Network access security is in general access specific, that is, the detailed solutions, algorithms, etc differ between accesses. Further details for different types of accesses are provided later in this chapter.

2. Network domain security

Mobile networks contain many network entities and reference points between these entities. The network domain security refers to the features that allow these network nodes to securely exchange data and protect against attacks on the network between the nodes.

3. User domain security

User domain security refers to the set of security features that secure the access to terminals. This can for example, be that the user needs to enter a PIN code before being able to access the terminal.

4. Application domain security

With application domain security is meant the security features used by applications such as HTTP (for web access) or IMS.

Application domain security is in general end-to-end between the application in the terminal and the peer entity providing the service. This is in contrast to the previous security features listed which provide hop by hop security, that is, they apply to a single link in the network only. If each link (and node) in the chain that requires security is protected, the whole end-to-end chain can be considered secure.

Since application level security traverses on top of the user plane transport provided by EPS, and as such is more or less transparent to EPS, it will not be discussed further in this book. For more information on IMS security, see for example Camarillo and Garcia-Martin (2008).

5. Visibility and configurability of security

This is the set of features that allows the user to learn whether a security feature is in operation or not and whether the use and provision of services should depend on the security feature. In most cases the security features are transparent to the user and the user is unaware that they are in operation. For some security features the user should however be informed about the operational status. For example, usage of encryption in E-UTRAN depends on operator configuration and it should be possible for the user to know whether it is used or not, for example, by a symbol on the terminal display. Configurability is the property that the user can configure whether the use or the provision of a service should depend on whether a security feature is in operation.

7.3 Network access security

7.3.1 Introduction

As mentioned previously, network access security is in many aspects specific to each access. Below we go into some details on access security in different types of accesses such as E-UTRAN, HRPD and a WLAN hotspot. These three examples represent well the different possibilities to get access to the EPS. We also describe additional aspects for the case DSMIPv6 is used.

Common in all cases is the use of USIM.

7.3.2 Access security in E-UTRAN

It was clear from the start of the standardization process that E-UTRAN should provide a security level, at least as high as that of UTRAN. Access security in

E-UTRAN therefore consists of different components, similar to those that can be found in UTRAN:

- Mutual authentication between UE and network.
- Key derivation to establish the keys for ciphering and integrity protection.
- Ciphering, integrity and replay protection of NAS signalling between UE and MME.
- Ciphering, integrity and replay protection of RRC signalling between UE and eNB.
- Ciphering of the user plane. The user plane is ciphered between UE and eNB.
- Use of temporary identities in order to avoid sending the permanent user identity (IMSI) over the radio link.

Figure 7.3.2.1 illustrates some of these components in the network.

Below we will discuss in detail how each of these components has been solved.

The authentication procedure in E-UTRAN is in many ways similar to the authentication procedure in GERAN and UTRAN but there are also differences. To understand the reason behind the differences, it is useful to first briefly look at the security features for GERAN and UTRAN systems. As with all security features in communication systems, what was considered sufficiently secure at one point in time may turn out not to be sufficient a few years later when attack methods and computing power have developed further. This is also true for 3GPP radio accesses. When GERAN was developed, some limitations were deliberately accepted. For example, mutual authentication is not performed in GERAN where it is only the network that authenticates the terminal. It was thought that there was no need for the UE to authenticate the network, since it was unlikely that anyone

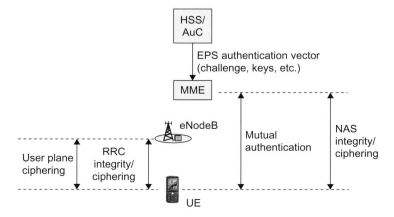

Figure 7.3.2.1   *Security features for E-UTRAN.*

would be able to set up a rogue GERAN network. When UTRAN/UMTS was developed, enhancements were made to avoid some of the limitations of GERAN. For example, mutual authentication was introduced. These new security procedures are one reason why a new type of SIM card was needed for UMTS; the so called UMTS SIM (or USIM for short). With the introduction of E-UTRAN, further improvement is taking place. One important aspect is, however, that it has been agreed that the use of USIM in the terminal shall be sufficient to access E-UTRAN, that is, no new type of SIM card shall be needed. The new features are instead supported by software in the terminal and the network.

Mutual authentication in E-UTRAN is based on the fact that both the USIM card and the network have access to the same secret key K. This is a permanent key that is stored on the USIM and in the HSS/AuC in the home operator's network. Once configured, the key K never leaves the USIM or the HSS/AuC. The key K is thus not used directly to protect any traffic and it is also not visible to the end-user. During the authentication procedure, other keys are generated from the key K in the terminal and in the network that are used for ciphering and integrity protection of user plane and control plane traffic. For example, one of the derived keys is used to protect the user plane, while another key is used to protect the NAS signalling. One reason why several keys are produced like this is to provide key separation and to protect the underlying shared secret K. In UTRAN and GERAN, the same keys are used for control signalling and user traffic, and hence this is also an enhancement compared to these earlier standards. This is, however, not the only key management enhancement as will be discussed below.

The mechanism for authentication as well as key generation in E-UTRAN is called EPS Authentication and Key Agreement (EPS AKA). Mutual authentication with EPS AKA is done in the same manner as for UMTS AKA, but as we will see when we go through the procedure, there are a few differences when it comes to key derivation.

EPS AKA is performed when the user attaches to EPS via E-UTRAN access. Once the MME knows the user's IMSI, the MME can request an EPS authentication vector (AV) from the HSS/AuC as shown in Figure 7.3.2.2. Based on the IMSI, the HSS/AuC looks up the key K and a sequence number (SQN) associated with that IMSI. The AuC steps (i.e. increases) the SQN and generates a random challenge (RAND). Taking these parameters and the master key K as input to cryptographic functions, the HSS/AuC generates the UMTS AV. This AV consists of five parameters: an expected result (XRES), a network authentication token (AUTN), two keys (CK and IK), as well as the RAND. This is illustrated in Figure 7.3.2.3. Readers familiar with UMTS will recognize this Authentication Vector as

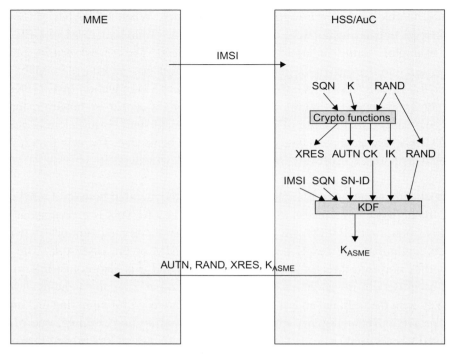

Figure 7.3.2.2 *MME fetching the EPS Authentication Vector from HSS/AuC.*

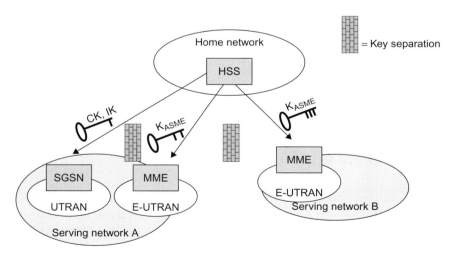

Figure 7.3.2.3 *Key separation between 3GPP accesses and serving networks.*

the parameters that the HSS/AuC would send to the SGSN for access authentication in UTRAN. For E-UTRAN, however, the CK and IK are not sent to the MME. Instead the HSS/AuC generates a new key, K_{ASME}, based on the CK and IK and other parameters such as the serving network identity (SN ID). The SN

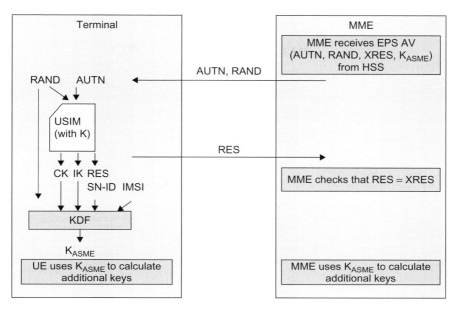

Figure 7.3.2.4 *EPS AKA between UE and MME.*

ID includes the Mobile Country Code (MCC) and Mobile Network Code (MNC) of the serving network. A reason for including SN ID is to provide a better key separation between different serving networks to ensure that a key derived for one serving network cannot be (mis-) used in a different serving network. Key separation is illustrated in Figure 7.3.2.3.

The K_{ASME} together with XRES, AUTN and RAND constitutes the EPS AV that is sent to MME. The CK and IK never leave the HSS/AuC when E-UTRAN is used. In order to distinguish the different AVs, the AUTN contains a special bit called the 'separation bit' indicating whether the AV shall be used for E-UTRAN or for UTRAN/ GERAN. A reason for going through this extra step with the new key K_{ASME}, instead of using CK and IK for ciphering and integrity protection like in UTRAN, is to provide a strong key separation towards legacy GERAN/UTRAN systems. For more details on the generation of the EPS AV, please see 3GPP TS 33.401 [33.401].

Mutual authentication in E-UTRAN is performed using the parameters RAND, AUTN and XRES. The MME keeps the K_{ASME} and XRES but forwards RAND and AUTN to the terminal shown in Figure 7.3.2.4. Both RAND and AUTN are sent to the USIM. AUTN is a parameter calculated by the HSS/AuC based on the secret key K and the SQN. The USIM now calculates its own version of AUTN using its own key K and SQN and compares it with the AUTN received from the MME. If they are consistent, the USIM has authenticated the network. Then the

USIM calculates a response RES using cryptographic functions with the key K and the challenge RAND as input parameters. The USIM also computes CK and IK in the same way as when UTRAN is used (it is after all a regular UMTS SIM card). When the terminal receives RES, CK and IK from the USIM, it sends the RES back to the MME. The MME authenticates the terminal by verifying that the RES is equal to XRES. This completes the mutual authentication. The UE then uses the CK and IK to compute K_{ASME} in the same way as HSS/AuC did. If everything has worked out, the UE and network has authenticated each other and both UE and MME now have the same key K_{ASME}. (Note that none of the keys K, CK, IK or K_{ASME} was ever sent between UE and network.)

Now all that remains is to calculate the keys to be used for protecting traffic. As mentioned above, the following type of traffic is protected between UE and E-UTRAN:

- NAS signalling between UE and MME
- RRC signalling between UE and eNB
- User plane traffic between UE and eNB.

Different keys are used for each set of procedures above, and also different encryption and integrity protection keys are used. The key K_{ASME} is used by UE and MME to derive the keys for encryption and integrity protection of NAS signalling (K_{NASenc} and K_{NASint}). In addition, the MME also derives a key that is sent to the eNB (the K_{eNB}). This key is used by the eNB to derive keys for encryption of the user plane (K_{UPenc}) as well as encryption and integrity protection of the RRC signalling between UE and eNB (K_{RRCenc} and K_{RRCint}). The UE derives the same keys as eNB. The 'family tree' of keys is typically referred to as a *key hierarchy*. The key hierarchy of E-UTRAN in EPS is illustrated in Figure 7.3.2.5.

Once the keys have been established in the UE and the network it is possible to start ciphering and integrity protection of the signalling and user data. The standard allows use of different cryptographic algorithms for this and the UE and the NW need to agree on which algorithm to use for a particular connection. For more details on which ciphering and integrity algorithms are supported with E-UTRAN, please see 3GPP TS 33.401 [33.401].

The final aspect that should be mentioned is the identity protection. In order to protect the permanent subscriber identity (i.e. IMSI) from being exposed in clear text over the radio interface, temporary identities are used whenever possible in a similar way to what is done in UTRAN. Please see the identities section in Chapter 6 for a description on how temporary identities are used in E-UTRAN.

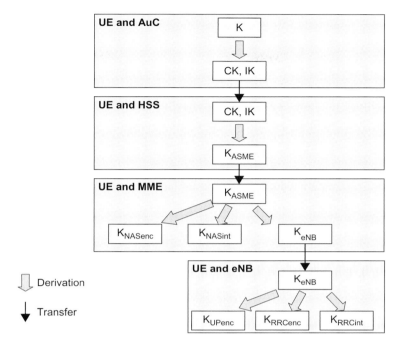

Figure 7.3.2.5 *Key hierarchy for E-UTRAN.*

A main enhancement in E-UTRAN as compared to UTRAN is, as was discussed above, the strong key separation between networks and key-usage. A few other enhancements are also worth brief mentioning:

- Larger key sizes. E-UTRAN supports not only 128-bit keys but can (in future deployments) also use 256-bit keys.
- Additional protection against compromised base stations. Due to the flattened architecture in E-UTRAN, additional measures were added to protect against a potentially compromised 'malicious' radio base station. One of the most important features is the added forward/backward security: each time the UE changes its point of attachment (due to mobility), or, when the UE changes from IDLE to ACTIVE, the air interface keys are updated according to a sophisticated procedure. This means that even in the unlikely event that the keys used so far have been compromised, security can be restored.

7.3.3 Interworking with GERAN/UTRAN

In this book we will not describe the security features applicable to GERAN and UTRAN in any detail. The interested reader is instead referred to books dedicated to GERAN and UTRAN. However, the interworking between GERAN/ UTRAN and E-UTRAN will be discussed below.

When a UE moves between GERAN/UTRAN and E-UTRAN, there are different possibilities in order to establish the security context to be used in the target access. One possibility would be to perform a new authentication and key agreement procedure every time the UE enters a new access. In order to reduce the delays during handover between GERAN/UTRAN and E-UTRAN, however, this may not be desirable. Instead, handovers can be based on *native* (or cached) or *mapped* security contexts. If the UE has previously been active in E-UTRAN access, then moved to GERAN/UTRAN and later returns to E-UTRAN, the UE and NW may have cached a native security context for E-UTRAN, including a native K_{ASME}, from the previous time the UE was in E-UTRAN. In this way a full AKA procedure in the target access is not needed during the inter-RAT handover. If a native context is not available, it is instead possible to map the security context used in the source access to a security context for the target access. This security context mapping is supported when moving between different 3GPP accesses. When mapping is performed, the UE and MME derive keys applicable to the target access (e.g. K_{ASME} for E-UTRAN) based on the keys used in the source access (e.g. CK, IK for UTRAN). The mapping is based on a cryptographic function, f, having the property that it protects the source context from the mapped target context. This assures that a mapped context cannot 'backwards' compromise the context from which it was mapped. An example of such a mapping is illustrated in Figure 7.3.3.1.

There are, however, a few important aspects related to such a mapping of security context since the protection is only one-way. If the source context has already been compromised, then the mapped context will inherit this property. Also, as we have already mentioned above, the level of security is not the same in all accesses. This is where key separation becomes important. Unless the different accesses are kept separated from a security point of view, the

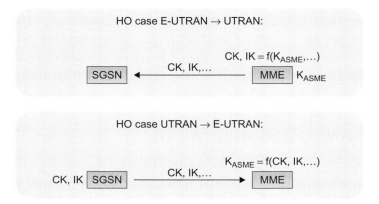

Figure 7.3.3.1 *Examples of mapping security context in handover between E-UTRAN and UTRAN.*

vulnerability of one access may spread into other accesses not susceptible to the same vulnerabilities. Therefore, in case a security context from for example, GERAN has been mapped to a security context for E-UTRAN, it is highly recommended that a full AKA run is performed as soon as possible after entering E-UTRAN to establish a fresh, native, E-UTRAN security context.

7.3.4 Access security in trusted non-3GPP accesses

One example of a cellular technology not specified by 3GPP, but that can be used to provide access to EPC, is evolved HRPD (eHRPD). The security features of HRPD that are specified by 3GPP2 are thus not under the control of 3GPP. Still, HRPD has the capabilities to provide a strong access control, mutual authentication as well as protection of signalling and user plane traffic sent over the HRPD radio link. Even though not specified by 3GPP, it is reasonable that these security features are sufficient for providing access to EPS. Typically, HRPD would be connected directly to EPC using the S2a or S2c reference points.

Access authentication in eHRPD is based on EAP-AKA; or, to be more precise, 3GPP has agreed to use a revision of EAP-AKA called EAP-AKA', but more about this below. On a high level, EAP-AKA (and EAP-AKA') is a method to perform AKA-based authentication over an access even if there is not native support for AKA in that particular access. This makes it possible to perform 3GPP-based access authentication using the same credentials – the shared secret key K located in USIM and HSS/AuC– as for 3GPP accesses. EAP-AKA runs between the UE and the 3GPP AAA Server, (Figure 7.3.4.1). In order to perform the AKA-based authentication, the 3GPP AAA Server downloads the Authentication Vector from HSS/AuC. For more details on EAP, please see the EAP protocol section in Chapter 11.

It should, however, be noted that the key hierarchy and details regarding key derivation differ somewhat between EPS AKA and EAP-AKA. With EAP-AKA,

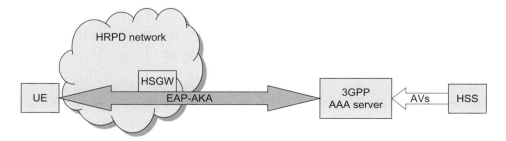

Figure 7.3.4.1 *EAP-AKA based access authentication for eHRPD.*

the Authentication Vector from HSS/AuC is a starting point for the authentication procedure, just like in EPS AKA. Then, a Master Key is derived by the UE and 3GPP AAA Server based on the CK and IK; in a way this is conceptually similar to how K_{ASME} is derived from CK and IK with EPS AKA. This Master Key is used to derive further keys, for example, keys that are used to protect user plane and control plane traffic in the eHRPD access. In this book we will not go into further details regarding security in the eHRPD access.

As already mentioned above, a revision of EAP-AKA called EAP-AKA' is used for access authentication in eHRPD and other trusted non-3GPP accesses. EAP-AKA is defined in RFC 4187 [4187] and EAP-AKA' in RFC 5448 [5448]. The revisions made in EAP-AKA' is that the procedure has been more aligned with EPS AKA. In practice this means that the serving network name is taken into account in the key derivation schemes. This even further strengthens the key separation as the keys used in one serving network cannot be used in another serving network.

There is no security context mapping when moving between 3GPP and eHRPD, or between 3GPP and non-3GPP access in general. However, despite the lack of context mapping, the optimized eHRPD interworking, described in the mobility section of Chapter 6, allows for optimizations of the security procedures during handover between E-UTRAN and eHRPD. When a UE is active in E-UTRAN access and the dedicated signalling connection with eHRPD (via S101) is setup, the UE can perform the EAP-AKA' procedure for eHRPD before actually handing over to eHRDP access. Once the handover takes place, the security context is already established in the target eHRPD access. Also in the other direction, that is, handover from eHRPD to E-UTRAN, the EPS AKA security procedures for E-UTRAN can be performed via the S101 signalling connection.

For privacy protection, EAP-AKA supports means to use temporary identities (so called pseudonyms) in a similar way as for E-UTRAN access. The EAP-AKA pseudonyms are, however, of a different format than the temporary identities used in 3GPP accesses.

7.3.5 Access security in untrusted non-3GPP access

Another example of an access that can be used to provide connectivity to EPS is WLAN. WLAN can be used in many scenarios, for example, in corporate environments, in the home or in public places such as airports and coffee shops. The level of security provided by a WLAN access also differs between deployments. For home and corporate use, WLAN security solutions such as WEP and WPA are often used. In public places it is, however, more common to turn WLAN security off completely. Instead access control is provided by the means of a web page where the user can enter a username and a password. The user may for example, have received a username and password for a temporary subscription at the same time he or she bought a cup of coffee. Once the user has entered the credentials on the web page, Internet access is provided. The WLAN access does in this case not provide any encryption or integrity protection of the user plane and is vulnerable to many types of attacks. In such deployments, WLAN access to EPS will therefore most likely be handled as an un-trusted non-3GPP access. In other deployments with WLAN security turned on, WLAN might be treated as a trusted access.

Now the user wants to get access to the services provided by his or her opera-tor via EPS. In general the coffee shop where the end-user is located and wants to connect does not have any agreement to connect to the EPC directly. Equally important, since no or very limited security is provided, providing direct access to EPC would make EPC vulnerable to attacks. The solution to this problem, as defined by EPS, is to set up an IPSec tunnel between the UE and a so called ePDG inside the operator's network when the access is un-trusted (Figure 7.3.5.1). The ePDG acts as a secure entry point into the EPC.

In order to setup the IPSec tunnel, the UE must first perform mutual authentica-tion towards the ePDG and operator network, as well as establish keys for the IPSec security association. This is done using the IKEv2 protocol. Once the UE

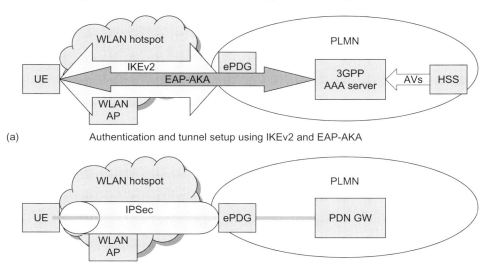

(a)   Authentication and tunnel setup using IKEv2 and EAP-AKA

(b)   User plane traversing IPSec tunnel between UE and ePDG

Figure 7.3.5.1 *(a) Authentication using IKEv2 and EAP-AKA between UE and ePDG. (b) User plane traffic is protected in IPSec tunnel between UE and ePDG.*

has connected to the WLAN and discovered the IP address of the ePDG (can be done using DNS) it starts the IKEv2 procedure. As part of IKEv2, public key-based authentication with certificates is used to authenticate the ePDG. The UE is on the other hand authenticated in a similar manner as for E-UTRAN, that is, based on the credentials on the USIM. Within IKEv2, EAP-AKA is run to perform the AKA-based authentication and key agreement. Therefore, the USIM-based authentication described in the previous sections can be performed also when the UE accesses over a generic WLAN hotspot. EAP-AKA is the same protocol as was described in Section 7.3.4. The difference is that now EAP-AKA is run as part of the IKEv2 procedure, while in Section 7.3.4 EAP-AKA' is run as part of the attach procedure in a trusted non-3GPP access such as eHRPD. Similar properties regarding key generation and privacy protections as mentioned about EAP-AKA' in Section 7.3.4 apply to the EAP-AKA in this scenario as well.

It should be noted that the WLAN hot spot has been used in this section as an example scenario where connection using IKEv2 and IPSec towards an ePDG is suitable. The ePDG may also be used for any access that can provide IP connectivity, for example, from the DSL connection at home, independent of security properties of the underlying access.

7.3.6   Special considerations for host-based mobility DSMIPv6

In the previous sections we have described the security features for two main scenarios, a user attaching to an access network that provides a high level of security, for example, E-UTRAN and eHRDP, and a user attaching to an access network where additional security protection is needed (IPSec tunnel over an unprotected WLAN access).

However, also the choice of mobility protocol has impact on the security features. As has been described in earlier chapters, specifically the mobility section of Chapter 6, there are two main methods for providing access in EPS, either using network-based mobility (GTP or PMIP) or host-based mobility (DSMIPv6 or MIPv4). When network-based mobility is used, the access security described in the previous chapters provides the needed security between UE and the EPC. However, when host-based mobility is used, there is a need to provide security also for the host-based mobility protocol between UE and PDN GW.

When DSMIPv6 is used, the DSMIPv6 signalling between UE and PDN GW is integrity protected using IPSec. In order to establish the IPSec security association for the DSMIPv6 signalling, the user is first authenticated using IKEv2 and

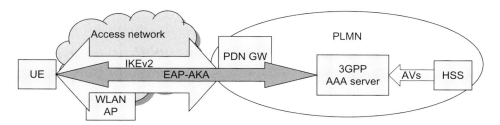

Figure 7.3.6.1 *IKEv2 and EAP-AKA based authentication and key agreement for DSMIPv6.*

EAP-AKA, as illustrated in Figure 7.3.6.1. This EAP-AKA based authentication for DSMIPv6, as well as IPSec protection of the DSMIPv6 signalling, is done in addition to any access level authentication and user plane protection that may be performed (as described in previous chapters). This means that the DSMIPv6 signalling may be protected twice, first using the general user plane protection on access level and then using IPSec between UE and PDN GW. Additionally, in case there is an ePDG on the path, there is an IPSec tunnel between UE and ePDG.

On a high level, the basic security features for MIPv4 are similar to those of DSMIPv6, that is, the MIPv4 signalling needs to be integrity protected to ensure that only authenticated UEs can send MIPv4 signalling messages to the PDN GW. However, the details of the security solution for MIPv4 are quite different compared to DSMIPv6. MIPv4 performs integrity protection of signalling using a special authentication element in the signalling messages. The messages are thus not protected using IPSec between UE and the network. For more details on the MIPv4 security solution, please see 3GPP TS 33.402 [33.402].

7.4 Network domain security

When GSM/GERAN was developed, no solution was specified for how to protect the traffic in the core network. This was perceived not to be a problem, since the GSM networks typically were controlled by a small number of large institutions. Furthermore, the original GSM networks were only running circuit-switched traffic. These networks used protocols and interfaces specific for circuit switched voice traffic and typically only accessible to large telecom operators. With the introduction of GPRS as well as IP transport in general, the signalling and user plane transport in 3GPP networks now runs over networks and protocols that are more open and accessible to others than the major institutions in the telecom community. This brings a need to provide enhanced protection also to traffic running over core network interfaces. For example, the core network interfaces may traverse third-party IP transport networks, or the interfaces may cross operator boundaries like in roaming cases. 3GPP has therefore developed specifications for how IP-based traffic is to be secured also in the core network and/or between a core network and some other (core) network. On the other hand, it can be noted that also today, if the core network interfaces run over trusted networks, for example, a transport network owned by the operator that is physically protected, there would be little need for this additional protection.

The specifications for how to protect the IP-based control plane traffic is called Network Domain Security for IP-based control planes (NDS/IP) and are available in 3GPP TS 33.210 [33.210]. This specification introduces the concept of security domains. The security domains are networks that are managed by a single administrative authority. Hence, the level of security and the available security services are expected to be the same within a security domain. An example of a security domain could be the network of a single telecom operator, but it is also possible that a single operator divides its network into multiple security domains. On the border of the security domains, the network operator places Security Gateways (SEGs) to protect the control plane traffic that passes in and out of the domain. All NDS/IP traffic from network entities of one security domain is routed via a SEG before exiting that domain towards another security domain. The traffic between the SEGs is protected using IPSec, or to be more precise, using IPSec Encapsulated Security Payload (ESP) in tunnel mode. The Internet Key Exchange (IKE) protocol, either IKEv1 or IKEv2, is used between the SEGs to set up the IPSec security associations. An example scenario is illustrated in Figure 7.4.1.

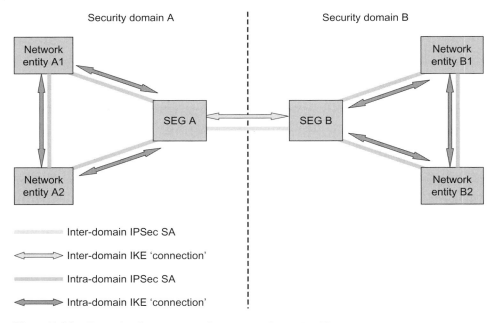

Figure 7.4.1   *Example of two security domains employing NDS/IP.*

A case of special relevance to EPS and E-UTRAN is the S1-U interface between EPC and the E-UTRAN. This interface needs to be properly protected (physically and/or by NDS/IP) since the user plane data protection would otherwise be terminated in the eNodeB, potentially exposing sensitive data on S1.

Also within a security domain, that is, between different network entities or between a network entity and a SEG, the operator may choose to protect the traffic using IPSec. The end-to-end path between two network entities in two security domains is thus protected in a hop-by-hop fashion.

Although NDS/IP was initially intended mainly for the protection of control plane signalling only, it is possible to use similar mechanisms to protect the user plane traffic. Indeed, for the aforementioned case of user data over S1, NDS/IP will be used in deployments that require it.

7.5 User domain security

The most common security feature in this user domain context is the secure access to the USIM. Access to the USIM shall be blocked until the USIM has authenticated the user. Authentication is in this case based on a shared secret (the PIN code) that is stored inside the USIM. When the user enters the PIN code on the terminal, it is passed on to the USIM. If the user provided the right PIN code, the USIM allows access from the terminal/user, for example, to perform the AKA-based access authentication.

7.6 Lawful intercept

Lawful Interception (LI) is one of the regulatory requirements operators must satisfy as legal obligation toward the Law Enforcement Agencies (LEA) and Government Authorities in most countries they are operating their businesses in. Within 3GPP standards, this is currently defined as: *Laws of individual nations and regional institutions (e.g. European Union), and sometimes licensing and operating conditions define a need to intercept telecommunications traffic and related information in modern telecommunications systems. It has to be noted that lawful interception shall always be done in accordance with the applicable national or regional laws and technical regulations* (as per 3GPP TS 33.106 'Lawful Interception Requirements' [33.106]). LI allows appropriate authorities to perform interception of communication traffic for specific user(s) and this includes activation (requires legal document such as a warrant), deactivation, interrogation as well as invocation procedures. A single user (i.e. interception subject) may be involved where interception is being performed by different LEAs. In such scenarios, it must be possible to maintain strict separation of these interception measures. The Intercept Function is only accessible by authorized personnel. As LI has regional jurisdiction, national regulations may define specific requirements on how to handle the user's location and interception across boundaries. As a necessary part of the mobile communications systems, handover is a basic process in EPS. Interception is also carried out when handover has taken place, when required by national regulations.

This subsection deals with this aspect on a brief and high level in order to complete the overall EPS functionalities; it is intended as a description of the 3GPP LI standards and not of any function implemented in the Ericsson or other vendors' nodes. The Legal Intercept function in itself does not put requirements on how a system should be built but rather requires that provisions be made for legal authorities to be able to get the necessary information from the networks via legal means, according to specific security requirements, without disruption of the normal mode of operations and without jeopardizing the privacy of communications not to be intercepted. Note that LI functions must operate without being detected by the person(s) whose information is being intercepted and other unauthorized person(s). As this is the standard practice for any communications networks already operating today around the world, EPS is no exception.

The process of collection of information is done by means of adding specific functions into the network entities where certain trigger conditions will then cause these network elements to send data in a secure manner to a specific network

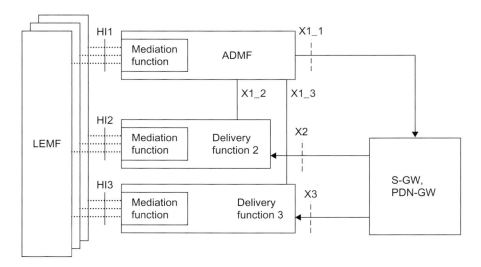

Figure 7.6.1   *High-level EPS-LI architecture for S-GW/PDN-GW.*

entity responsible for such role. Moreover, specific entities provide administration and delivery of intercepted data to Law Enforcement in the required format. As an example, Figure 7.6.1 shows the LI architecture for some of the EPS nodes:

Intercept-related information (also referred to as Events) are triggered by activities detected at the network element. Some events applicable to the MME are:

- Attach
- Detach
- Tracking Area Update
- UE requested PDN connectivity
- UE requested PDN disconnection.

Events are triggered in case of E-UTRAN access at the following user plane-related activities detected at the Serving GW and at the PDN GW:

- Bearer activation (valid for both Default and Dedicated bearer)
- Bearer Modification
- Bearer Deactivation
- Start of Intercept with bearer active
- UE requested bearer resource modification.

Depending on national regulations, intercept-related information collected may also be reported by the HSS.

Local regulations may allow for the operators to charge for the services rendered towards the Legal Interception requesting party, such as charging data collection process may include some or all of the following situations to be supported:

- Use of network resources
- Activation and deactivation of the target
- Every intercept invocation
- Flat rate.

This brief overview represents high-level functions supported in EPS in order to fulfil the legal interception requirements. This does not in any way show the complete possibilities or aspects of this function since it is deemed unrelated to the overall architecture aspects of the new system, but rather shown for completeness of the system in itself.

8
Quality of service, charging and policy control

8.1 Quality of service

8.1.1 General

Many mobile broadband operators aim at providing multiple services (Internet, voice, video) across their packet-switched access networks. These services will share the radio and core network resources with best effort services such as Internet browsing and e-mail download, and they all have different Quality of Service (QoS) requirements in terms of required bit rates as well as acceptable packet delays and packet loss rates. Furthermore, with mobile broadband subscriptions offering flat rate charging, high-bandwidth services such as file sharing become more common in cellular systems also. In such a multi-service scenario, it is important that EPS provides an efficient QoS solution that ensures that the user experience of each service running over the shared radio links is acceptable. Simply solving these issues through over-provisioning is not economical; the available radio spectrum is limited, and the cost of transmission capacity, including both spectrum allocations and backhaul links to potentially remote base stations, are important factors to an operator.

In addition to service differentiation, an important aspect is subscriber differentiation. The operator may provide differentiated treatment of the IP traffic for the same service depending on the type of subscription the user has. These subscriber groups can be defined in any way suitable to the operator, for example corporate vs. private subscribers, post-paid vs. pre-paid subscribers and roaming vs. non-roaming subscribers as illustrated in Figure 8.1.1.1.

The conclusion is that there is a need to standardize simple and effective QoS mechanisms for multi-vendor mobile broadband deployments. Such QoS mechanisms should allow the operator to enable service and subscriber differentiation and to control the performance experienced by the packet traffic of a certain service and subscriber group.

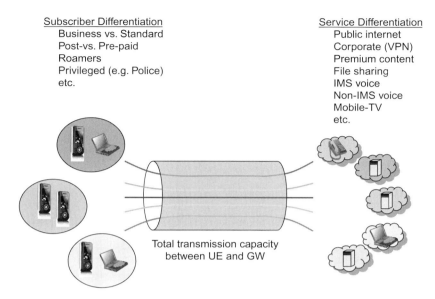

Figure 8.1.1.1   *Service and subscriber differentiation.*

8.1.2   QoS in E-UTRAN

8.1.2.1   General

Before going into details regarding the QoS parameters and mechanisms for E-UTRAN and EPS, we will put the EPS bearer QoS concept into a wider context.

The EPS only covers QoS requirements for the traffic within the EPS, that is, between UE and PDN GW. If the service extends beyond that, QoS is maintained by other mechanisms which, for example, depend on operator deployments and service level agreements (SLAs) between network operators. This book will not go further into those aspects.

The EPS bearer service was introduced in Chapter 6. The EPS bearer represents the level of granularity for QoS control in E-UTRAN/EPS and provides a logical transmission path with well-defined QoS properties between UE and the network. The QoS concepts of the EPS bearer is then mapped to the QoS concepts of the underlying transport. For example, over the E-UTRAN radio interfaces, the EPS bearer QoS characteristics are implemented using E-UTRAN specific traffic handling mechanisms. Each EPS bearer is transported over an E-UTRAN radio bearer with the corresponding QoS characteristics. In the 'backbone' network between eNB, Serving GW and PDN GW, the EPS bearer QoS may be mapped to IP transport layer QoS, for example, using DiffServ. In this book we will only very briefly touch upon the lower layer QoS mechanisms. An interested reader is referred to Dahlman (2008) for details regarding QoS mechanisms in the E-UTRAN radio layer.

8.1.2.2   Differences compared to QoS for pre-EPS GERAN/UTRAN

The QoS solutions for E-UTRAN have a few differences compared to the QoS solutions defined for GERAN/UTRAN. The two most prominent differences are described below.

Bearer control paradigm

The bearer control paradigm has changed. In GPRS Rel-6 and earlier, it was only the UE that could initiate a new bearer (i.e. a PDP context). The UE also controlled the traffic to bearer mapping information. Rel-7 was then amended with a NW-initiated procedure to establish bearers and traffic mapping information. EPS and E-UTRAN on the other hand implement a fully network-controlled bearer concept. The UE may request resources, but it is always the network that controls the EPS bearer state and the traffic to bearer mapping.

QoS parameters of a bearer

E-UTRAN also simplifies the QoS parameters that are associated with each bearer. As we will describe in more detail in later subsections, the EPS bearer is associated with two QoS parameters, a QoS class and an allocation and retention parameter. Certain EPS bearers also have associated bit rates resulting in a total of four QoS parameters for those EPS bearers.

Pre-EPS GPRS on the other hand defines a QoS concept for GERAN/UTRAN with 4 Traffic Classes and 13 different QoS Attributes. This QoS concept is often referred to as 'release 99 QoS' since it was introduced already for release 99 GPRS (completed in 2000). Each PDP context is assigned one of the four Traffic Classes together with values of the associated QoS Attributes. The QoS Attributes specify, for example, bit rates supported by the PDP context, the priority of the traffic, error rates, maximum transfer delay, etc. This has turned out to generate a complex system and many of the QoS Attributes are in practice not used. We will not go into any further details on Rel-99 QoS in this book. We will only discuss Rel-99 QoS on a high level to understand how GERAN/UTRAN is used together with EPS. The reader interested in more information on GRPS and Rel-99 QoS is instead referred to a book on GPRS.

Subscribed QoS parameters

In pre-EPS GPRS the subscribed QoS profile is the maximum QoS that could be allocated for each PDP context. A terminal activating multiple PDP contexts would thus have a subscribed QoS for each of them. In EPS however, the subscribed QoS stored in HSS only applies to the default bearer. There is no such thing as subscribed QoS for a dedicated bearer. Instead it is the PDN GW that determines the QoS of the dedicated bearer based on the authorized QoS received from the PCRF (Policy and Charging Rules Function). There is thus no need to have specific subscription parameters for dedicated bearers in the HSS. If the terminal is allowed to access a certain service, the PCRF will authorize the resources in the network.

Below we discuss the EPS bearer QoS in more detail. We then return to rel-99 QoS and GERAN/UTRAN to describe how QoS for GERAN/UTRAN works when GERAN/UTRAN is connected to the EPC.

8.1.2.3 QoS parameters of an EPS bearer

The basic aspects of the EPS bearer and its use for QoS purposes were already introduced in Section 6.2. In this section we discuss the QoS parameters of the EPS bearer in more detail.

Each EPS bearer has two QoS parameters associated with it: the QoS Class Identifier (QCI) and the Allocation and Retention Priority (ARP). As we will see below, the QCI determines what user plane treatment the IP packets transported on a given bearer should receive, while the ARP specifies the control plane treatment a bearer should receive. Some EPS bearers also have associated bit rate parameters to support allocation of a guaranteed bit rate (GBR) when establishing the bearer.

As we will see below, the bearer concept and the associated QoS mechanisms provide two essential features; traffic separation and resource-based admission control. The class-based QoS concept using QCI allows the network to separate between bearers carrying real-time and bearers carrying non-real time traffic. The network can then provide the appropriate forwarding treatment for each QoS class. For the support of services that require a certain GBR, the network reserves a GBR when establishing the corresponding bearer. These bearers are subject to admission control to ensure that sufficient resources are available before allowing the GBR bearer to be set up. These mechanisms are described further below.

QoS class identifier

The EPS uses a class-based QoS concept where each EPS bearer is assigned a QCI. This QCI is a number and the numerical value in itself does not represent any QoS property. The QCI is just a pointer, or reference, to node specific parameters, which defines what packet forwarding treatment a particular bearer should receive when processed in a node (e.g. scheduling weights, admission thresholds, queue management thresholds, link layer protocol configuration, etc.). The node specific parameters for each QCI have been pre-configured by the vendor designing the node or by the operator owning the node (e.g. eNodeB).

Allocation and retention priority

The ARP is used to indicate a priority for the allocation and retention of bearers. It is typically used by the network to decide whether a bearer establishment or modification can be accepted or needs to be rejected due to resource limitations.

In the 2G/3G core network, only three ARP values are supported by the packet core network. This was considered sufficient since emergency calls run over circuit-switched voice services only and the ARP mechanism in the packet-switched domain could thus be used for commercial purposes only. However, with the packet-only EPS, emergency services have to be supported also in the packet-switched domain. Therefore the ARP definition has been more aligned with the ARP used with circuit-switched services. For example, EPS supports 15 priority values for the ARP.

In situations where resources are scarce, the network can use the ARP to prioritize establishment and modification of bearers with a high ARP over bearers with a low ARP when performing admission control; note that bearers with high ARP are assigned low ARP *values*, and vice versa. For example, a VoIP call for emergency services should have a higher chance of being accepted than a regular VoIP call, and should thus be assigned a high ARP. ARP also supports pre-emption of bearers. In exceptional resource limitations, the network can use the ARP to decide which bearers to drop. This could, for example, occur at handover situations. Another exceptional circumstance is disaster situations where the ARP could be used to free up capacity by dropping low ARP bearers.

GBR and non-GBR bearers

One of the properties of a bearer is the bit rates it is associated with. We distinguish between two types of bearers; GBR bearers and non-GBR bearers. A GBR bearer has, in addition to the QoS parameters discussed above, also associated bit rate allocations: the GBR and Maximum Bit Rate (MBR). A non-GBR bearer does not have associated bit rate parameters.

A bearer with an associated GBR means that a certain amount of bandwidth is reserved for this bearer, independently of whether it is utilized or not. The GBR bearer thus always takes up resources over the radio link, even if no traffic is sent. The GBR bearer should in normal cases not experience any packet losses due to congestion in the network or radio link. This is ensured since GBR bearers are subject to admission control when they are set up. A GBR bearer is only allowed by the network if there are enough resources available. The MBR limits the bit rate that can be expected to be provided by a GBR bearer. Any traffic in excess of the MBR may be discarded by a rate shaping function. EPC currently only supports the case where the MBR and GBR are equal. In a future release there may be enhancements to allow a MBR which is larger than the GBR. One of the reason for such enhancements would be to better support rate adaptive codecs where a minimum bit rate is guaranteed by the network (the GBR) while additional bandwidth may be allowed if available (the MBR).

A non-GBR bearer does not have a fixed bandwidth allocated and there is thus no guarantee for how much traffic it can carry. The non-GBR bearer may therefore experience packet loss in case of congestion. The availability of radio resources for an existing non-GBR bearer thus depends on the total load of the cell as well as the QCI of the bearer. No transmission resources are reserved for non-GBR bearers. In EPS, non-GBR bearers are rate policed on aggregate level instead of on a per bearer (or PDP context) level. Consequently, even though the non-GBR bearers do not have any associated MBR, the operator may still police the utilized bandwidth of non-GBR bearers using the Aggregate Maximum Bit Rate (AMBR) as described further below.

Whether or not a bearer should be a GBR or a non-GBR bearer typically depends on what service is carried over that bearer. The GBR bearers are typically used for those services where it is better to block a service rather than degrade already admitted services in case resources are not available. Some services such as VoIP and streaming services benefit from a constant bandwidth and a GBR value may thus be needed to ensure a satisfactory user experience. If those resources are not available, it is better to block the service. Other services, such as Internet browsing, e-mail, chat programs, etc., normally do not require a constant fixed bandwidth. Those services would typically use non-GBR bearers. Whether to use GBR or non-GBR bearers for a certain service is up to operator configuration and can, for example, be controlled using the PCC framework. The choice depends to large extent on the expected traffic load compared to the available capacity.

Standardized QCI values and corresponding characteristics

Certain QCI values have been standardized to reference specific QoS characteristics. The QoS characteristics describe the packet forwarding treatment that the traffic for a bearer receives edge-to-edge between the UE and the GW in terms of certain performance characteristics such as priority, packet delay budget and packet error loss rate. The standardized characteristics are not signalled on any interface; they should instead be understood as guidelines for the pre-configuration of node-specific parameters for each QCI. For example, the radio base station would need to be configured to ensure that traffic belonging to a bearer with a certain standardized QCI receives the appropriate QoS treatment. The goal of standardizing a QCI with corresponding characteristics is to ensure that applications/services mapped to that QCI receive the same minimum level of QoS in multi-vendor network deployments and in case of roaming. The standardized QCI characteristics are defined in clause 6.1.7 in 3GPP TS 23.203 [23.203]. A simplified description can be found in Table 8.1.2.3.1.

The QCI values 1–4 are allocated for traffic that require dedicated resource allocation for a GBR, while values 5–9 are not associated with GBR requirements.

Table 8.1.2.3.1   *Standardized QCI characteristics.*

QCI	Resource Type	Priority	Packet Delay Budget	Packet Error Loss Rate	Example Services
1	GBR	2	100 ms	10^{-2}	Conversational Voice
2		4	150 ms	10^{-3}	Conversational video (live streaming)
3		3	50 ms	10^{-3}	Real-time gaming
4		5	300 ms	10^{-6}	Non-conversational video (buffered streaming)
5	Non-GBR	1	100 ms	10^{-6}	IMS signalling
6		6	300 ms	10^{-6}	Video (buffered streaming)TCP-based (e.g. www, e-mail, chat, ftp, p2p file sharing, progressive video, etc.)
7		7	100 ms	10^{-3}	Voice, video (live streaming) interactive gaming
8		8	300 ms	10^{-6}	Video (buffered streaming) TCP-based (e.g. www, e-mail, chat, ftp, p2p filesharing, progressive video, etc.)
9		9	300 ms	10^{-6}	Video (buffered streaming) TCP-based (e.g. www, e-mail, chat, ftp, p2p filesharing, progressive video, etc.)

Each standardized QCI is associated with a priority level, where priority level 1 is the highest priority level. The Packet Delay Budget can be described as an upper bound for the time that a packet may be delayed between the UE and the PCEF (Policy and Charging Enforcement Function). The Packet Error Loss Rate can in a simplified manner be described as an upper bound for the rate of non-congestion related packet losses.

Note that the description above gives a very simplified definition of the standardized QCIs, hiding many of the details. The purpose is to give the general reader a basic view of the topic. The interested reader should consult [23.203] for the complete definitions.

Apart from these standardized QCI, also non-standardized QCIs may be used. In this case it is the operators and/or vendors who define what node-specific parameters are used for a given QCI.

8.1.2.4   APN-AMBR and UE-AMBR

In addition to the bit rate parameters associated with each GBR bearer, EPS also defines the AMBR parameters that are associated with non-GBR bearers. These parameters are not specific to each non-GBR bearer, but rather define a total bit rate that a subscriber is allowed to consume for an aggregate of non-GBR bearers. The bit rate consumed by the GBR bearers does not count towards the AMBR. Two variants of the AMBR are defined; the APN-AMBR and the UE-AMBR.

One reason for why an aggregate rate policing of non-GBR bearers is preferable over a per-bearer policing is that network planning becomes easier. With a subscribed per-bearer MBR (as in GPRS) it is difficult to estimate the total bit rate the subscribers will use. Also, an AMBR can provide a more understandable subscription for the end-user compared to an MBR that is per bearer.

The APN-AMBR defines the total bit rate that is allowed to be used for all non-GBR bearers associated with a specific APN. This parameter is defined as part of a user's subscription but may be overridden by the PCRF. The APN-AMBR limits the total non-GBR traffic for an APN, independent of the number of PDN connections and non-GBR bearers that are opened for that APN. In other words, if a user has multiple PDN connections for the same APN, they all share the same APN-AMBR. For example, if an operator provides an APN for Internet access, the operator may then limit the total bandwidth for that APN and thus prevent the UE from increasing its accessible bandwidth by just opening new PDN connections to the same APN. This is different from

2G/3G core network where the subscribed QoS is defined per PDP context. The APN-AMBR is enforced by the PDN GW.

The UE-AMBR is defined per subscriber and defines the total bit rate allowed to be consumed for all non-GBR bearers of a UE. The subscription profile contains a subscribed UE-AMBR. However, the actual UE-AMBR value that is enforced by the network is set as the minimum of the subscribed UE-AMBR and the sum of the APN-AMBR of all active APNs (i.e. all APNs for which the UE has active PDN connections). The UE-AMBR is enforced by the eNB.

Different AMBR values are defined for uplink and downlink directions. There are thus in total four AMBR values defined: UL APN-AMBR, DL APN-AMBR, UL UE-AMBR and DL UE-AMBR.

The UE-AMBR and APN-AMBR are not dependent on each other and an operator may choose to apply either UE-AMBR or APN-AMBR (or both). The enforcement of the UE-AMBR and APN-AMBR are two tools for the operator to realize the business model. The UE-AMBR may be used to put upper limits on a subscription or to limit the total amount of traffic in the network. The APN-AMBR on the other hand is more related to, for example, Service Level Agreements with external PDNs or subscriptions related to specific APNs.

8.1.2.5   User plane handling

Some aspects of the user plane handling of EPS bearers were introduced already in Chapter 6. In particular it was there shown how packet filters in the UE and the GW are used to determine which IP flows shall be carried over a certain EPS bearer. Now, after the description of the mechanisms available for QoS control in E-UTRAN, it is again useful to look at user plane handling and how the QoS functions and QoS parameters described above are allocated to different nodes in the network. Figure 8.1.2.5.1 indicates different user plane QoS functions for E-UTRAN/EPS.

The UE and GW (PDN GW for GTP-based S5/S8 and Serving GW for PMIP-based S5/S8) carry out uplink and downlink packet filtering respectively in order to map the packet flows on to the intended bearer.

The GW and the eNB can implement functions related to admission control and preemption handling (i.e. congestion control) in order to allow these nodes to limit and control the load put on them. These functions can take the ARP value as an input in order to differentiate the treatment of different bearers in these functions, as described in the ARP section above.

The GW and eNB further implement functions related to rate policing. The goal of these functions is twofold: to protect the network from becoming overloaded

	UE	eNB	Transport network	PDN GW
Packet Filtering	**X** (uplink)			**X** (downlink)
GBR/ARP Admission		**X**		**X**
ARP Preemption		**X**		**X**
Rate Policing		**X**		**X**
Queue Management	**X**	**X**		
Uplink + Downlink Scheduling		**X**		
Configuring layer 1 and layer 2 protocols		**X**		
Map QCI to DSCP		**X**		**X**
Queue Management			**X**	
Uplink + Downlink Scheduling			**X**	

Functions operating per EPS bearer

Functions operating on transport layer, e.g. per DSCP

Figure 8.1.2.5.1 *Overview of user plane QoS functions for E-UTRAN/EPS. The functional allocation for the PDN GW is for GTP-based S5/S8. For PMIP-based S5/S8, the bearer-related functions are moved to the Serving GW.*

and to ensure that the services are sending data in accordance to the specified maximum bit rates (AMBR and MBR). For the non-GBR bearers, the PDN GW performs rate policing based on the APN-AMBR value(s) for both uplink and downlink traffic, while the eNB performs rate policing based on the UE-AMBR value for both uplink and downlink traffic. For GBR bearers, MBR policing is carried out in the GW for downlink traffic and in the eNB for up-link traffic.

In order to distribute radio network resources (radio and processing resources) between the established bearers, the eNB implements uplink and downlink scheduling functions. The scheduling function is to a large extent responsible for fulfilling the QoS characteristics associated to the different bearers.

The eNB is responsible for configuring the lower layer (layers 1 and 2) protocols of the radio connection of the bearer in accordance to the QoS characteristics associated with the bearer. Among others, this includes configuring the error-control protocols (modulation, coding, and link layer retransmissions) so that the QoS characteristics, packet delay budget and packet error loss are ful-filled. For more details on QoS handling in E-UTRAN, see Dahlman (2008).

On the transport level, that is, the basic IP transport between EPC network enti-ties (including intermediate transport entities in the packet cores such as regular IP routers), are not aware of the EPS bearers and instead queue management and packet forwarding treatment is done according to a transport layer mecha-nism such as DiffServ. The EPC entities map the QCI of the EPS bearer on to DiffServ Code Point (DSCP) values that are used by the transport network.

It should be noted that there are also service aware QoS control functions oper-ating on a finer granularity than the EPS bearer. These functions are defined as part of the Policy and Charging Control (PCC) architecture and are described in the PCC section of this chapter.

8.1.3   Interworking with GERAN/UTRAN

As mentioned above, E-UTRAN and the EPS have a different QoS control architecture compared to GERAN/UTRAN access – the pre-EPS QoS model is often referred to as the release 99 QoS, or Rel-99 QoS for short. When connect-ing GERAN/UTRAN accesses to EPS, via a S4-based SGSN, there are in theory two main alternatives for how to handle QoS in GERAN/UTRAN:

1. Implement the EPS QoS solutions also in GERAN/UTRAN access. This would, for example, imply that the QCI would be used for each PDP con-text, instead of the GERAN/UTRAN (Rel-99) QoS profile. This has the ben-efit that 3GPP family of accesses in EPS uses the same QoS parameters. It should however be noted that the GERAN/UTRAN radio interface has to be backwards compatible with the Rel-99 QoS scheme and PDP context procedures in order to allow pre-EPS terminals to connect. An EPS-based GERAN/UTRAN network would thus anyhow need to implement Rel-99 based QoS.
2. Keep the existing Rel-99 QoS and PDP context procedures for GERAN/ UTRAN. This would imply the least changes to the current GERAN/UTRAN radio interface, but there would be a need to specify a mapping to EPS-based QoS solutions and bearer procedures.

3GPP decided for the second alternative. The main motivation was that this was the simplest option that was also backwards compatible with pre-EPS terminals.

Please see the bearer section of Chapter 6 for a description of the interwork-ing between PDP context and EPS bearer procedures. In this section we look at the corresponding mapping between QoS parameters. Depending on scenario, different network entities need to perform mapping between QoS parameters associated with EPS bearer and the QoS parameters associated with the PDP context. The detailed mapping between the EPS QoS parameters and the rel-99 QoS parameters is defined in Annex E of 3GPP TS 23.401 [23.401] and briefly described below:

1. The ARP for a PDP context can take three possible values, while the ARP for an EPS bearer can take 15 possible values. The mapping between PDP context ARP values and EPS bearer ARP values is a one-to-many mapping. The exact scheme for which range of EPS bearer ARP values is mapped to an ARP value of a PDP context is not standardized but can be configured by each operator.
2. The GBR and MBR values are mapped one-to-one (for GBR bearers only).
3. The MBR for PDP contexts without GBR is mapped to/from APN-AMBR.
4. The mapping between the standardized QCIs and the rel-99 Traffic Class and QoS Attributes is described in Annex E of 3GPP TS 23.401. Only a subset of the Rel-99 QoS Attributes are specified by the mapping. The setting of the values of the other rel-99 QoS Attributes is not specified by the standard and is instead based on operator policy pre-configured in the SGSN.
5. In the first release of EPS (3GPP Rel-8), UE-AMBR is only used in E-UTRAN and has no counterpart when using 2G/3G. In the next release (Rel-9), UE-AMBR is applied also for 2G/3G accesses.

8.1.4 QoS aspects when interworking with other accesses

So far we have been considering QoS aspects related to the 3GPP family of accesses. EPS does however support interworking and mobility with other accesses as well, defined by other standardization bodies. Each such access may have its own set of QoS mechanisms and QoS parameters, as defined by the relevant standardization body. It is not possible within this book to go through each access and discuss the access-specific QoS solutions. There are, however, a few aspects that are either independent of access or related to the interworking between EPS and the access-specific QoS mechanism.

One QoS parameter that is common to all accesses is the APN-AMBR as described in the APN-AMBR section above. The APN-AMBR is enforced by the PND GW and can be enforced independent of which access the UE may be using.

Other access independent parameters are the QCI and ARP. As described above the QCI and ARP are parameters of the EPS bearer when using 3GPP family of accesses. As we will see in the next section on PCC, the PCC architecture also uses QCI and ARP, but as access independent parameters. When interworking with other accesses, these parameters are mapped by each individual access to access-specific parameters and mechanisms.

8.2   Policy and charging control

8.2.1   Introduction

PCC provides operators with advanced tools for service-aware QoS and charging control. In wireless networks, where the bandwidth is typically limited by the radio network, it is important to ensure an efficient utilization of the radio and transport network resources. Furthermore, different services have very different requirements on the QoS, which are needed for the packet transport. Since a network in general carries many different services for different users simultaneously, it is important to ensure that the services can co-exist and that each service is provided with an appropriate transport path.

PCC enables a centralized control to ensure that the service sessions are provided with the appropriate transport, for example, in terms of bandwidth and QoS treatment. The PCC architecture enables control of the media plane for both the IP Multimedia Subsystem (IMS) and non-IMS services. Furthermore, PCC also provides the means to control charging on a per-service basis.

When in 3GPP access, bearer procedures are available for QoS management in the access. While the EPS bearer and PDP context procedures are specific to the 3GPP family of accesses, corresponding QoS procedures exist for many other accesses as well. In this section we focus on how the operator can *control* those the QoS procedures and the charging mechanisms used for each service session.

When it comes to PCC, the term 'bearer' is used in a more generic fashion to denote an IP transmission path with well-defined characteristics (e.g. capacity, delay and bit error rate). This allows us to use the bearer terminology in an access agnostic fashion, independent of the details for how this transmission path is created or how QoS is managed for each access technology.

The term 'service session' is also important here. The bearer concept handles traffic aggregates, that is, all conformant traffic that is transported over the same bearer receives the same QoS treatment. This means that multiple service sessions transported over the same bearer will be treated as one aggregate. These bearer concepts still apply when PCC is used. As we will see however, PCC adds a 'service aware' QoS and charging control mechanism that in certain aspects is more fine grained; that is, it operates on a per-service session level rather than on a per-bearer level.

The PCC architecture for EPS is an evolution of the PCC architecture defined in 3GPP Rel-7. PCC has nevertheless evolved significantly from 3GPP Rel-7 in order to support new features in EPS, such as multiple access technologies, roaming and multi-access mobility. The goal in 3GPP has been to define an access-agnostic policy control framework, and as such, make it applicable to a number of accesses such as E-UTRAN, UTRAN, GERAN, HRPD and WiMAX. Furthermore, the introduction of a complete roaming model for PCC allows operators to have the same dynamic PCC, and provide the same access to services independently of whether a user is making this access through a gateway in their home or visited network.

It is also worth mentioning that standardization bodies standardizing other access technologies for fixed or wireless access have also created policy control specifications targeting their particular access technologies. When it comes to wireless accesses such as WiMAX and HRPD, an alignment towards a common policy control architecture based on 3GPP PCC has already been materialized with the EPS. For the fixed accesses, in particular related to the standardization work being done in ETSI TISPAN and in Broadband Forum (BBF) the work towards alignment and/or interworking has not come as far as for the wireless accesses.

8.2.2 The PCC Architecture

The basic aspects of the EPS architecture, including the PCC aspects, were introduced already in Chapter 3. In this section we give a more in-depth description as well as describe the basic concepts and functions of PCC.

The reference network architecture for PCC in EPS is shown in Figure 8.2.2.1. The functional entities that are part of the PCC architecture are briefly described below.

The Application Function (AF) interacts (or intervenes) with applications or services that require dynamic PCC. Typically the application level signalling for the service passes through, or is terminated, in the AF. The AF extracts session information from the application signalling and provides this to the PCRF over the Rx reference point. The AF can also subscribe to certain events that occur at the traffic plane level (i.e. events detected by either PCEF or BBERF (Bearer Binding and Event Reporting Function)). Those traffic plane events include events such as IP session termination or access technology type change. When

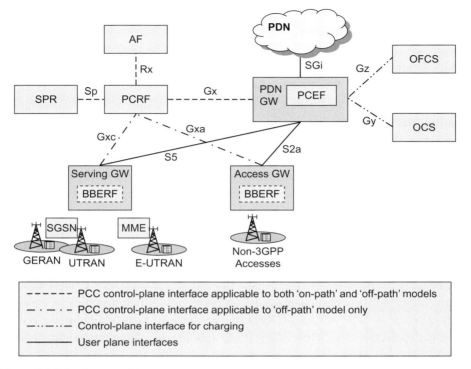

Figure 8.2.2.1   *3GPP Rel-8 PCC non-roaming architecture for EPS. Note that only a subset of the EPS reference points and EPS network entities are shown.*

the AF has subscribed to a traffic plane event, the PCRF will inform the AF of its occurrence. The term 'Application Function' is a generic term used by PCC for this entity, and in practice the AF functionality is contained within a specific network entity depending on the type of service. For the IMS, the AF corresponds to the P-CSCF. For a non-IMS service, the AF could, for example, be a *video streaming server*.

The Subscription Profile Repository (SPR) contains subscription information, such as user specific policies and data.

The Online Charging System (OCS) is a credit management system for pre-paid charging. The PCEF interacts with the OCS to check out credit and report credit status.

The Offline Charging System (OFCS) is used for offline charging. It receives charging events from the PCEF and generates Charging Data Records (CDRs) that can be transferred to the billing system.

The PCRF is the policy control function of PCC. It receives session information over Rx as well as information from the access network via the Gx. If a BBERF is used (see below), the PCRF also receives information via Gxa/Gxc reference points. (Gxb was defined between ePDG and PDN GW but is not used in the first release of EPS.) The PCRF may as well receive subscription information from the SPR. The PCRF takes the available information, as well as configured operator policies, into account and creates service-session level policy decisions. The decisions are then provided to the PCEF and the BBERF. Another task of the PCRF is to forward event reports between the BBERF, the PCEF and the AF.

The PCEF enforces policy decisions (e.g. gating, maximum bit rate policing) received from the PCRF and also provides the PCRF with user- and access-specific information over the Gx reference point. The PCEF may also perform measurements of user plane traffic (e.g. user plane traffic volume and/or time duration of a session). It reports usage of resources to the OFCS and interacts with the OCS for credit management.

In the PCC architecture for EPS, there are two main architecture alternatives; with and without BBERF in the Access GW (e.g. Serving GW or HSGW). In common language the two alternatives are referred to as 'off-path' and 'on-path' models respectively. The BBERF supports a subset of the functions supported by the PCEF. The details regarding the BBERF and the two architecture alternatives are discussed further below.

8.2.2.1   Multi-access and the off-path PCC model

As described in Section 6.4, EPS supports different mobility protocols depending on which access technology is used. For the 3GPP family of accesses (GERAN, UTRAN and E-UTRAN) the GTP or PMIPv6 may be used on the S5/S8 reference points. For connecting other accesses to EPC, it is possible to use PMIPv6, DSMIPv6, Mobile IPv4 (MIPv4) on S2a/b/c reference points. These different protocols have different properties when it comes to how the EPS bearers are implemented. These differences result in different requirements on PCC.

When GTP is used between the Serving GW and the PDN GW, the bearers are terminated in the PDN GW and the PDN GW can thus use the bearer procedures to control the EPS bearers. We refer to this model as the 'on-path' model because the QoS/bearer signalling takes place (using GTP) on the same 'path' as the user plane. In this model, the PCRF controls the QoS by providing the QoS policy information to the PCEF via the Gx reference point. The BBERF and Gxa/Gxc have no role here and are thus not used at all in the 'on-path' model.

When a Mobile IP-based protocol, such as PMIP or DSMIPv6, is used towards the PDN GW, the bearers and QoS reservation procedures are terminated closer to the (radio) access network and the PDN GW has thus no knowledge about bearers. For 3GPP family of accesses, the bearers only extend between the UE and the Serving GW. Between the Serving GW and the PDN GW there is no notion of EPS bearers. See the bearer section in Chapter 6 for illustrations. For other accesses, the bearers and QoS reservation procedures (if existing) extend between the UE and an 'access GW' in the access network. In this case, the PDN GW only handles mobility signalling towards the access network and the UE, not any QoS signalling. Therefore the PDN GW cannot control the QoS using bearer procedures and it is not sufficient for the PCRF to provide the QoS information to the PCEF. The PCRF has to provide the QoS info to the entity where the bearers are terminated. For this purpose, the BBERF and the Gxa/Gxc reference points are introduced.

When it comes to other functions of PCC, not related to where bearers are terminated, there is in most cases no difference between the 'on-path' and 'off-path' models. For example, the service aware charging functionality is always located in the PCEF. Further details on the functional content of PCEF and BBERF can be found in later subsections.

For EPS, the PCEF is always located in the PDN GW. The BBERF location, however, depends on the particular access technology. For example, for the 3GPP family of accesses, the BBERF (if applicable) is located in the Serving GW, whereas for eHRPD access the BBERF is located in the HSGW. Since the PDN GW is the mobility anchor for the UE, the same PCEF is kept during the whole IP session. The BBERF allocated for a UE may however change due to the mobility of the UE. For example, the Serving GW may change as the UE moves within the 3GPP accesses. The BBERF location will also change when the UE moves between 3GPP and other access technologies. Support for BBERF relocation is thus an inherent part of the off-path PCC architecture for EPS.

8.2.3   Basic PCC concepts

8.2.3.1   PCC
As the name suggests, the purpose of PCC is policy and charging control.

Policy control is a very generic term and in a network there are many different policies that could be implemented, for example, policies related to security, mobility, use of access technologies etc. When discussing policies, it is thus important to understand the context of those policies. When it comes to PCC, policy control refers to the two functions gating control and QoS control:

1. Gating control is the capability to block or to allow IP packets belonging to IP flow(s) for a certain service. The PCRF makes the gating decisions which are then enforced by the PCEF. The PCRF could, for example, make gating decisions based on session events (start/stop of service) reported by the AF via the Rx reference point.
2. QoS control allows the PCRF to provide the PCEF with the authorized QoS for the IP flow(s). The authorized QoS may, for example, include the authorized QoS class and the authorized bit rates. The PCEF or BBERF enforces the QoS control decisions by setting up the appropriate bearers. The PCEF also performs bit rate enforcement to ensure that a certain service session does not exceed its authorized QoS.

Charging Control includes means for both offline and online charging. The PCRF makes the decision on whether online or offline charging shall apply for a certain service session, and the PCEF enforces that decision by collecting charging data and interact with the charging systems. The PCRF also controls what measurement method applies, that is, whether data volume, duration, combined volume/duration or event-based measurement is used. Again it is the PCEF that enforces the decision by performing the appropriate measurements on the IP traffic passing through the PCEF.

With online charging, the charging information can affect, in real-time, the services being used and therefore a direct interaction of the charging mechanism with the control of network resource usage is required. The online credit management allows an operator to control access to services based on credit status. For example, there has to be enough credit left with the subscription in order for the service session to start or an ongoing service session to continue. The OCS may authorize access to individual services or to a group of services by granting credits for authorized IP flows. Usage of resources is granted in different forms. The OCS may, for example, grant credit in the form of certain amount of time, traffic volume or chargeable events. If a user is not authorized to access a certain service, for example, in case the pre-paid account is empty, then the OCS may deny credit requests and additionally instruct the PCEF to redirect the service request to a specified destination that allows the user to re-fill the subscription.

PCC also incorporates service-based offline charging. With offline charging, the charging information is collected by the network for later processing and billing.

Therefore, the charging information does not affect, in real-time, the service being used. Since billing is taking place after the service session has completed, for example, via a monthly bill, this functionality does not provide any means for access control in itself. Instead policy control must be used to restrict access and then service-specific usage may be reported using offline charging.

Online and offline charging may be used at the same time. For example, even for billed (offline charged) subscriptions, the online charging system may be used for functionality such as Advice of Charge. Conversely, for prepaid subscribers, the offline charging data generation may be used for accounting and statistics.

8.2.3.2   PCC decisions, the PCC rule and the QoS rule

The PCRF is the central entity in PCC making PCC decisions. The decisions can be based on input from a number of different sources, including:

*   Operator configuration in the PCRF that defines the policies applied to given services.
*   Subscription information/policies for a given user, received from the SPR.
*   Information about the service received from the AF.
*   Information from the access network about what access technology is used, etc.

The PCRF provides its decisions in the form of so called 'PCC rules'. The PCRF also provides a subset of the information in the so called 'QoS rules' to the BBERF if the 'off-path' model is used. In this section we first describe the main content of the PCC rules and then the subset of information contained in a QoS rule.

A PCC rule contains a set of information that is used by the PCEF and the charging systems. First of all it contains information (in a so called 'Service Data Flow (SDF) template') that allows the PCEF to identify the IP packets that belong to the service session. All IP packets matching the packet filters of a SDF template are designated a SDF. The filters in a SDF template contain a description of the IP flow and typically contain the source and destination IP addresses, the protocol type used in the data portion of the IP packet, as well as the source and destination port numbers. These five parameters are often referred to as the IP 5-tuple. It is also possible to specify other parameters from the IP headers in the SDF template. The PCC rule also contains the gating status (open/closed) as well as QoS and charging-related information for the SDF. The QoS information for a SDF includes the QCI, MBR, GBR and ARP. The definition of the QCI is the same as that described in the QoS section in Chapter 8 for the EPS bearer and the reader is referred to that section for a more thorough description of those parameters. However, one important aspect of the QoS parameters in the PCC rule is that they have a different scope than the QoS parameters of the EPS bearer. The QoS and charging parameters in the PCC rule apply to the SDF. More precisely, the QCI, MBR, GBR and ARP in the PCC rule apply to the IP flow described by the SDF template, while the QCI, MBR, GBR and ARP discussed in Chapter 8 applies for the EPS bearer. A single EPS bearer may be used to carry traffic described by multiple PCC rules, as long as the bearer provides the appropriate QoS for the service data flows of those PCC rules. We will further below discuss more on how PCC rules and SDFs are mapped to bearers. Table 8.2.3.2.1 lists

Table 8.2.3.2.1   *A subset of the elements that may be included in a dynamic PCC rule. Text copied from 23.203.*

Type of element	PCC rule element	Comment
Rule identification	Rule identifier	It is used between PCRF and PCEF for referencing PCC rules
Information related to Service Data Flow detection in PCEF and BBERF	Service Data Flow Template Precedence	List of packet filters for the detection of the service data flow Determines the order, in which the service data flow templates are applied at PCEF
Information related to policy control (i.e. gating and QoS control)	Gate status	Indicates whether a SDF may pass (gate open) or shall be discarded (gate closed)
	QoS Class Identifier (QCI)	Identifier that represents the packet forwarding behaviour of a flow.
	UL and DL Maximum bit rates	The maximum uplink (UL) and downlink (DL) bitrates authorized for the service data flow
	UL and DL Guaranteed bit rates	The guaranteed uplink (UL) and downlink (DL) bitrates authorized for the service data flow
	ARP	The Allocation and Retention Priority for the service data flow
Information related to charging control	Charging key	The charging system uses the charging key to determine the tariff to apply for the service data flow
	Charging method	Indicates the required charging method for the PCC rule Values: online, offline or no charging
	Measurement method	Indicates whether the SDF data volume, duration, combined volume/ duration or event shall be measured

a subset of the parameters that can be used in a PCC rule sent from PCRF to PCEF. For a full list of parameters, please see 3GPP TS 23.203 and 3GPP TS 29.212 [23.203, 29.212].

The same standardized QCI values and QCI characteristics outlined in Section 8.1 applies when QCI is used in the PCC rule. The standardized QCI and corresponding characteristics is independent of the UE's current access. The access network receiving the PCC rule will thus map the QCI value of the PCC rule on to any access-specific QoS parameters that apply in that access. This is further elaborated below.

The discussion so far has assumed a case where the PCRF provides the PCC rules to the PCEF using Gx. These rules, which are dynamically provided by the PCRF, are denoted 'dynamic PCC rules'. There is however also a possibility for the operator to configure PCC rules directly into the PCEF. Such rules are referred to as 'pre-defined PCC rules'. In this case the PCRF can instruct the PCEF to activate such pre-defined rules by referring to a PCC rule identifier. While the packet filters in a dynamic PCC rule are limited to the IP header parameters (the IP 5-tuple and other IP header parameters), filters of a PCC rule that is pre-defined in the PCEF may use parameters that extend the packet inspection beyond the IP 5-tuple. Such filters are sometimes referred to as Deep Packet Inspection (DPI) filters and they are typically used for charging control where more fine grained flow detection is desired. The definition of filters for pre-defined rules is not standardized by 3GPP.

As described above, in case the 'off-path' model applies the PCRF need to provide the QoS information to the BBERF via the Gxa/Gxc reference points. The QoS information provided to the BBERF is the same as is present in the corresponding PCC rule. However, since the BBERF only needs a subset of the information available in a PCC rule, the PCRF does not send the full PCC rule to the BBERF. Instead the PCRF generates a so called 'QoS rule' with information from the corresponding PCC rule. The QoS rules contain the information needed for the BBERF to ensure that bearer binding (see below) can be performed. The QoS rules thus contain the information needed to detect the SDF (i.e. SDF template and precedence) as well as the QoS parameters (e.g. QCI and bit rates). QoS rules do not contain any charging-related information.

8.2.3.3   Use cases

As a result of the interactions with the PCRF, the PCEF and BBERF perform several different functions. In this subsection we present two use cases in order to get an overview of the dynamics of PCC and how PCC interacts with the application level as well as the access network level. Some of the aspects brought up in the use cases will actually be discussed in more detail later. The intent with placing the use cases first is that a basic overview of the procedures described in the use cases should simplify the understanding of the PCC aspects being discussed in the later subsections.

The first use case is intended to illustrate a service session setup using 'on-path' PCC, network-initiated QoS control and online charging (Figure 8.2.3.3.1).

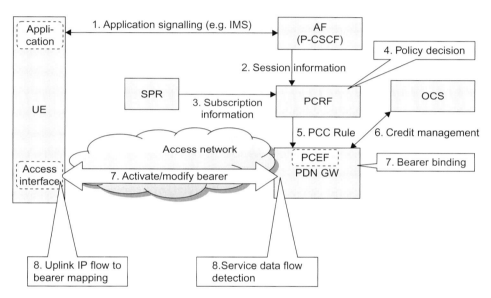

Figure 8.2.3.3.1 *High level use case for PCC in the EPS for the 'on-path' model and NW-initiated bearer procedures.*

The first use case is described below:

1. The subscriber initiates a service, for example, an IMS voice call, and performs end-to-end application session signalling that is intercepted by the AF (P-CSCF in the IMS case). In the IMS case, the application signalling is using the Session Initiation Protocol (SIP). A description of the service is provided as part of the application signalling. In IMS, the Session Description Protocol (SDP) is used to describe the sessions.

2. Based on service description information contained in the application signalling, the AF provides the PCRF with the service-related information over the Rx interface. The session information is mapped at the AF from a SDP (e.g. SIP/SDP for IMS) into information elements in the Rx messages to the

PCRF. This information typically includes QoS information (type of service, bit rate requirements) as well as traffic parameters (e.g. the IP 5-tuple) that allow identification of the IP flow(s) corresponding to this service session.

3. The PCRF may request subscription-related information from the SPR. (The PCRF may have requested subscription information earlier but it is shown at this step for illustrative purposes.)

4. The PCRF takes the session information, operator-defined service policies, subscription information and other data into account when building policy decisions. The policy decisions are formulated as PCC rules.

5. The PCC rules are sent by the PCRF to the PCEF. The PCEF will enforce the policy decision according to the received PCC rule. All user plane traffic for a given subscriber and IP connection pass through the network entity where the PCEF is located. For EPS, the PCEF is located in the PDN GW.

6. If the PCC rule specified that online charging shall be used for this PCC rule, the PCEF contacts the OCS via the Gy reference point to request credit according to the measurement method specified in the PCC rule.

7. The PDN GW (PCEF) installs the PCC rules and performs bearer binding to ensure that the traffic for this service receives appropriate QoS. This may result in the establishment of a new bearer, or a modification of an existing bearer. More details on bearer binding is provided further below.

8. The media for the service session is now being transported across the network and the PCEF performs SDF detection to detect the IP flow for this service. This IP flow is transported over the appropriate bearer. Further details on SDF detection can be found below.

The second example intends to illustrate the same basic use case but in a different network scenario using 'off-path' PCC, UE-initiated QoS control and offline charging. With UE-initiated QoS control, the UE and the network rely on UE-initiated triggers that start the bearer operations for this application. More discussion on UE- and network-initiated QoS control principles can be found below. Since offline charging is used, the PDN GW (PCEF) does not perform credit-based access control. The interactions with the charging system are therefore not shown in the use case.

It can be noted that the first three steps in the second use case are the same as in the first use case. These steps concern the application level signalling and the Rx signalling. This signalling is not dependent on access network properties, such as whether on-path or off-path PCC is used or whether UE-initiated or NW-initiated procedures are used. It is only the handling in the PCRF and in the access network that differ depending on PCC architecture model and UE/NW-trigger of bearer procedures.

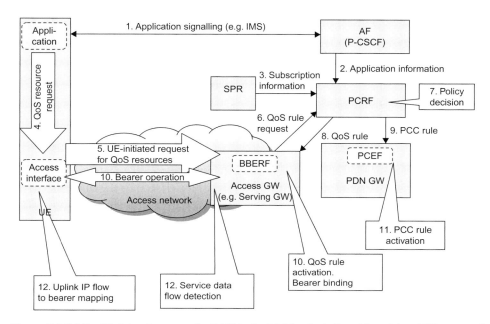

Figure 8.2.3.3.2 *High level use case for PCC in the EPS for the 'off-path' model and UE-initiated bearer procedures.*

The use case with 'off-path' model and UE-initiated procedures is described below (Figure 8.2.3.3.2). The first three steps are included in shortened form below. The full description can be found in the first use case.

1. The subscriber initiates a service, for example, an IMS voice call, and performs IMS session signalling via the AF.
2. Based on service description information contained in the application signalling, the AF provides the PCRF with the service-related information over the Rx interface.
3. The PCRF may request subscription-related information from the SPR.

The difference between UE-initiated and NW-initiated procedures will now become evident. In the first use case, the PCRF 'pushed' the rules to the PDN GW, and the PDN GW initiated bearer procedures to ensure that the service receives the appropriate QoS treatment. In the second use case, the PCRF instead waits until the request from the UE triggers a 'pull' of rules from the PCRF. This is described in the steps below:

4. The application in the UE makes a (internal) request for the access interface to request the QoS resources needed by the newly started application.

5. The UE sends a request to the network for QoS resources for this service. The UE includes the QoS class and packet filters associated with the service. The UE may also include a request for a certain GBR. The exact details regarding this request depend on which access technology the UE is using. For E-UTRAN, the UE would send a UE-requested bearer resource modification. For GERAN/UTRAN access, the UE would make a secondary PDP context activation or modification request. Other accesses may provide similar access-specific signalling.

6. Since off-path PCC is used, the BBERF initiates a PCRF interaction when receiving the request sent by the UE. (For comparison; in the on-path model, the request from the UE would have been forwarded from the Serving GW to the PDN GW, and the PDN GW would have sent a request for PCC rules to the PCRF.)

7. Similar to the first use case (step 4 in the first use case), the PCRF takes the session information, operator-defined service policies, subscription information and other data into account when building policy decisions. The policy decisions are formulated as PCC rules. Since off-path PCC is used, the PCRF also derives QoS rules based on the PCC rules.

8. The PCRF sends the QoS rules to the BBERF.

9. The PCRF sends the corresponding PCC rules to the PCEF.

10. The BBERF (e.g. Serving GW) installs the QoS rules and performs bearer binding to ensure that the traffic for this service receives appropriate QoS. This may result in the establishment of a new bearer, or the modification of an existing bearer.

11. The PDN GW (PCEF) installs the PCC rules. The PCEF performs gating, bit rate enforcement and service level charging as defined by the PCC rule.

12. The media for the service session is now being transported across the network. The UE uses uplink packet filters to determine which bearer shall carry the uplink traffic. Both the BBERF and PCEF perform SDF detection to detect the IP flow for this service. The BBERF forwards the downlink traffic over the appropriate bearer.

Note that the above two use case examples are not exhaustive in any sense. There are many other scenarios and configurations. For example, for services that do not provide an AF or Rx interface, it is still possible to use PCC. In that case step 2 of the second use case would be missing and the PCRF would authorize PCC/QoS rules based on pre-configured policies without access to dynamic session data.

8.2.3.4   Bearer binding

The PCC rule needs to be mapped to a corresponding bearer in the access network to ensure that the packets receive the appropriate QoS treatment. This mapping is one of the central components of PCC. The association between a PCC/QoS rule and a bearer is referred to as bearer binding. The bearer binding is done by the Bearer Binding Function (BBF) which is located either in the PCEF (for on-path) or in the BBERF (for off-path). When the PCEF (or BBERF) receives new or modified PCC or QoS rules, the BBF evaluates whether or not it is possible to use the existing bearers. If one of the existing bearers can be used, for example, if a bearer with the corresponding QCI and ARP already exists, the BBF may initiate bearer modification procedures to adjust the bit rates of that bearer. If it is not possible to use any existing bearer and NW-initiated bearer procedures are used, the BBF initiates the establishment of a suitable new bearer. In particular, if the PCC rule contains GBR parameters, the BBF must also ensure the availability of a GBR bearer which can accommodate the traffic for that PCC rule. If NW-initiated bearer procedures are used, the BBF triggers resource reservation in the access network to ensure that the authorized QoS of the PCC rule can be provided. Further details on the bearer concept can be found in Chapter 6.

For EPS and in case the UE is using 3GPP accesses, the BBF uses the EPS bearer procedures when activating the PCC/QoS rules. Other accesses interworking with EPS may have other, access specific, QoS signalling mechanisms. It is the task of the BBF to interact with the appropriate QoS procedures depending on access technology. In order to set up the right QoS resources in the access, the PCEF/BBERF not only need to invoke the appropriate QoS procedures but may also need to map the QoS parameters. In particular the BBF must map the QCI of the PCC/QoS rule, which is an access independent parameter, to access specific QoS parameters. For the 3GPP family of accesses in EPS this is simple since the QCI is used also as a QoS parameter for the EPS bearers. For other accesses, the mapping may involve a 'translation' from the QCI in the PCC/QoS rule to other access specific QoS parameters that are used in that particular access.

8.2.3.5 Service Data Flow detection

Once the service session is set up and the media traffic for the service is flowing, the PCEF and the BBERF use the packet filters of installed PCC and QoS rules to classify IP packets to authorized SDFs. This process is referred to as SDF detection. Each filter in the SDF filter is associated with a precedence value. The PCEF (or BBERF) matches the incoming packets against the available filters of the installed rules in order of precedence. The precedence is important if there is an overlap between filters in different PCC rules. One example of such overlap is a PCC rule which contains a wildcard filter that overlaps with more narrowly scoped filters in other PCC rules. In this case, the wildcard filter should be evaluated after the more narrowly scoped filters; otherwise the wildcard filter will

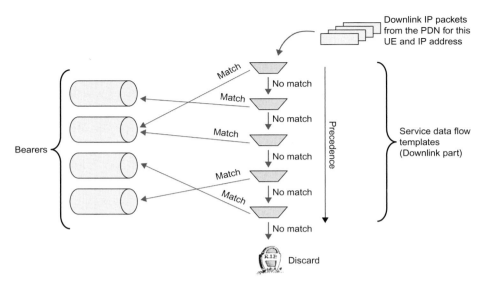

Figure 8.2.3.5.1   *Example of SDF detection and mapping to bearers for downlink traffic.*

cause a match before the PCEF/BBERF even tries the narrowly scoped filters. If a packet matches a filter, and the gate of the associated rule is open, then the packet may be forwarded to its destination. For the downlink part, the classification of an IP-packet to an SDF also determines which bearer should be used to transfer the packet (Figure 8.2.3.5.1). See also Section 6.3 for more details on bearers and how packet filters are used to direct packets on to the right bearer.

An additional aspect related to SDF detection occurs when DSMIPv6 is used as a mobility protocol. In that case the user plane traffic is tunnelled between UE and PDN GW, and thus also when passing through the BBERF; see Section 11.3 for more details on DSMIPv6. Since the packet filters of the PCC rule refer to un-tunnelled packet flows, the BBERF has to 'look inside' the DSMIPv6 tunnel to be able to apply the packet filters in the SDF template. This is something often referred to as 'tunnel look-through' and is illustrated in Figure 8.2.3.5.2. The outer tunnel header is determined when the DSMIPv6 tunnel is established by the UE and the PDN GW. Information about the tunnel header, that is, the outer header IP addresses, etc., is sent from the PDN GW to the BBERF via the PCRF so that the BBERF can apply the right packet filters for the tunnel.

8.2.3.6   Events and renewed policy decisions

When the PCRF makes a policy decision, information received from the access network may be used as input. For example, the PCRF may be informed about the current access technology used by the UE, or whether the user is in their home

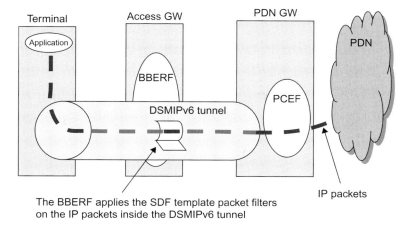

The BBERF applies the SDF template packet filters
on the IP packets inside the DSMIPv6 tunnel

Figure 8.2.3.5.2 *BBERF 'tunnel look-through' in case DSMIPv6 is used.*

network or is roaming. During the lifetime of a session, the conditions in the access network may however change. For example, the user may move between different access technologies or different geographical areas. There may also be situations where a certain authorized GBR can no longer be maintained over the radio link. In these cases, the PCRF may want to re-evaluate its policy decisions and provide new or updated rules to the PCEF (and BBERF, if applicable). The PCRF should thus be able to keep itself up-to-date about events taking place in the access network. To achieve this, procedures have been defined that allow the PCRF to notify the PCEF/BBERF about which events the PCRF is interested in. In PCC terminology we say that the PCRF subscribes to certain events, and that the PCEF/BBERF sets the corresponding event triggers. When an event occurs, and the corresponding event trigger is set, the PCEF/BBERF will report the event to the PCRF and allow the PCRF to revisit its previous policy decisions.

In the 'on-path' model, information about the access network (e.g. information regarding available QoS on radio link, etc.) is available in the PDN GW and the PCEF can thus report on any status change via the Gx reference point. As mentioned above, there is no need for the BBERF in this case. In the 'off-path' model, however, the PCRF will need to subscribe to events either in the PCEF or in the BBERF, depending on the nature of the event. With Mobile IP-based protocols, the access-specific bearers terminate in the BBERF instead of the PCEF. This implies that certain information about the access network is only available to the BBERF. Therefore, the BBERF detects such events and reports them over the Gxa/Gxc reference points. Other events, such as events related to multi-access mobility, are only known to the PCEF and thus reported by the PCEF also in the 'off-path' model.

In the 'off-path' model, the Gxa/Gxc and Gx interfaces will also be used for more generic parameter transfer. Some of the information provided by the BBERF is also needed in the PDN GW/PCEF. For example, the PDN GW may need to know which 3GPP radio technology is used (GERAN, UTRAN or E-UTRAN) to enable proper charging and this information is not necessarily provided via the PMIP-based S5/S8 reference point. It must then be provided by the BBERF to the PCEF (PDN GW), via the PCRF, as illustrated in Figure 8.2.3.6.1.

Also the AF may be interested in notifications about conditions in the access network, such as what access technology is used or the status on the connection with the UE. Therefore the AF may subscribe to notifications via the Rx reference point. In this case it is the PCRF that reports to the AF. The notifications over Rx are not directly related to renewed policy decisions in the PCRF, but event triggers play a role also here. The reason is that in case the AF subscribes to a notification over Rx, the PCRF will need to subscribe to a corresponding event via Gx or Gxa/Gxc interface.

Figure 8.2.3.6.1 shows a high-level summary of the information flow in a PCC architecture.

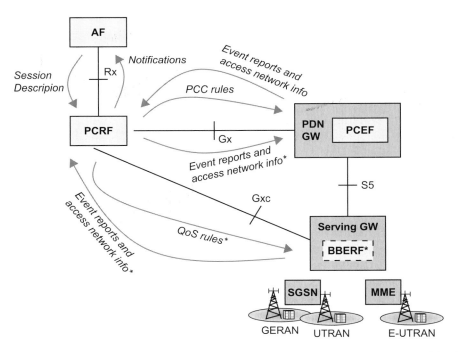

Figure 8.2.3.6.1   *High-level information flow. Items marked with * apply to off-path model only. Only 3GPP family of accesses is shown in the figure for simplicity.*

8.2.3.7 Functional allocations

Most functions of the PCEF are common to both 'on-path' and 'off-path' models. For example, service level charging, gate control, QoS enforcement and event reporting are done in the PCEF in both cases. However, as we have also seen above, the bearer-related functions and certain event reporting need to be performed by the BBERF in the 'off-path' case. Table 8.2.3.7.1 summarizes the allocation of different functions in the two architecture alternatives.

Table 8.2.3.7.1 *Allocation of functions in 'on-path' and 'off-path' models.*

	'on-path' (BBERF is not used)	'off-path' (BBERF is used)
Service level charging (flow based charging)	PCEF	PCEF
Service level gating control	PCEF	PCEF
Service level UL and DL bit rate enforcement	PCEF	PCEF
Bearer binding	PCEF	BBERF
Event reporting	PCEF	BBERF and PCEF

8.2.4 Network vs. terminal-initiated QoS control

As was already indicated in the two use cases above, there are two basic methods for initiating the QoS allocation in the access; either triggered by the UE or triggered by the network. We refer to these as the terminal-initiated and network-initiated QoS control paradigms. Below we look in more detail at a few general aspects of the two paradigms.

Originally, GPRS only supported the UE-initiated QoS control paradigm. To use UE-initiated procedures was very reasonable since there were actually no means to trigger resource reservation procedures from the network until policy control was introduced in 3GPP. However, in 3GPP Rel-7, when PCC was developed, a mechanism became available to trigger QoS resource reservation from the network based on application signalling. To support the network-initiated QoS control paradigm, the network-initiated secondary PDP context activation procedure was introduced in Rel-7 GPRS. For EPS, both network-initiated and terminal-initiated procedures are supported for GERAN/UTRAN and E-UTRAN. CDMA2000 systems specified by 3GPP2, including the eHRPD, support terminal-initiated procedures in general, while network-initiated procedures have been introduced for the eHRPD system.

With the terminal-initiated QoS control paradigm, it is the terminal that initiates the signal to set up a specific QoS towards the network. For the particular case of

E-UTRAN, the terminal would send a request for bearer resources to the network. The application in the terminal must know what QoS that it wants, and trigger the access interface part in the terminal (e.g. the E-UTRAN part) over a terminal-internal 'interface', or Application Programming Interface (API). This API is typically not standardized and may, for example, differ between terminal vendors and access technologies; the usage of the API is illustrated in step 4 of the second use case in Section 8.2.3.3. This means that in order to specify the QoS information for the access, the client applications need to be aware of the specifics of the QoS model of the access network. With this paradigm, there is no need for a PCRF to push QoS information to the network. A PCRF may however still be used to authorize the QoS resources requested by the terminal, as was illustrated in the second use case above. The terminal-initiated QoS control principles have been illustrated in Figure 8.2.4.1.

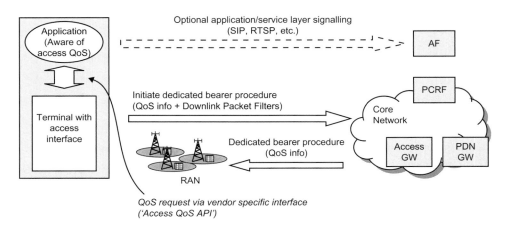

Figure 8.2.4.1   *Terminal-initiated QoS control.*

Using network-initiated QoS control, it is the network that initiates the signal to set up specific QoS towards the terminal and the radio network. For the particular case of E-UTRAN, it would be the network that initiates the dedicated bearer procedure. The trigger for this signal is received from other network nodes, typically an AF in combination with a PCRF. The signalling is sent over standardized references points such as Rx and Gx. This scenario is described by the first use case in Section 8.2.3.3 and has also been illustrated in Figure 8.2.4.2.

Using the network-initiated paradigm, the client application does not need to worry about the specifics of the QoS model of the access network. The application in the terminal can instead rely on the network to ensure that the access-specific QoS procedures are executed as needed. The application may however

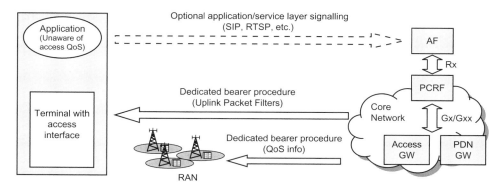

Figure 8.2.4.2   *Network-initiated QoS control.*

have access agnostic knowledge of the QoS that it wants to be provided with and make a request for this QoS via the application layer. For example, the QoS to be applied to the session may be negotiated with the network by means of application-layer signalling such as SIP and Real Time Streaming Protocol (RTSP) combined with the SDP. Note, however, that there is no access specific information in this signalling. This property of network-initiated QoS control is very attractive since it can be used to provide QoS to access agnostic client applications, for example, applications that are downloaded and installed by the user. This is not possible for terminal-initiated QoS control which requires access-specific client applications that need to be programmed towards a vendor-specific QoS API. The possibility to be 'access QoS agnostic' also enables QoS to be provided in the 'split terminal' case where the client applications reside in a node (e.g. a laptop or set top box) that is physically separated from the terminal. The signalling for a network-initiated QoS control was illustrated already in the first use case in Section 8.2.3.3.

A prerequisite for the network-initiated paradigm, is that the network is able to understand what QoS resources are needed for the service. However, in practice many services (e.g. Mobile-TV, IMS voice) are actually provided by the access network operator; possibly through agreements with third-party service operators, and are thus known to the operator. It is therefore reasonable that the operator also assigns the QoS level for the SDF associated with the service.

Due to the mentioned advantages with network-initiated QoS control paradigm, we consider it to be the most beneficial in cases where the operator controls and has full knowledge of the service. For services not known to the operator there is also a possibility to use the terminal-initiated QoS control paradigm. An example may be that the user accesses a streaming server (not known to the operator) on the Internet and the application in the terminal wants to set up premium QoS for that service. In this case terminal-initiated QoS could be used, assuming that it is allowed by the operator.

8.2.5   PCC and roaming

As already briefly mentioned in Chapter 3, PCC in 3GPP Rel-8 supports roaming for both on-path and off-path scenarios. When a user is roaming in a visited network, we distinguish between two main roaming cases: the 'Home Routed' case and the 'Visited Access' case. The last alternative is often also called 'Local Breakout' (LBO). In the Home Routed case, the user is connected through a PDN GW in the home network and all traffic for that IP connection is routed via the home network. In the Visited Access case, the user is connected via a PDN GW in the visited network and traffic is transported between the UE and the PDN without traversing a PDN GW in the home network. Since the PCEF is located in the PDN GW, this means that the PCEF may be located in the home or visited network. The BBERF (if applicable) is always located in the visited network when the user is roaming.

To support such roaming scenarios, two different architecture alternatives would be possible:

* One alternative would be that the PCRF in the home network directly controls the PCEF and/or BBERF in the visited network via Gx and/or Gxa/Gxc interface.
* Another alternative would be to introduce a reference point between the PCRF in the home network and a PCRF in the visited network. The Gx/Gxa/Gxc interface would then go between the visited PCRF and the PCEF/BBERF in the visited network.

One main principle when developing the PCC architecture for these roaming scenarios has been that no policy control entity is allowed to directly control a policy enforcement entity in another operator's network. The interaction should always go via a policy control entity in the same network as the policy enforcement entity. Therefore it was decided to go for the second alternative above, introducing a new reference point, S9, between two PCRFs, one in the home network and one in the visited network. These two PCRFs are denoted Home PCRF (H-PCRF) and Visited PCRF (V-PCRF) respectively. The two roaming scenarios and the associated PCC architecture are illustrated in Figure 8.2.5.1. It should be noted that in the Visited Access case, the AF may be associated with either the home or the visited network.

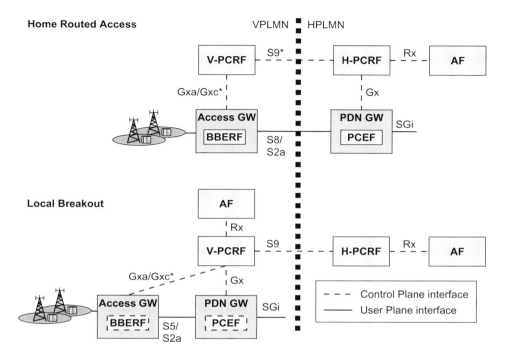

Figure 8.2.5.1 *PCC architecture for home routed and visited access (local breakout) roaming cases. Items marked with * are only applicable to off-path model.*

Control of allowed services and the authorization of resources are always handled by the H-PCRF. Since the home operator provides this control in both roaming and non-roaming scenarios a consistent user experience is possible. For the roaming scenarios when S9 is used, the V-PCRF may accept or reject, but not change, policy decisions coming from the home network. This allows the visited operator to control the usage of the resources in its (radio) access network.

In the Home Routed roaming scenario the PCEF is located in, and controlled by, the home operator. The PCEF connects to the H-PCRF via Gx and online charging can be performed via Gy to the OCS in a similar way as for non-roaming scenarios.

If the on-path model is used, that is, the roaming interface for home routed traffic is based on GTP, there is no need for a BBERF or Gxa/Gxc, and consequently no V-PCRF or S9 either. All QoS signalling with the visited network is taking place over S8 using GTP. There is no need to invoke a PCRF in the visited operator's network. This roaming model is basically the same as that existing for pre-EPS GPRS.

If the off-path model is used, that is, if the roaming interface for home routed traffic is using a Mobile IP-based protocol; the S9 reference point is needed. The BBERF in the visited network is connected via a V-PCRF to the H-PCRF over the S9 reference point. For this case the H-PCRF is responsible for controlling the BBERF in the visited network, via the V-PCRF. Consequently the H-PCRF provides policy decisions (QoS rules) to the BBERF in the visited network, via the V-PCRF.

In the Visited Access case a PDN connection is established via a PDN GW in the visited operator's network. If GTP is used towards the PDN GW in the visited network, then the PDN GW (PCEF) is connected via a V-PCRF to the H-PCRF via S9. On the other hand, if a Mobile IP-based protocol is used towards the PDN GW in the visited network, then the S9 reference point and also the role of the V-PCRF become more complex. The reason is that both Gx and Gxa/Gxc procedures are handled within the same S9 session. The V-PCRF must be able to split and combine messages between S9 on one side and Gx and Gxa/Gxc on the other side.

In the Visited Access case the V-PCRF and the S9 reference point are used independent on whether the visited network is using the on-path or off-path architecture. It would therefore be desirable that the S9 interface would also be independent of the PCC model. Naturally this is not possible to achieve in the Home Routed case since the S9 only exists when off-path is used. In the Visited Access case however, it is to a large extent possible to 'hide' the Gxa/Gxc aspects in the visited network from the S9 interface. This has been one of the goals when designing the S9 protocol. For additional details, see Chapter 10 and 3GPP TS 29.125 [29.125].

For the Visited Access scenario it is possible to use AFs connected via Rx to the V-PCRF. In this situation, Rx signalling is forwarded via the V-PCRF to the H-PCRF using S9.

8.3 Charging

As operators invest in new infrastructure and persuade end-users to enjoy the benefits of the newly deployed networks, the revenue generating options become a key factor for the business cycle. How the end-users/subscribers are actually charged and how billing information is packaged towards them is very much according to individual operator's business model and competitive environment they are operating in. From the EPS point of view, the system needs to enable collection of enough information related to different aspects of the usage for individual user so the operator has the flexibility to determine his own variant of billing as well as packaging towards the end-users. It has become increasingly important in today's competitive business environment for the operators to be able to provide lucrative and competitive option packages towards their potential customers to lure businesses away from other operators. The process of collecting information related to charging can provide tools and means for the operators to make this possible.

For EPS, the existing charging models as well as mechanism apply except for the circuit-switched domain charging aspects of course.

The 3GPP charging infrastructure principles and mechanism did not change due to EPS, but rather EPS entities have been included within this infrastructure. Figure 8.3.1 shows the overall high-level charging system reference model.

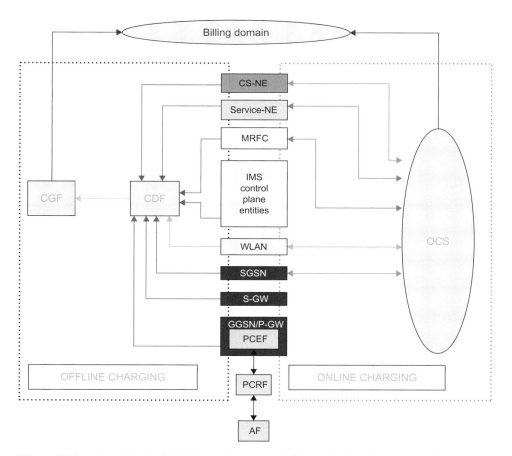

Figure 8.3.1 *Overall logical high-level reference model for collection of charging information.*

The two main charging mechanisms provided by the model are Offline and Online charging, though the terms Online and Offline are not necessarily related to how the end-users are billed at all. These two mechanisms are the means of how the charging-related data are collected and transported to the billing system for further processing as per individual customer's billing options and for settling accounting relations between operators and between operators and subscribers. Offline charging facilitates collection of charging-related data concurrently as the resources are being used in a specific network's various elements provisioned to support such collection of information on an individual basis and then may be sent towards the billing system/domain according to the operator's configuration. Whereas Online charging requires the network to actually get authorization of network resource usage before such usage can occur and the OCS (Online Charging System) is the authorization entity which either grants or rejects the request made by the appropriate network element. In order to do this, the network needs to assemble the relevant charging- and resource usage-related information (known as charging event) and send it to the OCS in real-time, thus allowing OCS to make the appropriate level of authorization which may be limited in its scope such as the volume of data or time of usage, etc. Depending on the level of authorization, the network may need to get a re-authorization performed for any additional resource usage. Note that information collected by the charging system can be used in various manners, such as it may provide statistical measure of network resource usage at certain time of day, usage behaviour, application usage, etc.

In case of Offline charging, various network elements can have a distributed collection role, which would allow for more detailed information availability. Alternatively they may have a centralized role, with a limited collection of events capability as per the entity's role within the network. This role is specified by the Charging Trigger function (CTF) that causes the entities to collect charging events, for example, appropriate charging-related data, and provides them to the Charging Data Function (CDF).

The CDF sends these data towards Charging Gateway Function, which is responsible for sending the data to the billing system. The heart of the Offline system is the CTF which keeps account of usage information regarding the services being delivered to the end-user based on either:

- Signalling information for sessions and service events by the users of the network or
- The handling of user traffic for these events and sessions. The Off line charging system can be illustrated in a simplified manner as shown in Figure 8.3.2.

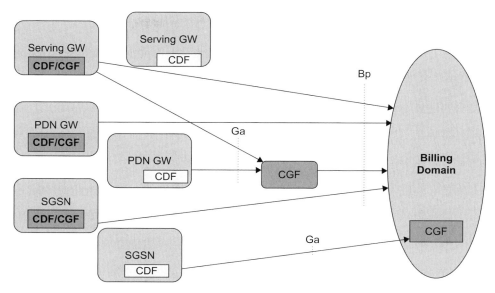

Figure 8.3.2 *Off line Charging entities.*

The information must be made available in real-time with data that uniquely associates and correlates the user's consumption of network resources and/or services.

Note that even though it is not required for the various types of information to be sent in a synchronous manner, the overall charging event must be able to receive and process all relevant data for a specific service/session in real-time in order to provide accurate, billable data towards end-users. So all offline processing of Charging Data Records (CDR) towards the end-user's billing is performed after the usage of the network resources is complete. The billing domain is responsible for generation and handling of the settlement/billing process offline.

In case of Online charging, the CTF and OCS and several other network entities form the charging function. The OCS contains functions like Online Charging Function (OCF), Rating Function (RF) and Account Balance Management Function (ABMF) in order to handle the online charging process. The OCF is the entity that connects to the network elements responsible for providing the charging data, that is, it supports the CTF function. Even though the CTF performs very similar functions for both the Offline and Online charging mechanisms, online charging requires additional handling for authorization prior to the resource usage and thus also requires additional functions from the CTF which are crucial for the real-time Online process. Some of these functions are:

- The charging events are sent to the OCF in order to retrieve authorization for the chargeable event/network resource usage requested by the user.

- The CTF must be able to delay the actual resource usage until permission has been granted by the OCS.
- The CTF must be able to track the availability of resource usage permission ('quota supervision') during the network resource usage.
- The CTF must be able to enforce termination of the end-user's network resource usage when permission by the OCS is not granted or expires.

The OCF supports two methods of charging:

1. Session-Based Charging Function providing online charging for the network/ user's sessions, examples of such services could be the PS resource usages for IMS sessions.
2. Event-Based Charging Function (i.e. content charging) provides online charging in support of application server or service, such as SIP AS or MMS.

The Rating Function is responsible for providing the OCF with the actual value of the network resource usage/service usage, which may be either monetary or non-monetary information. This is determined by information provided by the OCF and CTF; the actual rating or determining the value of the usage is very much operator specific and can be quite wide ranging, some basic examples of such rating could be as follows:

- Rating of data volume (e.g. based on charging initiated by an access network entity, that is, on the bearer level).
- Rating of session/connection time (e.g. based on charging initiated by an IMS level application).
- Rating of service events on the service level (e.g. based on charging of web content or MMS).

The ABMF is responsible for the subscriber's account balance within OCS.

In the case of Online charging, the network resource usage must be authorized and thus a subscriber must have a pre-paid account in the OCS in order for the Online pre-network resource usage authorization to be performed. The two methods used to achieve this are known as Direct Debiting and Unit Reservation. As their names imply, in case of Direct Debiting, the user is immediately debited the amount of resource usage needed for that specific service/session, where as in case of Unit Reservation a predetermined unit is reserved for the usage and the user is then allowed to use that amount, or less, for that service/session. When resource usage has been completed (i.e. session terminated, or the service is completed, etc.), the actual amount of resource usage (i.e. the used units) must be returned by the network entity responsible for monitoring the usage, to the

OCS so that over-reserved amounts can be re-credited to the subscriber account, ensuring that the correct amount gets debited.

Note that PCC makes it possible to have quite detailed charging mechanisms and allows for the possibility of operators having granular control over the subscriber's usage of the network resources. PCC also allows operators to offer various flexible charging and policy schemes towards their subscribers. More details can be found in Section 8.2 on policy control and charging.

For Online charging, the PCEF function as specified in the PCC section interacts with the OCS as specified above and provides the Online charging functions in the PS Domain. Note that the GGSN in GPRS provides the PCEF functions and related support for PCC for Online charging. It also provides the necessary configuration options as specified here for PDN GW in case of Offline charging; GGSN aspects are not described within the context of EPS. Key functions related to charging data collection triggers in the EPS can be described on a high level as follows; it should however be noted that much of the standardization for charging is still underway at the time of writing. We have therefore provided a high-level overview and indicated which parts are still under consideration. Readers are therefore advised to check the relevant charging specifications for more details.

Mobility management-related events relevant for charging, for example, Inter-RAT handover, user's activity/inactivity during established sessions and roaming/non-roaming status, etc., are collected at the SGSN. The authors of this book believe that similar functions are also needed in the MME. However, at the time of writing this book, the standardization process in 3GPP had not been fully completed and the functions mentioned are currently not provided by the MME.

IP-CAN bearers and related functions may be collected at the SGSN/MME (again, this is still under consideration in 3GPP) and Serving GW/PDN GW. Individual service data flows within an IP-CAN bearer according to PCC may be collected at PDN GW but only for the GTP protocol variant. At the time of writing, for the case of PMIP, the data is collected on a per-PDN level but is still undergoing work in the standards body within 3GPP.

At the time of writing, the MBMS and Location Services EPS have remained work in progress and thus unspecified, though charging for these functions is supported in GPRS.

While the SGSN/MME and Serving GW charging data collection is more related to the radio access type used, the PDN GW collects for external network-related data relevant to a subscriber. A subscriber's subscription-related data can also be relevant for charging, an example of such data would be the APN used, more on APN can be found on session management subsection in Chapter 6. A unique charging Identity is assigned; uplink and downlink data volume, date and time are collected for individual subscriber for the purpose of charging.

One example of items that may be contained in a CDR record generated by a PDN GW, as depicted in 3GPP TS 32.251, is illustrated in the table below:

Table 8.3.1   *Example of items that may be contained in CDRs generated by the PDN GW.*

Field	Description
Served IMSI	IMSI of the served party.
Served MN NAI	Mobile Node Identifier in NAI format (based on IMSI), if available.
P-GW Address used	The control plane IP address of the PDN GW used.
Charging ID	IP CAN bearer identifier used to identify this IP CAN bearer in different records created by PCNs.
PDN Connection Id	This field holds the PDN connection identifier to identify different records belonging to same PDN connection.
Serving node Address	List of SGSN/Serving GW control plane IP addresses used during this record.
Serving node Type	List of serving node types in control plane. The serving node types listed here map to the serving node addresses listed in the field "Serving node Address" in sequence.
PGW PLMN Identifier	PLMN identifier (Mobile Country Code and Mobile Network Code) of the PDN GW.
Access Point Name Network Identifier	The logical name of the connected access point to the external packet data network (network identifier part of APN).
PDP/PDN Type	PDP type or PDN type (i.e. IPv4, IPv6 or IPv4v6).
Served PDP/PDN Address	IP address allocated for the PDP context / PDN connection, i.e. IPv4 or IPv6.
Dynamic Address Flag	Indicates whether served IP address is dynamic, which is allocated initial attach and UE requested PDN connectivity. This field is missing if the IP address is static.
List of Service Data	Consists of a set of containers, which are added when specific trigger conditions are met. Each container identifies the configured counts (volume separated for uplink and downlink, elapsed time or number of events) per rating group or combination of the rating group.

Table 8.3.1 *(Continued)*

Field	Description
Record Opening Time	Time stamp when IP CAN bearer is activated in this PDN GW or record opening time on subsequent partial records.
MS Time Zone	This field contains the MS Time Zone the MS is currently located, if available.
Duration	Duration of this record in the PDN GW.
Cause for Record Closing	The reason for the release of record from this PDN GW.
Record Sequence Number	Partial record sequence number, only present in case of partial records.
Record Extensions	A set of network operator/manufacturer specific extensions to the record. Conditioned upon the existence of an extension.
Local Record Sequence Number	Consecutive record number created by this node. The number is allocated sequentially including all CDR types.
Served MSISDN	The primary MSISDN of the subscriber.
User Location Information	This field contains the User Location Information of the UE
Serving node PLMN Identifier	Serving node PLMN Identifier (Mobile Country Code and Mobile Network Code) used during this record.
RAT Type	This field indicates the Radio Access Technology (RAT) type currently used by the Mobile Station, when available.
Start Time	This field holds the time when User IP session starts, available in the CDR for the first bearer in an IP session.
Stop Time	This field holds the time when User IP session is terminated, available in the CDR for the last bearer in an IP session.

Explicit identifiers are used in each of the domains: CS, PS, IMS and Applications that are involved in a specific session; this is because charging data is collected in several network entities and also is needed in order to settle inter-operator resource usage when a user is roaming. In EPS, the Charging Identity for EPC and the PDN GW Identity makes up this identification.

Various levels of correlation are supported in order to achieve a complete charging information profile for the individual usage for each subscriber.

Intra-level correlation aggregates the charging events belonging to the same charging session, for example, over a time period, and implies the generation of interim charging records. When an end-user has accessed a service during the same session from different radio accesses or while roaming, the network entities involved in charging perform correlation of the data. Inter-level correlation combines charging events generated by the different CTFs in different 3GPP domains and inter-network correlation for the IMS requires generation and transmission of specific identifiers across operators' networks. An example of the different level of correlation could be, an end-user in E-UTRAN access connected via EPC using MMTEL service towards an end-user connected to another operator in another network. All three types of correlation would then be generated (if supported and required) by the different operators and the specific domains.

How can all these be configured for a subscriber? Subscriber charging provides the means to configure the end-user's charging information into the network. Charging data collected by the different PLMNs involved (e.g. HPLMN, Interrogating PLMN and VPLMN) and may be used by the subscriber's home operator, dependant upon the deployment and user's roaming status, to determine the network usage and the services, either basic or supplemental. There may also be the possibility to use external Service Providers for billing.

For those subscribers handled by Service Providers, the billing information is utilized for both wholesale (Network Operator to Service Provider) and retail (Service Provider to subscriber) billing. In such cases, the charging data collected from the network entities may also be sent to the Service Provider for further processing after the Home PLMN operator has processed the information as may be desired. Figure 8.3.3 illustrates the different business relationships from the perspective of charging and billing – for the purposes of this book, the circuit-switched aspects are excluded.

The entities and their roles can be described as follows:

Users: Retail users that are charged by their Mobile Network Operator or third-party service provider. Normally users have subscriptions or similar relationship with either or both parties.

Third-Party Service Providers: charged wholesale by the Mobile Network Operator. Responsible for providing users the billing and other charging-related customer care type services towards the users for services rendered.

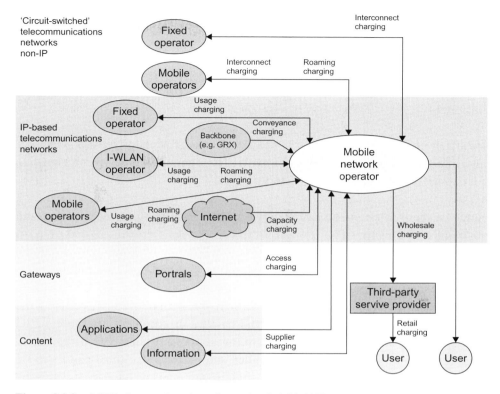

Figure 8.3.3 *3GPP view on charging relationship [TS 22.115].*

Other telecommunications operators: interconnect charging between Mobile Network Operator and non-IP 'circuit-switched' Network Operators for call traffic carried; usage charging between Mobile Network Operator and IP-based Network Operators for session traffic carried. For the purposes of the current section, this group is of no interest.

Other mobile operators: roaming charging between these entities, this may require different mechanisms for IP-based types from the traditional 'circuit-switched' types. Also, where mobile operators need to pass traffic to one another, there will be interconnect charging for non-IP 'circuit switched' types; usage charging for IP-based types.

I-WLAN operators: where I-WLAN operators need to pass traffic to mobile operators or mobile operators to I-WLAN operators, there may be roaming and usage charging.

IP backbone carriers: conveyance charging Mobile Network Operators for traffic carried.

Third-Party content and application suppliers: supplier charging between Mobile Network Operators and Value-Added Service Providers for information exchanged.

Third-Party Portals: access charging between Mobile Network Operators and this entity.

Internet: charge for capacity of connection between Mobile Network Operator and Internet. An Operator pays a provider for a connection based on capacity, for example, annual charge for a 2Mbit/s 'pipe'.

Some of these roles can be easily derived when evaluating the deployment and roaming scenarios for EPS.

EPS is an all-IP based network where Diameter is used for all charging data collection related functions within the EPS. Though support for CAP, GTP' as well as TAP remain due to existing widely deployed and used billing systems by the operators as well as for interworking and backwards compatibility with 2G/3G networks. The interfaces towards the billing system rely heavily on the operators' business model and billing principles and may involve third-party service providers which affect the end-users directly and may be visible towards the end-users if changed drastically and these are not easily feasible.

As charging data can not only convey operator's business aspects, but it may also reveal sensitive information regarding individual subscriber, transfer of such information towards external entities outside of an operator's secure domain must also be done in a secure manner so that information may not be disclosed to unauthorized personnel or entities. Integrity of charging information must be maintained; privacy and secrecy is integral and must be provided by the serving operator and validation of the content and receipt of charging information by the serving Network Operator must be possible. Note that by serving Network Operator, we mean both home and visited as well as any intermediary Network Operators/third-party providers involved in the transaction and handling of the charging/billing.

9
Selection functions

9.1 Architecture overview for selection functions

EPS is an all-IP based system, and thus naturally uses DNS-based mechanism as well as other IETF-defined node discovery mechanisms in order to find appropriate network entities within and among operators' networks. Even though it is based on IP and Internet-driven technology, EPS has specific requirements stemming from both existing GPRS networks deployed today as well as the specific nature of the networks the operators manage and share resources with and type of services they render. In addition, the nature of the 3GPP Packet Core network Selection Functions is also very much dictated by the fundamental principles of mobility, roaming and security and whether the different network elements should or should not be visible from external networks (e.g. the Internet). In addition, operators also want to be able to manage their existing and evolved packet core through single selection mechanism and include the terminal's access network capabilities as part of the selection criteria.

The method for how the network entities would be selected for a certain terminal also dictate how the DNS as well as other selection mechanism have been developed for EPS, specially selection of entities such as MME, Serving GW, PCRF and so on. Such core network entities shall not be reachable from external networks or entities which are not governed by the roaming agreements and roaming interconnect networks known as GRX and IPX. Another important factor in the selection of the key network nodes like the Serving GW and PDN GW is the role of certain information elements such as the Access Point Name (APN) identifying the target PDN, the protocol type towards the PDN GW (e.g. PMIP or GTP) and also in certain cases the terminal Identity and the geographic location of some of these entities.

Even though certain adverse network conditions such as overload or even complete failure of a selected MME may also influence the selection of such entities, this is expected to be a rare situation and thus not discussed further.

The main network entities that require special means of selection are: MME, SGSN, Serving GW and PDN GW. In addition, PCRF selection plays an important role when PCC is used.

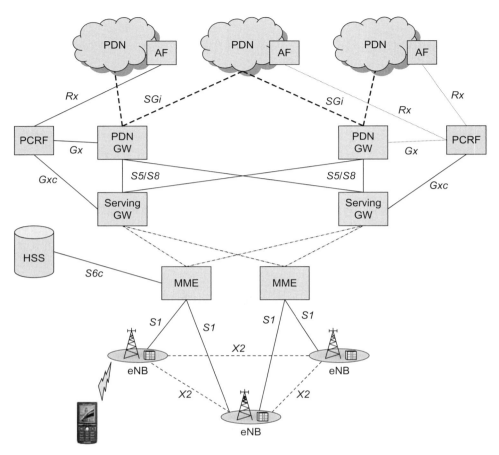

Figure 9.1.1 *High-level hierarchical overview of selection entities. Note that for simplicity not all interfaces are shown in the figure.*

Figure 9.1.1 shows a topological connection between the nodes that are involved in the selection process of a user's network attachment procedure in E-UTRAN, these entities remain connected/active for the duration of that specific user's attachment to the network and removes/disconnects the association only when the user's session/IP connectivity is removed/closed/disconnected. Additional nodes are involved and participate in the selection process when 2G/3G and non-3GPP accesses are part of the network.

9.2   Selection of MME, SGSN, serving GW and PDN GW

9.2.1   Selection procedure at a glance

Let us first illustrate a simplified Initial Attach procedure where an MME, Serving GW and PDN GW need to be selected for the user. The actual message

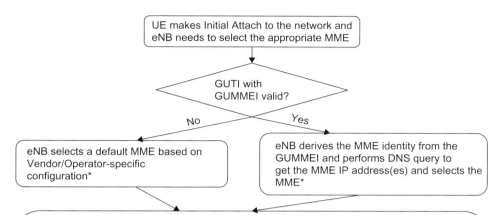

The selected MME now needs to select the appropriate S-GW and P-GW for the user. The key information for the selection is first to use the APN provided by either the UE or the one available in the user's subscriber record from HSS. Note that entities like MME has two roles, one is the entity responsible for selecting the GWs according to 3GPP-defined (as well as vendor-specific and operator-configured proprietary input parameters) and then as a DNS Resolver as per IETF RFCs used during this process. Initial selection criteria for GW selection needs to choose whether: (1) GWs must be collocated; (2) The topological/geographical closeness; (3) Protocol option on the S5/S8 (GTP or PMIP, note that a GW selected may also support both protocols); (4) NAPTR 'Preference' and SRV 'Weight'.

For S-GW selection, the TAI is used as specified in TS 23.003, since TAI provides information about where the UE is being served. Then further selection may be made with S-GW areas that can be specified in the SRV record of the DNS**

For P-GW selection, the APN, the collocation status, topology closeness, protocol option as well as the preference and weight of SRV records may be used. In its simplest form, the APN would be able to identify a set of valid P-GWs uniquely for that user, MME can then select one from this list.**

So in a nutshell, the process is as follows:
– Both S-GW and P-GW selection procedures executed up to getting the list of candidate host names in the form: **[topon|topoff].<1_label>.<X_labels_nodename>**. The host names are provided as output from a DNS S-NAPTR procedure.

– A pair of S-GW and P-GW is selected from the lists (based on criteria configured in the MME)

– If the IP addresses were not provided together with the candidate host names, the IP addresses are then resolved using DNS with A/AAAA query procedures.

If the selected pair is not reachable, the MME may choose other candidates from the list.

*Note that eNB may take into account such information as the weight factor and overload conditions of the MME in question as well as vendor and operator specific information that may be configured and may select a different MME from the MME pool as appropriate. **Note that the S-GW/P-GW selection criteria may include additional parameters available to the MME based on vendor specific implementation and/or proprietary operator configuration data.

Figure 9.2.1.1 *High Level Flows for Selection Function for EPC nodes.*

flows for Initial Attach are shown in Chapter 12, but in the flow chart shown in Figure 9.2.1.1, we only show the order of events on a high level.

3GPP systems including EPS provide connectivity towards multiple PDNs with independent and unique IP connection for each PDN (see Section 6.2 for more details). Each PDN connection established for a new APN requires an additional PDN GW selection process in the MME. In somewhat simplified terms, the DNS is used to store the mapping between APN and PDN GWs. Since for each user (i.e. UE) there can only be a single Serving GW at any given time in the network, selection of the additional PDN GW must use the current Serving GW information as part of its input data for this selection process. With an APN-FQDN created from the APN as the first query parameter to the DNS procedures, the MME will receive a list of candidate PDN GWs. Based on whether the GW collocation is preferred or not, the MME uses the Serving GW already selected as the next criterion in the selection process. When collocation is desired, the MME selects the PDN GW collocated with the already selected Serving GW, if possible. When collocation is not set, then the protocol used on the S5/S8 (interface (GTP or PMIP) as well as other preference criteria set by the SRV and NAPTR records in the DNS system may provide the next selection criterion. Otherwise MME may select a random PDN GW from the candidates list.

Until now, we have focused on the selection process for the case when a user establishes connectivity in the EPS network. In addition to that, during the mobility procedure such as during handover between 3GPP accesses (for more details, see Section 6.4), an MME or SGSN may need to be changed (e.g. in case of MME pool change) and in such a case, the current serving MME needs to select a new candidate MME to transfer the session to. This selection is performed by the serving node and is based on the Tracking Area Identity (TAI) Fully Qualified Domain Names (FQDN), where the TAI is from the target Cell ID that the target eNB has provided to the serving MME. In case of SGSN selection, either the Routing Area ID of the GERAN or UTRAN is used or, in case of UTRAN, the RNC ID can be used in the same way as the TAI to construct the FQDN to find the right SGSN (pool) with the Rel-8 SGSNs. Note that in case of pre-Rel-8 SGSN, existing selection mechanism must be maintained (see TS29.303 [29.303] for more details) which is simpler and do not necessarily mandate DNS usage.

In the case of Serving GW relocation, which should be a rare procedure and only occur if the Serving GW service area changes, the MME uses the TAI as well as the protocol supported on the S5/S8 reference point (GTPv2 or PMIPv6) in its DNS query to get the appropriate candidate Serving GW list.

Note that the MME is responsible for keeping track of all relevant information for Serving GW as well as each PDN GW per PDN basis, this includes host name, IP address, selected port (if not according to standard) and selected protocol type. The MME needs to be configured with the variant of protocol that is used over the S5/S8 reference point. To support roaming cases, the MME is configured on a per HPLMN basis with the S8 protocol variant (PMIPv6 or GTP).

The HSS may also be provisioned with a specific PDN GW IP address and/or FQDN per APN basis and MME will have this information available during Initial Attach procedure and then this input shall be used to select the P-GW for that user. In particular, to support handovers between 3GPP and non-3GPP accesses where the same PDN GW as was used in source access needs to be allocated in target access, the HSS contains information about the currently used PDN GW. This PDN GW information is provided to the node performing PDN GW selection (e.g. MME for E-UTRAN).

9.2.2 Use of DNS infrastructure

Before going into the details of the individual node selection, it is important to understand the 3GPP structure and requirements on the DNS and its usage. The GPRS system as well as the IP Multimedia Subsystem (IMS) in 3GPP has used DNS extensively, including DNS NAPTR and SRV records developed within the IETF. The EPS continues this and expands DNS usage to include node selection.

The DNS parameters and procedures for EPS are specified in 3GPP TS 23.003 [23.003] and TS 29.303. These specifications, and the DNS related IETF RFCs such as RFC 2181 [RFC 2181], RFC 1035 [RFC 1035], RFC 3958 [RFC 3958] and RFC 2606 [RFC 2606] can be consulted for further details.

Note that the use of DNS infrastructure in order to discover the nodes is not mandatory as operators may choose to use O&M functions and other provisioning methods to achieve the selection process. Use of DNS and its supporting services definitely improves the selection process as various criteria can be used an input to the process and also makes it easy and flexible for inter-operator operations.

In EPS, the DNS is used to store information on mapping between APN, protocol (GTP or PMIP) and PDN GW, as well as mapping between TAI and Serving GW. The DNS can also be configured so that the DNS records provide information on collocated nodes as well as topological/geographical closeness between different nodes. In order to utilize the DNS infrastructure as well as IP-networks, rules are defined for how to construct the FQDN used for DNS procedures. While GGSN selection in GPRS systems use only address

(A/AAAA) records in DNS, the EPS DNS procedures support also SRV and NAPTR records and are able to provide the more powerful features already mentioned above, e.g. to support collocated nodes and topological closeness . In particular the Straightforward-NAPTR (S-NAPTR) procedure is used, with a suitable FQDN as input, to retrieve a candidate list of host names and corresponding service (e.g. protocol), port and IPv4/IPv6 addresses. For more detailed information on S-NAPTR, see RFC 3958 [3958] and TS 29.303 [29.303]. Below we briefly describe some of the definitions and rules that are needed for the selection functions.

9.2.2.1   Home network realm/domain

The home Network Realm/Domain need to be in the form of an Internet domain name, for example, operator.com, as specified in IETF RFC 1035 [RFC 1035].

The Home Network Realm/Domain need to be derived from the IMSI as described in the following steps:

1. Take the first 5 or 6 digits of the IMSI, depending on whether a 2 or 3 digit MNC is used and separate them into Mobile Country Code (MCC) and Mobile Network Code (MNC); if the MNC is 2 digits then a zero shall be added at the beginning.
2. Use the MCC and MNC derived in step 1 to create the 'mnc<MNC>. mcc<MCC>.3gppnetwork.org' domain name.
3. Add the label 'epc' to the beginning of the domain name.

An example of a Home Network Realm/Domain is:

IMSI in use: 234150999999999;

Where:

MCC = 234;
MNC = 15;
MSIN = 0999999999;

Which gives the Home Network Realm/Domain name: epc.mnc015.mcc234. 3gppnetwork.org.

9.2.2.2   DNS sub-domain for operator usage in EPC

The EPC nodes DNS sub-domain (DNS zone) is derived from the MNC and the MCC by adding the label 'node' to the beginning of the Home Network Realm/ Domain and is constructed as:

node.epc.mnc<MNC>.mcc<MCC>.3gppnetwork.org

This DNS sub-domain is formally placed into the operator's control. 3GPP never takes this DNS sub-domain back or any zone cut/sub-domain within it for any purpose. As a result the operator can safely provision any DNS records under this sub-domain without worrying about future 3GPP standards encroaching on the DNS names within this zone.

9.2.2.3   Access point name (APN)
Overall structure
The APN is composed of two parts as follows:

1. The APN Network Identifier (APN-NI); this defines the PDN to which the UE requests connectivity and optionally a requested service by the UE. This part of the APN is mandatory.
2. The APN Operator Identifier (APN-OI); this defines in which PLMN the PDN GW (or GGSN for GPRS) is located. This part of the APN is optional.

The APN is constructed by placing the APN Operator Identifier after the APN Network Identifier.

APN network identifier
In order to guarantee uniqueness of APN Network Identifiers within or between PLMNs, an APN Network Identifier containing more than one label shall correspond to an Internet domain name. This name should only be allocated by the PLMN if that PLMN belongs to an organization which has officially reserved this name in the Internet domain. Other types of APN Network Identifiers are not guaranteed to be unique within or between PLMNs.

An APN Network Identifier may be used to access a service associated with a PDN GW. This may be achieved by defining an APN which in addition to being usable to select a PDN GW, is also locally interpreted to be a request for a specific service by the PDN GW. An example would be a unique APN for IMS services.

APN operator identifier
The APN Operator Identifier is a 'domain name' of the operator. It contains two labels that uniquely identify the operator (PLMN) as well as a third label that can be common for all operators. For each PLMN there is a default APN-OI that is built up using the MNC and the MCC as well as the label 'gprs'. The result of the APN Operator Identifier Realm/Domain will be:

mnc<MNC>.mcc<MCC>.gprs

In the roaming case, the UE may utilize the services of the VPLMN. In this case, the APN Operator Identifier need to be constructed the same way, but replacing the Home operator's MNC and MCC with that of the VPLMN.

APN-FQDN

Note that when the selection function in EPC resolves an APN in DNS, the APN is not used as is. Instead an APN-FQDN is created by inserting the labels 'apn. epc' between the APN-NI and the default APN-OI and replacing '.gprs' with '3gppnetwork.org'. This results in an APN-FQDN with the following format:

<APN-NI>.apn.epc.mnc<MNC>.mcc<MCC>.3gppnetwork.org

Note that in existing GPRS network, the suffix '.gprs' is used in the DNS instead of '3gppnetwork.org'. However, with EPS the suffix '3gppnetwork.org' was chosen for the APN resolutions in DNS. One reason is that the usage of the top level domain '.org' aligns better with the existing top level domains of the DNS infrastructure while '.gprs' does not exist outside of the private 3GPP operator networks. Another reason is that the domain name suffix '3gppnetwork.org' is already used for IMS.

Service and protocol service names for 3GPP

To perform node selection in EPS using DNS, the selection function creates an FQDN (e.g. an APN-FQDN) that is provided in the DNS query to find the host names and IP addresses of target nodes such as PDN GW. Such a target node may, however, be multi-homed (i.e. have more than one IP address) and may use different IP addresses for different protocols (e.g. for PMIP and GTP on S5/S8 interface). In this case, the node selection function needs to resolve the allowed interfaces supporting a certain service (e.g. interface and protocol type) using DNS NAPTR procedure using certain 'Service Parameters' describing the service. More detailed description can be found in Section 6.5 of IETF RFC 3958 [RFC 3958] and in 3GPP TS 29.303 [29.303].

Table 9.2.2.3.1 lists 'Service Parameters' to be used in the procedures specified in 3GPP TS 29.303 [29.303].

9.2.3   MME selection

MME selection function is located in the eNB only. The architecture supports multiple eNBs connected to multiple MMEs as well as Serving GWs.

When an UE attempts to attach to E-UTRAN (see Session Management section in Chapter 6 for details), it provides the eNB with parameters; that is, GUTI, (see Section 6.6 on Identities for more details) that would facilitate selection

Table 9.2.2.3.1 *List of 'app-service' and 'app-protocol' names.*

Description	IETF RFC 3958 Section 6.5 'app-service' name	IETF RFC 3958 Section 6.5 'app-protocol' name
PGW and interface types supported by the PGW	x-3gpp-pgw	x-s5-gtp, x-s5-pmip, x-s8-gtp, x-s8-pmip, x-s2a-pmip, x-s2a-mipv4, x-s2b-pmip
SGW and interface types supported by the SGW	x-3gpp-sgw	x-s5-gtp, x-s5-pmip, x-s8-gtp, x-s8-pmip, x-s11, x-s12, x-s4, x-s1-u, x-s2a-pmip, x-s2b-pmip
GGSN	x-3gpp-ggsn	x-gn, x-gp
SGSN	x-3gpp-sgsn	x-gn, x-gp, x-s4, x-s3
MME and interface types supported by the MME	x-3gpp-mme	x-s10, x-s11, x-s3, x-s6a, x-s1-mme

The formats follow the experimental format as specified in IETF RFC 3958 [RFC 3958]. For example, to find the S8 PMIP interfaces on a PGW the Service Parameter of '3gpp-pgw:x-s8-pmip' would be used as input in the procedures defined in IETF RFC 3958.

of the appropriate MME based on how the GUTI is constructed. eNB is also informed of the load status of the MMEs within a pool (see Pool information in Section 6.7) via S1-AP signalling and thus able to provide additional information to the eNBs connected to the MME pool. This allows for efficient selection of MME within a pool and also allows triggers towards the UE (using Tracking Area Update or S1 Release procedures) to reconnect to a different MME within a pool when required. The process of MME selection is designed to be efficient from the UE movement point of view and have been developed to reduce MME change when serving within certain operating boundaries.

The eNB is responsible for the selection of an appropriate MME at UE attachment when no routing to an MME can be determined from the information provided by the UE.

It is important to understand how MME would be represented in the operators' IP network, in this case how the FQDN for the MME or MME Pool would be constructed, since identifying the correct MME for an UE is dependant on it.

An MME within an operator's network is identified using an MME Group ID (MMEGI), and an MME Code (MMEC), which is then made available to the eNB via GUTI and/or other parameters that identify the UE in the network (see Identifiers Section 6.6.).

A sub-domain name is derived from the MNC and MCC by adding the label 'mme' to the beginning of the Home Network Realm/Domain.

The MME node FQDN is constructed as:

Mmec<MMEC>.mmegi<MMEGI>.mme.epc.mnc<MNC>. mcc<MCC>.3gppnetwork.org

An MME pool FQDN is constructed as:

mmegi<MMEGI>.mme.epc.mnc<MNC>.mcc<MCC>. 3gppnetwork.org

When an MME or SGSN is selecting another MME the TAI FQDN is used towards the DNS to find the appropriate MME (in a pool) and then the information is analysed using A/AAAA queries to get the actual MME address details.

The TAI consists of a TAC, MNC and MCC.

A sub-domain name is derived from the MNC and MCC by adding the label 'tac' to the beginning of the Home Network Realm/Domain.

The TAI FQDN is constructed as:

tac-lb<TAC-low-byte>.tac-hb<TAC-high-byte>.tac.epc. mnc<MNC>.mcc<MCC>.3gppnetwork.org

9.2.4   SGSN selection function for EPS

As described in the overview section, the SGSN selection in case of EPS is relevant for Inter-Radio-Access-Technology Handover (Inter-RAT HO) within 3GPP access. In such case, the serving node is responsible to select the target SGSN using either RAI or RNC-ID (in case of UTRAN access only).

A specific SGSN within an operator's network is identified using the RAI FQDN and the Network Resource Identifier (NRI) which uniquely identifies the core network assigned to the terminal. This identifier can be used by a target MME or SGSN node to connect to the source SGSN node.

The SGSN FQDN is constructed as:

nri-sgsn<NRI>.rac<RAC>.lac<LAC>.rac.epc.mnc<MNC>. mcc<MCC>.3gppnetwork.org

Routing area identity (RAI) – EPC

The RAI consists of a RAC, LAC, MNC and MCC.

A sub-domain name for use by core network nodes based on RAI need to be derived from the MNC and MCC by adding the label 'rac' to the beginning of the Home Network Realm/Domain.

The RAI FQDN is constructed in the similar manner as the TAI and would be represented as:

**rac\<RAC>.lac\<LAC>.rac.epc.mnc\<MNC>.mcc\<MCC>.
3gppnetwork.org**

9.2.5 GW selection overview

GW Selection includes the Serving GW as well as PDN GW selection process which is performed by the MME. Compared to GPRS, selection of Serving GWs and PDN GWs have become more complex. It is very tightly coupled to the various protocol choices on certain reference points (e.g. S5/S8) as well as certain network configuration such as Local Breakout, in addition to maintaining existing GPRS GGSN selection function parameters like APN.

A related aspect is GW selection by an SGSN that support both Gn/Gp and S4 interfaces and thus may select between GGSN and PDN GW. In this case it is possible for the SGSN to use the information on whether the UE supports E-UTRAN or not (as indicated in the UE capability information sent by the UE) as input to select either a GGSN or a Serving GW and PDN GW. UEs not supporting E-UTRAN may, for example, be allocated a GGSN while UEs supporting E-UTRAN would need to be handled by a PDN GW in order to allow a potential handover to E-UTRAN.

9.2.6 PDN GW selection function

In case of 3GPP access, the PDN GW selection function is located in the MME and SGSN, whereas for non-3GPP access the PDN GW selection function is located depending on the protocol used in the appropriate reference points (i.e. S2a, S2b and S2c).

1. For PMIPv6, it is the MAG functional entity that performs the selection function (an example of such case is the 3GPP2 HSGW entity which is also known as Access Gateway in a generic manner in 3GPP specifications).

2. For MIPv4 FA mode on S2a, the entity requesting the PDN GW is the entity that plays the role of the FA.

3. For DSMIPv6, the UE needs to be provided with an appropriate PDN GW address (to be used as Home Agent address for DSMIPv6). Different methods are available in EPS. When connecting over an access supporting transfer of Protocol Configurations Options (PCO), the PDN GW address may be returned to the UE in such a PCO field. When connecting towards an ePDG, the PDN GW address may be returned to the UE in the IKEv2 signalling with ePDG. If none of these methods are available, then the UE may use DHCP mechanisms or querying a DNS server based on the PDN information (using APN) may be used to select the PDN GW.

The PDN GW selection function uses subscriber information provided by the HSS and possibly additional criteria. The PDN subscription contexts provided contains information such as the APN(s) the user has subscribed to and (for roaming cases) whether this APN(s) is allowed to be connected to using a PDN GW located in the VPLMN, or if the user has to connect via a PDN GW located in the HPLMN.

In the case of non-3GPP accesses using PMIPv6 or MIPv4, the PDN Gateway selection function interacts with the 3GPP AAA Server or 3GPP AAA Proxy and uses subscriber information provided by the HSS to the 3GPP AAA Server. To support separate PDN GW addresses at a PDN GW for different mobility protocols (PMIP, MIPv4 or GTP), the PDN GW Selection function takes mobility protocol type into account when deriving PDN GW address by using the Domain Name Service function, in a similar way as is done by the MME/SGSN.

In case of 3GPP access, the PDN GW selection function interacts with HSS to retrieve subscriber information and uses information such as whether the Serving GW and PDN GW must be collocated; the topological/geographical closeness (described in TS 29.303 [29.303]) and protocol option on the S5/S8 (GTP or PMIP, note that a GW selected may also support both protocols).

Note that collocation of Serving GW and PDN GW is not part of the 3GPP standards specifications for non-3GPP accesses since the use of Serving GW for non-3GPP access is only relevant to a specific case known as Chained S2a/S2b-S8. This chained scenario is not further described in this book. Interested readers should refer to TS 23.402 [23.402] for more details.

In the case a static PDN GW is configured by the operator to be used for that user, the PDN GW is selected by either having the APN configured to map to a given PDN GW (i.e. the APN resolves to a single PDN GW only), or the PDN GW identity provided by the HSS explicitly shows the use of static PDN GW.

The APN may also be provided by the UE. In this case, this UE-provided APN is used to derive the APN-FQDN as long as the subscription allows for it.

If the user is roaming and the HSS provides a PDN subscription context that allows for allocation of a PDN GW from the visited PLMN for this APN, the PDN GW selection function selects a PDN GW from the visited PLMN. If a visited PDN GW cannot be found or if the subscription does not allow for allocation of a PDN GW from the visited PLMN, then the APN is used to select a PDN GW from the HPLMN.

If the UE is requesting a connection to an existing PDN via the same APN, the selection function must select the same PDN GW as it used previously to establish the PDN connection to this APN.

To support the PDN GW selection using the S-NAPTR procedures, the authoritative DNS server responsible for the 'apn.epc.mnc<MNC>. mcc<MCC>.3gppnetwork.org' domain must provision NAPTR records for the given APN and corresponding PDN GWs under the APN-FQDN:

<APN-NI>.apn.epc.mnc<MNC>.mcc<MCC>.3gppnetwork.org

Some example S-NAPTR procedures are included here for illustration purposes. For detailed information on 3GPP usage of various IETF specifications and their adaptation for 3GPP network usage, TS 29.303 [29.303] is an excellent source. For 3GPP access:

When non-roaming, the Initial attach the S-NAPTR procedure to get the list of 'candidate' PDNGW shall use 'Service Parameters' of:

'x-3gpp-pgw:x-s5-gtp', 'x-3gpp-pgw:x-s5-pmip' depending on the protocol supported.

When roaming, the Serving GW is in the VPLMN and if the APN to be selected is in the HPLMN, then the selection function for the S-NAPTR procedure shall use 'Service Parameters' of:

'x-3gpp-pgw:x-s8-gtp', 'x-3gpp-pgw:x-s8-pmip' depending on the protocol supported.

When non-roaming and additional PDN connection is to be established, the Serving GW is already selected (i.e. the UE has an existing PDN connection), and the selection function shall use the S-NAPTR procedure 'Service Parameters':

'x-3gpp-pgw:x-s5-gtp', 'x-3gpp-pgw:x-s5-pmip' depending on the protocol supported.

For non-3GPP access:

When PMIP is used for Initial attach and both roaming and non-roaming cases, the selection function shall use the S-NAPTR procedure with 'Service Parameters' of:

'x-3gpp-pgw:x-s2a-pmip', 'x-3gpp-pgw:x-s2b-pmip'

The additional information is received from 3GPP AAA server/Proxy regarding the appropriate protocol is used over S8 (i.e. GTP vs. PMIP).

9.2.7   Serving GW selection function

In case of 3GPP access, the Serving GW is a mandatory node, whereas in case of non-3GPP access the Serving GW is used only is special case of chained scenario. This chained scenario is not further described in this book. Please refer to TS 23.402 [23.404] if interested to learn more about this feature.

The Serving GW selection function selects an available Serving GW to serve an UE. The selection is based on network topology, that is, the selected Serving GW serves the UE's location (derived from the TAI) and in case of overlapping Serving GW service areas, the selection may prefer Serving GWs with service areas that reduce the probability of changing the Serving GW as the UE moves around the network.

Due to the possibility of using either GTP or PMIPv6 over the S5 and S8 interfaces as well as possible multiple PDN connections involving HPLMN and VPLMN (in this case the VPLMN provides the network configuration required for Local Breakout where the PDN GW in the VPLMN needs to be selected) when roaming, Serving GWs may need to support both protocols for a single UE connected to different PDNs. This may, for example, be needed in case a UE has two PDN connections, one with PDN GW in Visited PLMN and one with PDN GW allocated in the Home PLMN. In this case PMIPv6 may be used on S5 between the Serving GW and the PDN GW in the Visited PLMN, but GTP could be used for the other PDN connection with PDN GW in the Home PLMN.

Again we show a few examples of S-NAPTR procedures. The detailed information can be found in TS 29.303. For 3GPP accesses:

When non-roaming and TAU procedure requires S-GW change, the S-GW selection function shall use the S-NAPTR procedure with the TAI FQDN to get the list of 'candidate' SGW 'Service Parameters' of

'x-3gpp-sgw:x-s5-gtp' and/or 'x-3gpp-sgw:x-s5-pmip'

When roaming, the SGW selection is needed due to TAU procedure in the VPLMN, then the selection function for the S-NAPTR procedure shall use the TAI FQDN and 'Service Parameters' of

'x-3gpp-sgw:x-s8-gtp' or 'x-3gpp-sgw:x-s8-pmip'

and make use of the operator's preference for the roaming protocol (GTP/PMIP).

When non-roaming and additional PDN connection, the SGW is already selected (i.e. the UE has an existing PDN connection), the selection function shall use the S-NAPTR procedure with the APN-FQDN and 'Service Parameters' of

'x-3gpp-pgw:x-s5-gtp', 'x-3gpp-pgw:x-s5-pmip'

For non-3GPP access when the Serving GW acts as a local mobility anchor for non-3GPP access S8-S2a/b chained roaming, Serving GW selection is not specified and thus may be left to the vendor-specific implementation choice of a operator's networks.

9.2.8 Handover (non-3GPP access) and PDN GW selection

Once the selection of PDN GW has occurred, the *PDN Gateway identity* is registered in the HSS so that it can be provided to the target access in case of an inter-access handover. The *PDN GW identity* registered in HSS can be either an IP address or a FQDN. Registering the PDN GW IP address is suitable in case the PDN GW only has a single IP address or can use the same IP address for all the mobility protocols it supports (or if it only supports one mobility protocol). Registering the FQDN is more flexible in the sense that it allows PDN GW to use multiple IP addresses depending on mobility protocol. So if a UE hands over from one access where PMIPv6 is used to another access where GTP is used, the PDN GW selection function could use the FQDN to derive different PDN GW IP addresses depending on mobility protocol.

In case the terminal activated the PDN connection in a non-3GPP access, it is the PDN GW that registers the PDN GW identity and its association with a UE and the APN in the HSS. For 3GPP access types, the MME/SGSN updates the HSS with the selected PDN GW identity. Once registered in the HSS, the HSS (possibly via 3GPP AAA Server or Proxy) can then provide the association of the PDN Gateway identity and the related APN for the UE later on. It can be noted that the PDN GW information is provided to target access only in case mobility is performed using a network based mobility protocols in the target access (PMIP and GTP). For handover to a non-3GPP access using DSMIPv6, it is instead the UE that knows the address of the PDN GW (Home Agent).

When the UE is moving between 3GPP and non-3GPP accesses, PDN Gateway selection information for the subscribed PDNs the UE has not yet connected to is returned to the target access system like during Initial Attach. For the PDNs the UE has already connected to, the PDN GW information is transferred as below:

1. If a UE hands over to a non-3GPP access using PMIPv6 in target access and it already has assigned PDN Gateway(s) due to a previous attach in a 3GPP access, the HSS provides the PDN GW identity for each of the already allocated PDN Gateway(s) with the corresponding PDN information to the 3GPP AAA server. The AAA server forwards the information to the PDN GW selection function in the target access.
2. If a UE hands over to a 3GPP access and it already has an assigned PDN Gateway(s) due to a previous attach in a non-3GPP access, the HSS provides the PDN GW identity for each of the already allocated PDN Gateway(s) with the corresponding PDN information to the MME.

9.3 PCRF selection

The PCRF and its role in the EPS are described in detail in Section 8.2. For the purpose of PCRF discovery, what is pertinent is that a PDN GW and associated AF may be served by one or more PCRF nodes in the HPLMN. When roaming and need to support local breakout scenarios, one or more PCRF nodes in the VPLMN may be serving the PDN GW and the associated AF. There may be either one PCRF per PLMN allocated for all PDN connections of a UE, or the different PDN connections may be handled by different PCRFs. Which option to use is up to operator configuration. When DSMIPv6 is used however, the only option is to use one PCRF for all PDN connections of a UE.

The Rx, Gx, Gxa/Gxc and S9 sessions for the same IP-CAN sessions must be handled by the same PCRF. The PCRF must also be able to link the different sessions (Rx session, Gx session, S9 session, etc.) for the same IP-CAN session. This applies also to roaming cases where one VPCRF handles the PCC sessions (Gx, Rx, Gxa/Gxc, etc.) belonging to one PDN connection of the UE. Therefore the selection of PCRF and relating an individual UE's different PCC sessions over the multiple PCRF interfaces for a UE IP-CAN session are very much related. This means that the information carried over these different interfaces must be able to be correlated for the purpose of providing the overall PCC functions towards individual users.

In order to ensure that the same PCRF is selected for all sessions (Gx, Rx, etc) of an IP-CAN session, a logical functional entity called DRA (Diameter Routing Agent) is used during the selection/discovery of the PCRF. 3GPP defines the DRA to be: *The DRA (Diameter Routing Agent) is a functional element that ensures that all Diameter sessions established over the Gx, S9, Gxx and Rx reference points for a certain IP-CAN session reach the same PCRF when multiple and separately addressable PCRFs have been deployed in a Diameter realm. The DRA is not required in a network that deploys a single PCRF per Diameter realm.* as per 3GPP TS 29.213 [29.213].

DRA complies with standard Diameter functionality as specified by IETF where the Diameter protocol and its applications are specified. DRA also needs to (both Redirect and Proxy DRA) advertise the specific applications supported by it which includes information like support of Gx, Gxx, Rx, S9 as well as necessary and relevant Authentication and vendor-related information for 3GPP.

When operating in the Redirect mode, the DRA Client shall use the value within the Redirect Host AVP of the redirect response in order to obtain the PCRF identity (see Diameter protocol details in Chapter 11).

When in the Proxy mode, the DRA shall select a PCRF only if one has not been already selected for that UE previously and if the request is an IP-CAN session establishment of gateway control session establishment. After a successful establishment of the PCRF for that UE, this PCRF must be used until all IP-CAN sessions for that UE has been terminated and removed from DRA.

When DRA is deployed in a network for PCRF's realm, the DRA is to be the first contact point from the clients for the sessions.

In roaming scenario, home routed or local breakout, if the DRA is deployed, the vPCRF is selected by the DRA located in the visited PLMN, and the hPCRF is selected by the DRA located in the home PLMN.

In case of roaming, depending on which PLMN the PCRF being selected and in what network configuration (i.e. Local Breakout or not, Home routed or both), the Serving GW, the non-3GPP Access GW and PDN GW may be the selection function trigger towards DRA. The DRA keeps status of assigned PCRF for a certain UE and IPCAN session. It is assumed that there is a single logical DRA serving a Diameter realm (Figure 9.3.1).

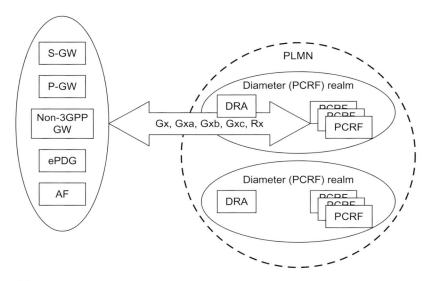

Figure 9.3.1 *PCRF selection and discovery using DRA.*

The parameters available for the DRA to be able to determine the already allocated PCRF depend on the reference point over which the DRA is contacted. Different information is available on Rx compared to for example, Gx and Gxa/Gxc. Concluding this chapter, we have illustrated in a simplified manner various network node selection criteria as well as methods used in 3GPP.

Part IV
The Nuts and Bolts of EPC

10

EPS network entities and interfaces

In this chapter we will describe in more detail the different network entities, interfaces, protocols and procedures used in EPS. This chapter can be used as a reference for readers wanting to see more details than preceding chapters. Before jumping into the actual descriptions however, it is useful to take a look at what we actual mean by a network entity, an interface, a protocol and a procedure.

A network entity in EPS is what sometimes called a 'logical entity'. This means that in the 3GPP standard it is a logically separate entity with a well-defined functionality. There are also well-defined interfaces between different network entities. This does not, however, imply that actual physical 'boxes' implemented by vendors and deployed in real networks have to correspond one-to-one with the network entities in the standard. Vendors may implement a network entity as a standalone product, or may choose to combine different network entities in the same product. It can for example be beneficial to combine a Serving GW and a PDN GW in the same node in order to reduce the number of physical in order to nodes that the user plane has to traverse in non-roaming cases.

Although 3GPP uses the term *reference point* to denote an association between two logical network entities we have chosen the more commonly used term *interface* in this book. There are some differences in the formal definition of a reference point and an interface but for the purpose of this book the difference has no practical implications. Interfaces in 3GPP typically have a prefix letter and one or two additional letters. In GPRS, most interfaces start with the letter 'G' while in EPS most interfaces start with the letter 'S'.

A protocol is defined in TS 29.905 [29.905] as 'A formal statement of the procedures that are adopted to ensure communication between two or more functions within the same layer of a hierarchy of functions'; this definition is taken from ITU-T, document I.112 [I.112]. This could possibly be described in other words as a well-defined set of rules for sending information between two network entities. These rules typically cover transmission, message and data formatting, error handling, etc. EPS uses protocols defined by the Internet Engineering Task Force (IETF) as well as protocols defined by 3GPP. When IETF protocols are used, such as Diameter or IPSec, 3GPP specifies how these protocols are applied in the 3GPP architecture and what protocol options and amendments shall be used or shall not be used. In some cases new IETF RFCs are created to specify protocol amendments as required by 3GPP.

It is important to note that there is not a one-to-one mapping between protocols and interfaces. The same protocol may be used on multiple interfaces, and the other way around, multiple protocols may be used on a single interface. One obvious example of the latter is of course that different protocols are used on different layers in the protocol stack. Another example is the S5/S8 interface that supports two protocol alternatives, Proxy Mobile IP (PMIP) and GPRS Tunnelling Protocol (GTP), to implement similar functions. It is also possible to use different protocols to implement different functions on an interface. For example, on the S2c interface IKEv2 is used to establish a security association (SA) while DSMIPv6 is used for mobility purposes.

EPS is an 'all IP' system where all protocols are transported over IP networks. This is different from the original GPRS standard where some interfaces supports protocols based on for example ATM and Signalling System No 7 (SS7).

Even though there are numerous interfaces in the EPS architecture, the protocols supported over those interfaces can be grouped into a relatively small number of groups. Figure 10.1 illustrates a few of the most significant protocols and interfaces in EPS, grouped by protocol type. It should be noted that in some cases multiple protocols are supported over a given interfaces.

The procedures, that is the message flows, define how commands and messages are transferred between the network entities in order to implement a function (e.g. a handover between two base stations). It may be worth pointing out that the messages and information content shown in the procedures described in Chapter 12 do not necessarily correspond one-to-one with the actual protocol messages. Messages with different names in the message flows may for example be implemented using the same protocol message (or the other way around). Another example is that different logical information elements may be combined into a single parameter when defining the actual protocol fields.

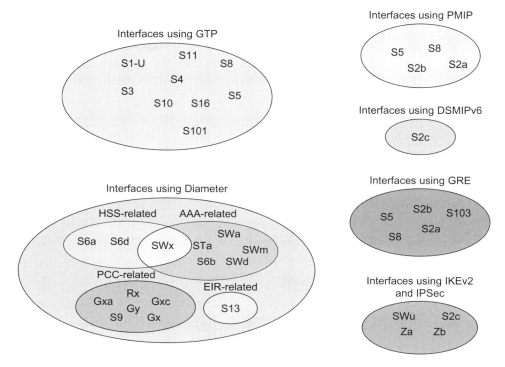

Figure 10.1 *Key protocols and interfaces in EPS.*

This chapter begins with a brief overview of the EPS network entities. Then the EPS interfaces are described; with an overview of the functions supported and the protocols used over each interface. Chapter 11 describes the protocols aiming to give a basic overview of what protocols are used in EPS and their basic properties. Chapter 12 provides a brief introduction to some of the procedures used in EPS. It should be noted that it is not feasible to make a complete description of all procedures that exist in EPS. Instead we have chosen a few procedures that we think will give a good overview for the intended readers without losing any conceptual information of the key functions supported by EPS. Interested readers may also consult the 3GPP technical specifications 23.060 [23.060], 23.401 [23.401] and 23.402 [23.402] for complete descriptions.

10.1 Network entities

The network architecture of SAE is comprised of a few different network entities; each network entity has a distinct role in the architecture. This section covers the roles of the different nodes; the eNodeB, the Mobility Management Entity (MME), the Serving GW, the PDN GW and the PCRF.

10.1.1   eNodeB

The eNodeB provides the radio interface and performs radio resource management for Long-Term Evolution (LTE) including radio bearer control, radio admission control and scheduling of uplink and downlink radio resources for individual UEs. The eNodeB also supports IP header compression and encryption of the user plane data. eNodeBs are interconnected to one another via an interface named X2; this interface has several uses, e.g. handover. eNodeBs are also connected to the EPC via the S1 interface, which is split up into the user plane and the control plane. The control plane interface is referred to as S1-MME and terminates in the MME. The S1-U interface, meanwhile, terminates at the Serving GW and handles user plane traffic. The S1 interface supports pooling, that is a many-to-many relation between the eNodeBs and the MMEs and also between the eNodeBs and the Serving GW. The S1 interface also supports network sharing. This allows operators to share the radio network, that is the eNodeBs, while maintaining their own EPC networks.

10.1.2   Mobility management entity

From a Core Network perspective, the MME is the main node for control of the LTE access network. It selects the Serving GW for a UE during the initial attach and also during handover, if necessary, between LTE networks. It is responsible for the tracking and paging procedures for UEs in idle mode and also the activation and de-activation of bearers on behalf of a UE. The MME through interaction with the HSS is responsible for authenticating the end-user. For UEs that are roaming, the MME terminates the S6a interface towards the UEs home HSS. The MME also ensures that the UE has authorization to use (camp on) an Operator's PLMN and also enforces any roaming restrictions that the UE may have.

In addition, the MME provides control plane functionality for mobility between LTE and 2G/3G access networks. The S3 interface terminates at the MME from the SGSN.

An MME is selected by the MME selection function. Selection is based on network topology, dependent on which MME serves the particular location that a UE is in. If several MMEs serve a particular area, the selection procedure is based on a few different criteria, for example selecting an MME that reduces the need to change it later or perhaps based on load balancing needs. A full description of the MME selection function is covered in Section 9.2.3.

The MME is also responsible for Non-Access Stratum (NAS) signalling, which terminates at the MME; the MME also acts as the termination point in the network for the security of NAS signalling, handling the ciphering protection and management of security keys.

Lawful Intercept related to signalling also handled by the MME.

10.1.3 Serving GW

The Serving GW performs several functions for both the GTP-based and PMIP-based network architectures. The Serving GW terminates the interface towards E-UTRAN; every UE that attaches to an EPS is associated with a single Serving GW. In the same was as the MME, the Serving GW is selected for the UE based on network topology and UE location. The Domain Name Service (DNS) may be used to resolve a DNS string of possible Serving GW addresses which serve the UE's location. The selection of Serving GW may be affected by a few criteria; firstly, a Serving GW may be selected based on the fact that its service area may reduce the necessity to change the Serving GW at a later time. Secondly, Serving GW selection may be based on the need for load balancing between different Serving GWs. A full description of the selection procedure is covered in Chapter 9.

Once a UE is associated with a Serving GW, it handles the forwarding of end-user data packets and also acts as a local anchor point when required for inter-eNodeB handover. During handover from LTE to other 3GPP access technologies (inter-RAT handover for other 3GPP access technologies), the Serving GW terminates the S4 interface and provides a connection for the transfer of user traffic from 2G/3G network systems and the PDN GW. During both the inter-NodeB and inter-RAT handovers, the Serving GW sends one or more 'end-markers' to the source eNodeB, SGSN, or RNC in order to assist the re-ordering function in the eNodeB.

When a UE is in idle state, the Serving GW will terminate the downlink (DL) path for data. If new packets arrive, the Serving GW triggers paging towards the UE. As part of this, the Serving GW manages and stores information relevant to the UE; for example parameters of the IP bearer service or internal network routing information.

The Serving GW is also responsible for the reproduction of user traffic in the case of lawful intercept.

10.1.4 PDN GW

The PDN GW provides connectivity to external PDNs for the UE, functioning as the entry and exit point for the UE data traffic. A UE may be connected to more than one PDN GW if it needs to access more than one PDN. The PDN GW also allocates an IP address to the UE. These PDN GW functions apply to both the GTP-based and the PMIP-based versions of the SAE architecture. This is covered in more detail in Section 6.1.

In its role as a gateway, the PDN GW may perform deep packet inspection, or packet filtering on a per-user basis. The PDN GW also performs service level gating control and rate enforcement through rate policing and shaping. From a QoS perspective, the PDN GW also marks the uplink and downlink packets with, for example, the DiffServ Code Point. This is covered in more detail in Section 8.1 and 8.2.

Another key role of the PDN GW is to act as the anchor for mobility between 3GPP and non-3GPP technologies such as WiMAX and 3GPP2 (CDMA/HRPD).

10.1.5   Policy and charging rules function

The Policy and Charging Rules Function (PCRF) is the policy and charging control element of the SAE architecture and encompasses policy control decision and flow-based charging control functionalities. This means that it provides network-based control related to service data flow detection, gating, QoS and flow-based charging towards the Policy and Charging Enforcement Function (PCEF). It should be noted, however, that the PCRF is not responsible for credit management.

The PCRF receives service information from the Application Function (AF) and decides how the data flow for a particular service will be handled by the PCEF. The PCRF also ensure that the user plane traffic mapping and treatment is in accordance with the subscription profile associated with an end-user. The PCRF functions are described in more detail in Section 8.2.

10.2   Control plane between UE, eNodeB and MME

10.2.1   S1-MME

10.2.1.1   General
The E-UTRAN-Uu interface is defined between the UE and the eNodeB and the S1-MME interface is defined between the eNodeB and the MME, as shown in Figure 10.2.1.1.1.

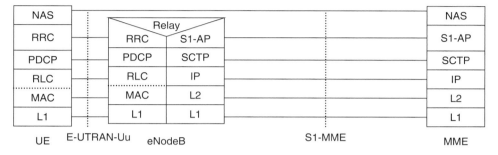

Figure 10.2.1.1.1 *S1-MME interface.*

Figure 10.2.1.3.1 *Protocol stack for E-UTRAN-Uu and S1-MME.*

10.2.1.2 Interface functionality

The S1-MME interface provides support for functionality such as paging, handover, UE context management, E-RAB management and transparent transport of messages between MME and UE.

10.2.1.3 Protocol

The protocol stack for E-UTRAN-Uu and S1-MME is depicted in Figure 10.2.1.3.1.

As can be seen in Figure 10.2.1.3.1, the NAS protocols run directly between the UE and the MME while the eNodeB just acts as a transparent relay. The protocol layers below NAS on E-UTRAN-Uu and S1-MME are called access stratum (AS).

The AS protocols on E-UTRAN-Uu (RRC, PDCP, RLC, MAC and the physical LTE layer) implements the Radio Resource Management and supports the NAS protocols by transporting the NAS messages across the E-UTRAN-Uu interface. Similarly the AS protocols on S1-MME (S1-AP, SCTP, IP, etc.) implements functionality such as paging, handover, UE context management, E-RAB management and transparent transport of messages between MME and eNodeB. The S1-MME interface is defined in 3GPP TS 36.410 [36.410].

The NAS layer consist of an EPS mobility management (EMM) protocol and an EPS session management (ESM) protocol. The EMM protocol provides procedures for the control of mobility and security for the NAS protocols. The ESM protocol provides procedures for the handling of EPS bearer contexts. Together with the bearer control provided by the access stratum, this protocol is used for the control of user plane bearers. The NAS protocols are defined in 3GPP TS 24.301 [24.301].

10.3 GTP-based interfaces

10.3.1 Control plane

For more detailed procedures and functions of GPTv2-C interfaces can be found in 3GPP technical specification 23.401 [23.401] and 23.060 [23.060].

The detailed messages and parameters for control plane GTPv2-C protocol are defined in 3GPP TS 29.274 [29.274]. At the time of writing, this specification is still under development, thus may not have a complete set of functions. Note that both IPv4 and IPv6 are supported on the transport layer.

The following table shows some messages supported over various interfaces using GTPv2-C.

Message name	Entity involved	Interfaces
• Create Session Request/Response	• SGSN/MME to Serving GW to PDN GW (response in opposite path)	• S4/S11, S5/S8
• Modify Bearer Request/Response	• SGSN/MME to Serving GW to PDN GW (response in opposite path)	• S4/S11, S5/S8
• Create Bearer Request/Response	• PDN GW to Serving GW to SGSN/ MME (response in opposite path)	• S5/S8, S4/S11
• Context Request/Response	• MME to MME	• S10
• Context Request/Response	• SGSN to/from MME	• S3
• Context Request/Response	• SGSN to/from SGSN	• S16
• Create Forwarding Tunnel Request	• MME to Serving GW	• S11
• Downlink Data Notification	• Serving GW to SGSN/MME	• S4/S11
• Detach Notification	• SGSN to MME, MME to SGSN	• S3

10.3.2 MME ↔ MME (S10)

The S10 interface is defined between two MMEs and this interface exclusively uses GTPv2-C and for LTE access only. The main function over this protocol is to transfer the contexts for individual terminals attached to the EPC network and thus sent on a per UE basis. This interface is primarily used when MME is relocated. Figure 10.3.2.1 shows the protocol stack for S10 interface.

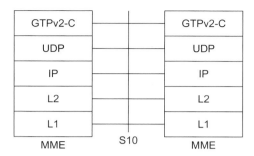

Figure 10.3.2.1 *S10 interface.*

10.3.3 MME ↔ Serving GW (S11)

The S11 interface is defined between MME and the Serving GW, this interface exclusively uses GTPv2-C and for LTE access only. Due to the separation of the control and user plane functions between MME and Serving GW, the S11 interface is used to create a new session (i.e. to establish the necessary resources for the session) and then manage these sessions (i.e. modify, delete and change) any sessions for a terminal (for each PDN connection) that has established connection within EPS.

The S11 interface is always triggered by some events either directly from the NAS level signalling from the terminal, such as a device attaching to the EPS network, adding new bearers to an existing session or creating a connection towards a new PDN, handover cases, or it may be triggered during network-initiated procedures such as PDN GW-initiated bearer modification procedures. As such the S11 interface keeps the control and user plane procedures in sync for a terminal during the period that the terminal is seen active/attached in the EPS. In case of handover, the S11 interface is used to relocate the Serving GW when appropriate, establish direct or indirect forwarding tunnel for user plane traffic and manage the user data traffic flow.

Figure 10.3.3.1 shows the protocol stack across S11 interface.

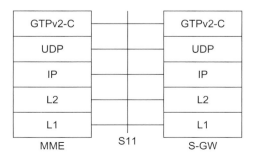

Figure 10.3.3.1 *S11 interface.*

Note that some of the interactions (e.g. signalling between MME and Serving GW) also need to be performed between Serving GW and PDN GW over S5/S8 interface as described below, though in such case, depending on the protocol choice over S5/S8, the messages are either continued over GTP or transferred over PMIP.

The S11 interface shares common functions with interfaces S4 as well, which can be seen in later parts of this section.

10.3.4 Serving GW ↔ PDN GW (S5/S8)

The S5/S8 interface is defined between the Serving GW and the PDN GW. The S5 interface is used in non-roaming scenarios where the Serving GW is located in the home network, or in roaming scenarios with both Serving GW and PDN GW located in the visited network. The latter scenario is also referred to as Local Breakout. The S8 interface is the roaming variant of S5 and is used in roaming scenarios with the Serving GW in the visited network and the PDN GW in the home network.

When the GTP variant of the protocol is used, the S5/S8 interface provides the functionality associated with creation/deletion/modification/change of bearers for individual user connected to EPS. These functions are performed on a per PDN connection for each terminal. The Serving GW provides the local anchor for all bearers for a single terminal and manages them towards the PDN GW. Figure 10.3.4.1 shows the protocol stack over S5/S8 interface. Section 10.4 contains information on the PMIP variant of S5/S8.

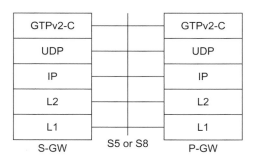

Figure 10.3.4.1 *S5 or S8 interface (GTP variant).*

10.3.5 SGSN ↔ MME (S3)

The S3 interface is between S4-based SGSN (this SGSN is enhanced to support EPS mobility compared to the Gn/Gp-based SGSN as described in Chapter 3) and MME in order to support handover to/from 2G/3G radio access network for 3GPP accesses. The functions supported include transfer of the information related to the terminal that is being handed over, handover/relocation messages and thus the messages are for individual terminal basis. This interface exclusively supports GTPv2-C and the protocol stack is shown in Figure 10.3.5.1.

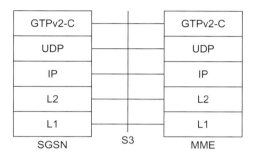

Figure 10.3.5.1 *S3 interface.*

10.3.6 SGSN ↔ Serving GW (S4)

The S4 interface is between SGSN supporting 2G/3G radio access and Serving GW, it has equivalent functions as the S11 interface but for 2G/3G radio access networks and supports related procedures to support legacy terminals connecting via EPS as well. This interface supports exclusively GTPv2-C and provides procedures to enable user plane tunnel between SGSN and Serving GW in case the 3G network has not enabled direct tunnel for user plane traffic from RNC to/ from Serving GW. Figure 10.3.6.1 shows the protocol stack for S4 interface.

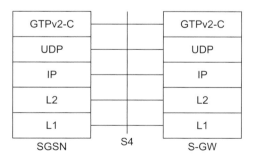

Figure 10.3.6.1 *S4 interface.*

10.3.7 SGSN ↔ SGSN (S16)

The S16 interface is defined between two SGSNs and this interface exclusively uses GTPv2-C and for 2G/3G accesses only when on an EPS network. The main function of this protocol is to transfer the contexts for individual terminals attached to the EPC network and are thus sent on a per UE basis as in the case of S10 interface. Figure 10.3.7.1 shows the protocol stack for S16 interface.

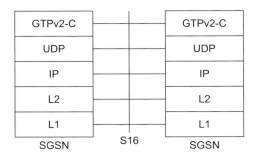

Figure 10.3.7.1   *S16 interface.*

10.3.8   User plane

The user plane protocols use GTPv1-U as defined in 3GPP TS29.281 [29.281]. This runs over X2-U, S1-U, S4 user plane, S5/S8 user plane and S12 user plane interfaces for GTP variant of the EPS architecture. GTPv1-U is also used for 2G/3G Packet Core (also known as GPRS) over Gn, Gp and Iu-U interfaces. Note that both IPv4 and IPv6 are supported on the transport layer of IP.

10.3.9   eNodeB ↔ Serving GW (S1-U)

The S1-U is the user plane interface carrying user data traffic between the eNodeB and Serving GW received from the terminal. The protocol stack for S1-U is shown in Figure 10.3.9.1.

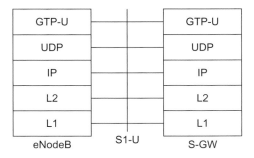

Figure 10.3.9.1   *S1-U interface.*

10.3.10   UE ↔ eNodeB ↔ Serving GW ↔ PDN GW (GTP-U)

Figure 10.3.10.1 shows end-to-end user plane traffic in case of LTE access and using GTPv1-U protocol stack (over S1-U-S5/S8 GTP variant). As can

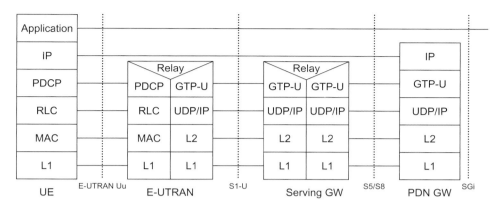

Figure 10.3.10.1   *User plane for LTE access (GTP-based S5/S8).*

be seen in Figure 10.3.10.1, the protocol stack on SGi is Application protocol over IP.

10.3.11   UE ↔ BSS ↔ SGSN ↔ Serving GW ↔ PDN GW (GTP-U)

Figure 10.3.11.1 shows end-to-end user plane traffic for the case of 2G access over EPC using the GTPv1-U protocol stack (over S4 and S5/S8 GTP variant). As can be seen in Figure 10.3.11.1, the protocol stack on SGi is Application protocol over IP.

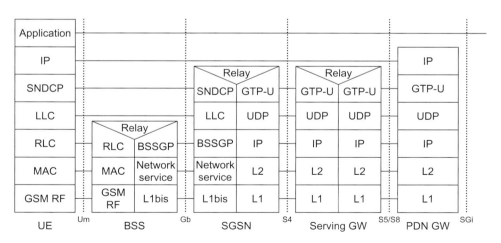

Figure 10.3.11.1   *User plane for 2G access (GTP-based S5/S8)*

10.3.12   UE ↔ UTRAN ↔ Serving GW ↔ PDN GW (GTP-U)

Figure 10.3.12.1 shows end-to-end user plane traffic in case of 3G access over EPC and using GTPv1-U protocol stack (over Iu-U-S5/S8 GTP variant) and

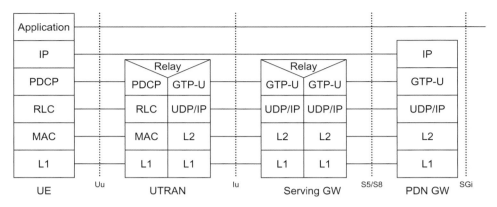

Figure 10.3.12.1   *User plane for 3G access with 'direct tunnel' (GTP-based S5/S8).*

where the SGSN is no longer in the user plane path (direct tunnel established between RNC and Serving GW). As can be seen in Figure 10.3.12.1, the protocol stack on SGi is Application protocol over IP.

10.3.13   *UE ↔ UTRAN ↔ SGSN ↔ Serving GW ↔ PDN GW (GTP-U)*

Figure 10.3.13.1 shows end-to-end user plane traffic in case of 3G access over EPC and using GTPv1-U protocol stack (over Iu-U-S5/S8 GTP variant) and where the SGSN is in the user plane path (direct tunnel not used). As can be seen in Figure 10.3.13.1, the protocol stack on SGi is Application protocol over IP.

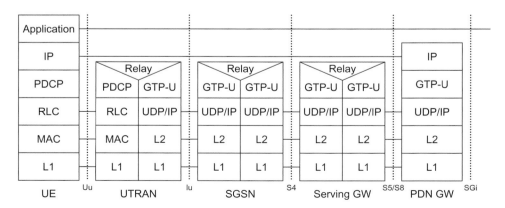

Figure 10.3.13.1   *User plane for 3G access without 'direct tunnel' (GTP-based S5/S8).*

10.4 PMIP-based interfaces

10.4.1 Serving GW – PDN GW (S5/S8)

10.4.1.1 General

The S5/S8 interface is defined between the Serving GW and the PDN GW. The S5 interface is used in non-roaming scenarios where the Serving GW is located in the home network, or in roaming scenarios with both Serving GW and PDN GW located in the visited network, see Figure 10.4.1.1.1. The latter scenario is also referred to as Local Breakout. The S8 interface is the roaming variant of S5 and is used in roaming scenarios with the Serving GW in the visited network and the PDN GW in the home network.

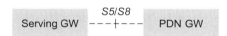

Figure 10.4.1.1.1   *S5/S8 interface.*

10.4.1.2 Interface functionality

The S5/S8 interface provides user plane tunnelling and tunnel management between the Serving GW and the PDN GW. It is used for Serving GW relocation due to UE mobility and in case the Serving GW needs to connect to a non-collocated PDN GW for the required PDN connectivity.

10.4.1.3 Protocol

Two protocol alternatives have been specified for the S5/S8 interface; a PMIP-based variant and a GTP-based variant. The protocol stacks for the GTP-based alternative is described in Section 10.3. The protocol stacks for the PMIP-based variant of S5/S8 is shown in Figure 10.4.1.3.1 PMIP-S5-CP (control plane) and Figure 10.4.1.3.2 PMIP-S5-UP (user plane). The Serving GW acts as a Mobile Access Gateway (MAG) for PMIPv6 and the PDN GW acts as an LMA. The

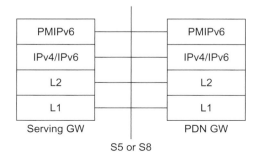

Figure 10.4.1.3.1   *Control plane protocol stack for the S5/S8 interface (PMIPv6 variant).*

user plane is tunnelled using Generic Routing Encapsulation (GRE). The key field extensions of GRE are used [draft-ietf-netlmm-grekey-option] where the key field value is used to identify a PDN connection.

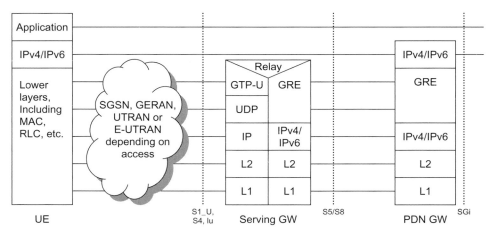

Figure 10.4.1.3.2   *User plane protocol stack for the S5/S8 interface (PMIPv6 variant).*

The definition of the PMIP-based S5/S8 interface and its functionality is given in 3GPP TS 23.402 [23.402]. The PMIPv6-based protocol for S5/S8 is defined in TS 29.275 [29.275].

10.4.2   Trusted non-3GPP IP access – PDN GW (S2a)

10.4.2.1   General

The S2a interface is defined between an access GW in the trusted non-3GPP access network and the PDN GW. The S2a interface is defined for both roaming and non-roaming scenarios, see Figure 10.4.2.1.1.

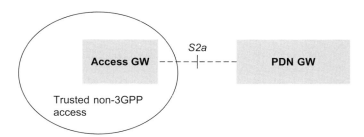

Figure 10.4.2.1.1   *S2a interface.*

10.4.2.2   Interface functionality

The S2a interface provides user plane tunnelling and tunnel management between the trusted non-3GPP access and the PDN GW. It provides mobility support for mobility within the trusted non-3GPP access and between different accesses.

10.4.2.3 Protocol

The protocol over the S2a interface is based on PMIPv6. The protocol stack for S2a is shown in Figure 10.4.2.3.1 (control plane) and Figure 10.4.2.3.2 (user plane). An Access GW in the trusted non-3GPP access acts as MAG for PMIPv6 and the PDN GW acts as LMA. The user plane is tunnelled using GRE. The key field extensions of GRE are used [draft-ietf-netlmm-grekey-option] where the key field value is used to identify a PDN connection.

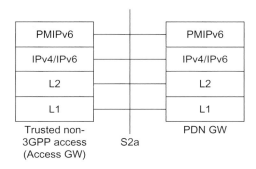

Figure 10.4.2.3.1 *Control plane protocol stack for the S2a interface.*

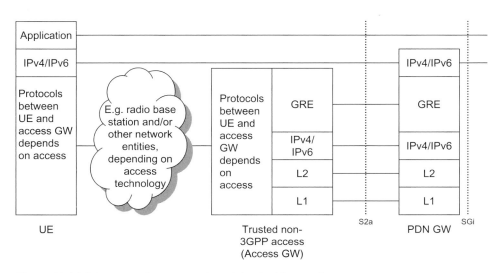

Figure 10.4.2.3.2 *User plane protocol stack for the S2a interface.*

The definition of the S2a interface and its functionality is given in 3GPP TS 23.402 [23.402]. The PMIPv6-based protocol for S2a is defined in TS 29.275 [29.275].

10.4.3   ePDG – PDN GW (S2b)

10.4.3.1   General

The S2b interface is defined between the ePDG and the PDN GW. The S2b interface is defined for both roaming and non-roaming scenarios, see Figure 10.4.3.1.1.

Figure 10.4.3.1.1   *S2b interface.*

10.4.3.2   Interface functionality

The S2b interface provides user plane tunnelling and tunnel management between the ePDG and the PDN GW.

10.4.3.3   Protocol

The protocol over the S2b interface is based on PMIPv6. The protocol stacks for S2b is shown in Figure 10.4.3.3.1 (control plane) and Figure 10.4.3.3.2 (user plane). An Access GW in the trusted non-3GPP access acts as an MAG for PMIPv6 and the PDN GW acts as an LMA. The user plane is tunnelled using GRE. The key field extensions of GRE are used [draft-ietf-netlmm-grekey-option] where the key field value is used to identify a PDN connection.

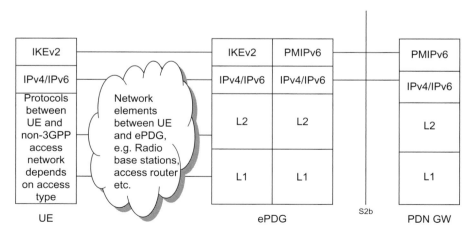

Figure 10.4.3.3.1   *Control plane protocol stack for the S2b interface.*

The definition of the S2b interface and its functionality is given in 3GPP TS 23.402 [23.402]. The PMIPv6-based protocol for S2b is defined in TS 29.275 [29.275].

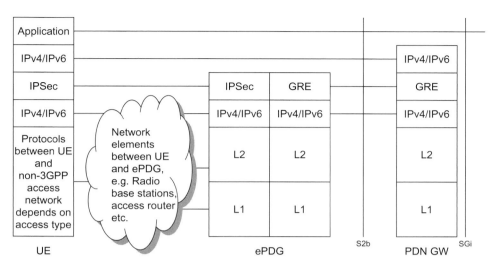

Figure 10.4.3.3.2 *User plane protocol stack for the S2b interface.*

10.5 DSMIPv6-based interfaces

10.5.1 UE – PDN GW (S2c)

10.5.1.1 General
The S2c interface is defined between the UE and the PDN GW. The S2c interface is defined for both roaming and non-roaming scenarios, see Figure 10.5.1.1.1.

Figure 10.5.1.1.1 *S2c interface.*

10.5.1.2 Interface functionality
The S2c interface provides the user plane with related control and mobility support between UE and the PDN GW. This interface is implemented over trusted and/or untrusted non-3GPP Access and/or 3GPP access. Special considerations apply for S2c when the UE is attached over a 3GPP access (see below).

10.5.1.3 Protocol
The protocol over the S2c interface is based on DSMIPv6. The protocol stacks when using S2c over a trusted non-3GPP access is shown in Figure 10.5.1.3.1 while the protocol stack for using S2c over an untrusted non-3GPP access is shown in Figure 10.5.1.3.2. The user plane is tunnelled using IP-in-IPv4/IPv6 encapsulation or using IP-in-UDP-in-IPv4 encapsulation to support NAT traversal.

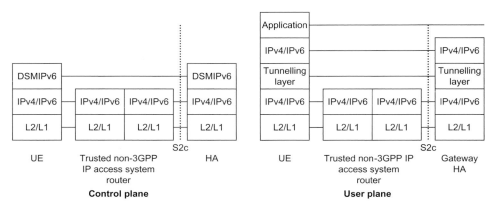

Figure 10.5.1.3.1 *Protocol stacks for the control plane and user plane when using S2c over a trusted non-3GPP access.*

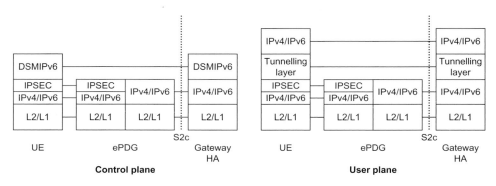

Figure 10.5.1.3.2 *Protocol stacks for the control plane and user plane when using S2c over an un-trusted non-3GPP access.*

When the UE is connected via a 3GPP access, the UE is considered to be on its home link (in DSMIPv6 sense). There is thus no S2c user plane tunnelling over 3GPP accesses. S2c is used only for DSMIPv6 bootstrapping and DSMIPv6 De-Registration (Binding Update with Lifetime equals zero) when the UE is connected via 3GPP access. More information on these and other DSMIPv6 aspects can be found in Section 11.3.

The definition of the S2c interface and its functionality is given in 3GPP TS 23.402 [23.402]. The DSMIPv6-based protocol for S2c is defined in TS 24.303 [24.303].

10.6 HSS-related interfaces and protocols

10.6.1 General

In this section we describe the interface S6a between the HSS and the MME as well as interface S6d between HSS and SGSN. The interface between the 3GPP AAA server and the HSS is described in Section 10.7.

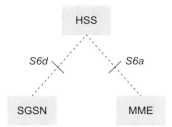

Figure 10.6.2.1.1 *S6a and S6d interfaces.*

10.6.2 MME – HSS (S6a) and SGSN – HSS (S6d)

10.6.2.1 General

The interface S6a is defined between the HSS and the MME and the interface S6d is defined between HSS and SGSN. The S6d interface is used in EPS and thus applies to the S4-based SGSN only. For a Gn/Gp-based SGSN, the Gr interface between SGSN and HSS/HLR applies, see Figure 10.6.2.1.1.

10.6.2.2 Interface functionality

The interfaces between HSS and SGSN as well as between HSS and MME are used for multiple purposes. It allows the MME/SGSN and HSS to:

- Exchange location information: As described in Section 6.4 and 6.5, the MME/SGSN currently serving the UE is notifying the HSS about the MME/SGSN identity. In some cases, for example if the UE attaches to a new MME/SGSN, the MME/SGSN downloads information from HSS about the MME or SGSN that previously served to UE.
- Authorize a user to access the EPS: The HSS holds the subscription data including for example the allowed Access Point Names (APNs) and other information related to the user's authorized services. The subscription profile is downloaded to the MME/SGSN and used when granting a user access to the EPS.
- Exchange authentication information: As described in Chapter 7, the HSS provides authentication data (the EPS Authentication vector) to the MME/SGSN when the user is being authenticated.
- Download and handle changes in the subscriber data stored in the server: When the subscriber data in the HSS is modified, for example in case the subscription is withdrawn, or access to certain APNs is withdrawn, the updated subscription data is downloaded to the MME/SGSN currently serving the UE. Based on the updated subscription data, the MME/SGSN may modify the ongoing session, or even detach the UE completely.
- Upload the PDN GW identity and APN being used for a specific PDN connection. The information about the currently active PDN connections (PDN

GW identity and APN) is stored in the HSS to support mobility with non-3GPP accesses.

- Download the PDN GW identity and APN pairs being stored in HSS for already ongoing PDN connection. This occurs for example during the handover from a non-3GPP access to a 3GPP access.

10.6.2.3 Protocol

The same protocol is used on both S6a and S6d. This S6a/S6d interface protocol is based on Diameter and is defined as a vendor specific Diameter application, where the vendor is 3GPP. The S6a/S6d Diameter application is based on the Diameter base protocol but defines new Diameter commands and attribute-value pairs (AVPs) to implement the functions described in the previous section. Diameter messages over the S6a, S6d interfaces shall use the Stream Control Transmission Protocol (SCTP) [2960] as a transport protocol. The protocol stack is illustrated in Figure 10.6.2.3.1.

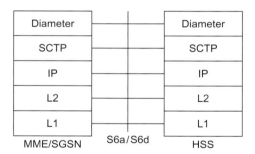

Figure 10.6.2.3.1 *Protocol stack for S6a/S6d.*

The protocol over the S6a/S6d interface, including the S6a/S6d Diameter application is defined in 3GPP TS 29.272 [29.272].

10.7 AAA-related interfaces

10.7.1 General

The network nodes specific to 3GPP accesses, such as the MME and SGSN, connect directly to HSS. Network entities related to other accesses, for example as described by 3GPP2, are however typically interfacing an AAA server instead. Therefore a 3GPP AAA server is used in the EPC architecture to interface entities related to for example CDMA accesses and other accesses outside the 3GPP family. In order to access for example subscription data and other data available in HSS, the 3GPP AAA server interfaces the HSS via the SWx interface.

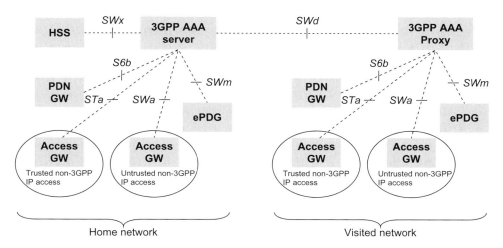

Figure 10.7.1.1 *AAA-related interfaces.*

The different AAA-related interfaces covered in this section are illustrated in Figure 10.7.1.1. The S6b, STa, SWa and SWm may either connect to the 3GPP AAA server or the 3GPP AAA proxy depending on whether it is a roaming or non-roaming scenario. For the roaming scenario, S6b may connect to the AAA server or AAA proxy depending whether the PDN GW is located in the visited network or in the home network. See Chapter 3 for a more complete description of the architecture alternatives.

10.7.2 AAA server – HSS (SWx)

10.7.2.1 General
The SWx interface is defined between the HSS and the 3GPP AAA server, see Figure 10.7.2.1.1.

10.7.2.2 Interface functionality
The SWx interface is used for subscriber profile management as well as for non-3GPP access-related location management.

Non-3GPP access location management procedures on SWx include the following functionality:

- AAA server registration: The 3GPP AAA server registers the current 3GPP AAA server address in the HSS for a given user when a new subscriber has been authenticated by the 3GPP AAA server.
- Upload PDN GW identity and APN: The 3GPP AAA server informs the HSS about the current PDN GW identity and APN being used for a given

Figure 10.7.2.1.1   *SWx interface.*

UE, or that a certain PDN GW and APN pair is no longer used. This occurs for example when a PDN connection is established or closed. This corresponds to the similar functionality as is available on S6a/S6d when the UE is in 3GPP access.

- PDN GW identity and APN download: The 3GPP AAA server downloads the PDN GW identity and APN information being stored in HSS for already ongoing PDN connections for a given UE. This is for the case when the UE has already been assigned PDN GW(s) due to a previous attach in a 3GPP access (when the UE is handed over from a 3GPP access to a non-3GPP access).
- AAA-initiated de-registration: The 3GPP AAA server may de-register the currently registered 3GPP AAA server in the HSS for a given user and purge any related non-3GPP user status data in the HSS. This occurs if the UE for some reason has been disconnected from the non-3GPP access.
- HSS-initiated de-registration: The HSS may initiate a de-registration procedure to purge the UE from the 3GPP AAA server. This happen when the user's subscription has been cancelled or for other operator-determined reasons. As a result, the 3GPP AAA server should de-activate any UE tunnel in the PDN GW and/or detach the UE from the access network.

The subscriber profile management procedures over SWx include the following functionality:

- Subscriber profile push: The HSS may decide to send the subscriber profile to a registered 3GPP AAA server. This occurs for example when the subscriber profile has been modified in the HSS and the 3GPP AAA server needs to be updated.
- Subscriber profile request: The 3GPP AAA server may also request the user profile data from the HSS. This procedure is invoked when for some reason the subscription profile of a subscriber is lost or needs to be updated.

10.7.2.3   Protocol

The SWx interface protocol is based on Diameter and is defined as a vendor specific Diameter application, where the vendor is 3GPP. The SWx Diameter application has its own Diameter application identifier but is re-using Diameter commands from the Diameter Cx/Dx application which is a 3GPP vendor specific application defined for IMS. New AVPs are however defined for the SWx

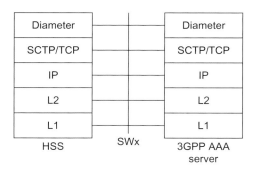

Figure 10.7.2.3.1 *Protocol stack for SWx.*

application to implement the functions described above. The protocol stack is illustrated in Figure 10.7.2.3.1.

The definition of the interface and its functionality is given in 3GPP TS 23.402. The specification of the SWx Diameter application can be found in 3GPP TS 29.273.

10.7.3 Trusted non-3GPP access – 3GPP AAA server/proxy (STa)

10.7.3.1 General

The STa interface is defined between the trusted non-3GPP IP access and the 3GPP AAA server in the non-roaming case. In the roaming case it is defined between the trusted non-3GPP IP access and the 3GPP AAA proxy, see Figure 10.7.3.1.1.

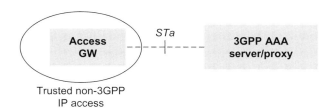

Figure 10.7.3.1.1 *STa interface.*

10.7.3.2 Interface functionality

The STa interface includes the following functionality to:

- Authenticate and authorize a user when the user attaches to a trusted non-3GPP IP Access.
- Transport subscription data such as the APN-AMBR and default QoS profile from the 3GPP AAA server to the trusted non-3GPP IP access. The 3GPP AAA server has in turn received the subscription data from the HSS.

- Transport mobility parameters that are needed for the S2a interface, that is when PMIPv6 or Mobile IPv4 is used to connect the UE to the EPC. In particular this information may include the PDN GW identity(s) and APN(s) currently allocated to a UE during a previous attach in a 3GPP access.
- Transport mobility parameters related to the S2c interface, that is when the UE is attaching to EPC using DSMIPv6. In particular the Home Agent IP address or Fully Qualified Domain Name (FQDN) may be sent from the 3GPP AAA server to the gateway of the trusted non-3GPP access for Home Agent discovery based on DHCPv6.
- To transport information about IP Mobility Mode Selection. This includes both information from the trusted non-3GPP IP access to the 3GPP AAA server/proxy about the mobility features supported by the non-3GPP access (e.g. if the Access GW in the non-3GPP IP access supports MAG functionality for PMIP) as well as information from the 3GPP AAA server/proxy to the Access GW regarding the selected mobility mechanism.

10.7.3.3   Protocol

This STa interface protocol is based on Diameter and is defined as a vendor specific Diameter application, where the vendor is 3GPP. The STa Diameter application is based on Diameter base protocol and also includes commands from the following specifications:

- The Diameter Network Access Server (NAS) Application which is a Diameter application used for AAA services in the Network Access Server (NAS) environment [4005].
- The Diameter EAP application which is a Diameter application to support EAP transport over Diameter [RFC 4072]. The EAP methods EAP-AKA and EAP-AKA'may be used as described in Chapter 7.
- Extensions relevant for PMIPv6 defined in [draft-ietf-dime-pmip6] and extensions relevant for DSMIPv6 defined in RFC 5447 [5447].

The protocol stack for STa is illustrated in Figure 10.7.3.3.1.

The definition of the interface and its functionality is given in 3GPP TS 23.402 [23.402]. The STa Diameter application is defined in TS 29.273 [29.273].

10.7.4   Untrusted non-3GPP IP access – 3GPP AAA server/proxy (SWa)

10.7.4.1   General

The SWa interface is defined between the untrusted non-3GPP IP access and the 3GPP AAA server (non-roaming case) or 3GPP AAA proxy (roaming case), see Figure 10.7.4.1.1.

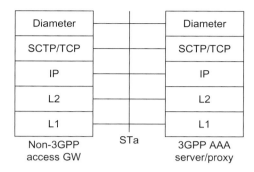

Figure 10.7.3.3.1 *Protocol stack for Sta.*

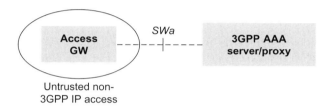

Figure 10.7.4.1.1 *SWa interface.*

10.7.4.2 Interface functionality

The SWa interface is used for 3GPP-based access authentication and authorization with an untrusted non-3GPP access. It also supports reporting of accounting information generated by the access network.

As described in Section 7.3 access authentication in untrusted non-3GPP IP access is optional. The reason is that when accessing an untrusted non-3GPP access, the UE will anyway be authenticated and authorized for EPS access using the tunnel procedures with an ePDG (using SWu and SWm interfaces). Use of SWa is therefore optional when a UE accesses an untrusted non-3GPP IP access.

10.7.4.3 Protocol

This SWa interface protocol is based on Diameter and is defined as a vendor specific Diameter application, where the vendor is 3GPP. It is using the Diameter base protocol and includes commands from the following two applications:

- The Diameter EAP application which is a Diameter application to support EAP transport over Diameter [4072]. The EAP methods EAP-AKA and EAP-AKA' may be used as described in Chapter 7.
- The Diameter Network Access Server (NAS) application which is a Diameter application used for AAA services in the Network Access Server (NAS) environment [4005], see Figure 10.7.4.3.1.

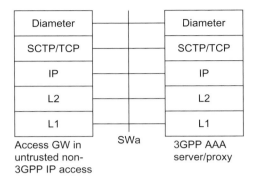

Figure 10.7.4.3.1   *Protocol stack for SWa.*

The definition of the interface and its functionality is given in 3GPP TS 23.402 [23.402]. The SWa protocol is defined in TS 29.273 [29.273].

10.7.5   ePDG – 3GPP AAA server/proxy (SWm)

10.7.5.1   General

The SWm interface is defined between the ePDG and the 3GPP AAA server or between the ePDG and the 3GPP AAA proxy, see Figure 10.7.5.1.1.

10.7.5.2   Interface functionality

The SWm interface includes the following functionality to:

- Authenticate and authorize a user at tunnel setup on the SWu interface (i.e. between UE and ePDG).
- Transport subscription profile data from the 3GPP AAA server to the ePDG. The 3GPP AAA server has in turn received the subscription profile data from the HSS.
- Transport mobility parameters that are needed for the S2b interface, that is when PMIPv6 is used to connect the UE to the EPC. In particular this information may include the PDN GW identity(s) and APN(s) currently allocated to a UE during a previous attach in a 3GPP access.
- Transport mobility parameters related to the S2c interface, that is when the UE is attaching to EPC using DSMIPv6. In particular the Home Agent IP address or FQDN may be sent from the 3GPP AAA server to the gateway of the trusted non-3GPP access for Home Agent discovery based on IKEv2.
- To transport information about IP Mobility Mode Selection. This includes both information from the ePDG to the 3GPP AAA server/proxy about the mobility features supported by the ePDG (e.g. if the ePDG supports MAG

Figure 10.7.5.1.1 *SWm interface.*

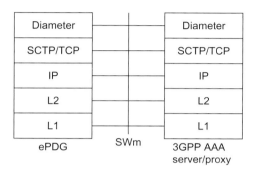

Figure 10.7.5.3.1 *Protocol stack for SWm.*

functionality for PMIP) as well as information from the 3GPP AAA server/proxy to the ePDG regarding the selected mobility mechanism.

- To transport session termination indications and requests. This includes both a session termination indication from the ePDG to the 3GPP AAA server/proxy in case the session with the UE has been terminated as well as session termination requests from the 3GPP AAA server/proxy to request the ePDG to terminate a given session.

10.7.5.3   Protocol

This SWm interface protocol is based on Diameter and is defined as a vendor specific Diameter application, where the vendor is 3GPP. The SWm Diameter application is based on Diameter base protocol and includes commands from the following specifications:

- The Diameter Network Access Server (NAS) Application which is a Diameter application used for AAA services in the Network Access Server (NAS) environment [4005].
- The Diameter EAP application which is a Diameter application to support EAP transport over Diameter [4072]. The EAP methods EAP-AKA and EAP-AKA' may be used as described in Chapter 7.
- Extensions relevant for PMIPv6 defined in [draft-ietf-dime-pmip6] and extensions relevant for DSMIPv6 defined in RFC 5447 [5447].

The protocol stack for SWm is illustrated in Figure 10.7.5.3.1.

The definition of the interface and its functionality is given in 3GPP TS 23.402 [23.402]. The SWm Diameter application is defined in TS 29.273 [29.273].

10.7.6   PDN GW – 3GPP AAA server/proxy (S6b)

10.7.6.1   General

The S6b interface is defined between the PDN GW and the 3GPP AAA server (for non-roaming case, or roaming with home routed traffic to PDN GW in home network) and between the PDN GW and the 3GPP AAA proxy (for roaming case with PDN GW in the visited network, i.e. local breakout), see Figure 10.7.6.1.1.

Figure 10.7.6.1.1   *S6b interface.*

10.7.6.2   Interface functionality

The S6b interface is not utilized when the UE is attached in GERAN, UTRAN or E-UTRAN. In these cases it is instead the S6a/S6d interfaces that provide the needed functionality as described in Section 10.6. When a UE attaches over another access not in the 3GPP family of accesses, S6b provides the functionality described below.

The S6b interface is used to inform the 3GPP AAA server/proxy about current PDN GW identity and APN being used for a given UE, or that a certain PDN GW and APN pair is no longer used. This occurs for example when a PDN connection is established or closed. (The information is then forwarded to the HSS via the SWx interface.) S6b may be used to retrieve specific subscription-related parameters such as a subscribed QoS profile for non-3GPP accesses.

The above functionality of S6b is common for all mobility protocols that can be used when a UE attaches over a non-3GPP access, that is PMIPv6-based S2a or S2b, Mobile IPv4-based S2a or DSMIPv6-based S2c.

Other main functions of the S6b interface depend on what mobility protocol is used when connecting a UE to the EPC.

When the UE attaches to the EPC using the DSMIPv6-based S2c interface, the S6b interface is also used to authenticate and authorize the UE. It is also used to indicate to the PDN GW that a PDN GW re-allocation shall be performed (see TS 23.402 for more details on the PDN GW re-allocation procedure and the scenarios when it is used). When S2c is used the S6b interface can furthermore be

used to transport a session termination indication from the 3GPP AAA server/
proxy to the PDN GW, to trigger a termination of a PDN connection (the 3GPP
AAA server/proxy may receive a trigger for sending the termination indication
from the HSS via the SWx interface).

When the UE attaches using the Mobile IPv4-based S2a interface, S6b is also
used to authenticate and authorize the Mobile IPv4 Registration Request message that was sent by the UE.

10.7.6.3   Protocol

This S6b interface protocol is based on Diameter and is defined as a vendor specific Diameter application, where the vendor is 3GPP. The S6b Diameter application is based on Diameter base protocol with the following additions:

- The Diameter Network Access Server (NAS) Application which is a Diameter
  application used for AAA services in the Network Access Server (NAS) environment [4005].
- Extensions relevant for PMIPv6 defined in [draft-ietf-dime-pmip6] and extensions relevant for DSMIPv6 defined in RFC 5447 [5447].
- The Diameter EAP application which is a Diameter application to support
  EAP transport over Diameter [4072]. The EAP method EAP-AKA is be used
  as described in Chapter 7.

The protocol stack for S6b is illustrated in Figure 10.7.6.3.1.

The definition of the interface and its functionality is given in 3GPP TS 23.402
[23.402]. The S6b Diameter application is defined in TS 29.273 [29.273].

10.7.7   3GPP AAA proxy – 3GPP AAA server/proxy (SWd)

10.7.7.1   General

The SWd interface is defined between the 3GPP AAA proxy and the 3GPP AAA
server. The SWd interface is used in roaming scenarios where the 3GPP AAA
proxy is located in the visited network and the 3GPP AAA server is located in
the home network.

The 3GPP AAA proxy acts as a Diameter proxy agent and forwards Diameter commands between the Diameter client and the Diameter server, see Figure 10.7.7.1.1.

10.7.7.2   Interface functionality

The prime purpose of the protocols crossing this interface is to transport AAA
signalling between home and visited networks in a secure manner. The SWd

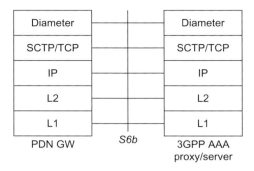

Figure 10.7.6.3.1   *Protocol stack for S6b.*

Figure 10.7.7.1.1   *SWd interface.*

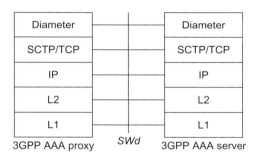

Figure 10.7.7.3.1   *Protocol stack for SWd.*

interface may be used in connection with any of the interfaces SWa, STa, SWm and S6b depending on the particular roaming scenario. The functionality of those interfaces applies to SWd as well.

10.7.7.3 Protocol

The SWd interface uses the Diameter applications and extensions that are used on the SWa, STa, SWm and S6b interfaces. There is thus no separate Diameter application defined for the SWd interface, instead the SWa, STa, SWm and S6b applications are proxied onto SWd. The same Diameter commands as defined for the SWa, STa, SWm and S6b interfaces are used on SWd as well, depending on the specific roaming scenario.

The protocol stack for SWd is illustrated in Figure 10.7.7.3.1.

The definition of the interface and its functionality is given in 3GPP TS 23.402 [23.402]. The SWd protocol is defined in TS 29.273 [29.273].

10.8 PCC-related interfaces

10.8.1 General

The PCRF-related interfaces include Gx, Gxa, Gxc, Rx, S9 and Sp. See Section 8.2 for more details on PCC and where these interfaces are located in the architecture. Below we will walk through the different interfaces related to PCC and briefly describe each of them.

10.8.2 PCEF – PCRF (Gx)

10.8.2.1 General

The Gx interface is defined between the PCEF (PDN GW) and the PCRF, see Figure 10.8.2.1.1.

Figure 10.8.2.1.1 *Gx interface.*

10.8.2.2 Interface functionality

The main purpose of the Gx interface is to support PCC rule handling and event handling for PCC.

PCC rule handling over the Gx interface includes the installation, modification and removal of PCC rules. All these three operations can be made upon any request coming from the PCEF or due to some internal decision in the PCRF.

The event handling procedures allows the PCRF to subscribe to those events it is interested in. The PCEF then reports the occurrence of an event to the PCRF.

For more details on PCC rule handling and event reporting, see Section 8.2.

10.8.2.3 Protocol

This Gx protocol is based on Diameter and is defined as a vendor specific Diameter application, where the vendor is 3GPP. 3GPP Rel-8 is re-using the Gx Diameter application that was defined for Gx in 3GPP Rel-7. Only minor updates

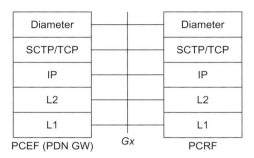

Figure 10.8.2.3.1 *Protocol stack for Gx.*

have been done to Gx in Rel-8. The Gx Diameter application is based on the Diameter base protocol and is also incorporating commands and AVPs from the Diameter Credit Control Application (DCCA) defined in RFC 4006 [4006].

The protocol stack for Gx is illustrated in Figure 10.8.2.3.1.

The definition of the Gx interface and its functionality is given in 3GPP TS 23.203 [23.203]. The Gx Diameter application is defined in TS 29.212 [29.212].

10.8.3 BBERF – PCRF (Gxa/Gxc)

10.8.3.1   General

The Gxa and Gxc interfaces are located between the PCRF and the BBERF. Gxc applies when the BBERF is located in the Serving GW and Gxa applies when the BBERF is located in an Access GW in a trusted non-3GPP IP access, see Figure 10.8.3.1.1.

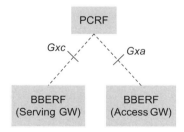

Figure 10.8.3.1.1 *Gxa/Gxc interfaces.*

10.8.3.2   Interface functionality

The main purpose of the Gxa and Gxc interfaces is to support QoS rule and event handling for PCC. This is similar to the Gx interface with the difference that the Gx interface handles PCC rules instead while Gxa and Gxc handle QoS rules.

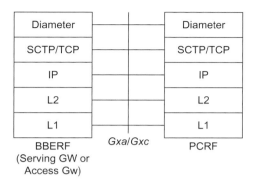

Figure 10.8.3.3.1 *Protocol stack for Gxa/Gxc.*

For more details on QoS rule handling and event reporting, see Section 8.2.

10.8.3.3   Protocol

The protocol over the Gxa and Gxc interfaces is based on Diameter. A new Diameter vendor specific application, the Gxx Diameter application, has been defined and is used on both Gxa and Gxc. The Gxx Diameter application is, similar to the Gx application, based on the Diameter base protocol and incorporating commands and AVPs from the DCCA defined in RFC 4006 [4006].

The protocol stack for Gxa and Gxc is illustrated in Figure 10.8.3.3.1.

The definition of the Gxa/Gxc interface and its functionality is given in 3GPP TS 23.203 [23.203]. The Gxx Diameter application is defined in TS 29.212 [29.212].

10.8.4   PCRF – AF (Rx)

10.8.4.1   General

The Rx interface is defined between the PCRF and the AF, see Figure 10.8.4.1.1.

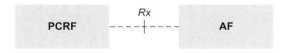

Figure 10.8.4.1.1 *Rx interface.*

10.8.4.2   Interface functionality

The main purpose of the Rx interface is to transfer session information from the AF to the PCRF. Rx is also used by the AF to subscribe to notifications about traffic plane events, for example that an IP session has been closed or that the

UE has handed over to a different access technology. The PCRF will notify the AF of the occurrence of a subscribed traffic plane event.

For more details on Rx procedures, see Section 8.2.

10.8.4.3 Protocol

The protocol over the Rx interface is based on Diameter. 3GPP Rel-8 is re-using the Rx Diameter application that was defined for Rx in 3GPP Rel-7. The Rx Diameter application is based on the Diameter base protocol and is also incorporating commands from the Diameter Network Access Server (NAS) Application defined in RFC 4005 [4005]. It can be noted however that the concept of a NAS (Network Access Server) is not used with Rx, it is merely the Diameter NAS application commands that are re-used for the Rx protocol, not its functional framework.

The protocol stack for Rx is illustrated in Figure 10.8.4.3.1.

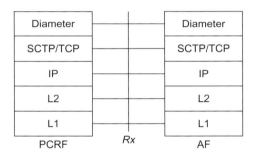

Figure 10.8.4.3.1 *Protocol stack for Rx.*

The definition of the Rx interface and its functionality is given in 3GPP TS 23.203 [23.203]. The Rx Diameter application is defined in TS 29.214 [29.214].

10.8.5 PCRF – PCRF (S9)

10.8.5.1 General

The S9 interface is defined between the PCRF in the home network (H-PCRF) and a PCRF in the visited network (V-PCRF). S9 is an inter-operator interface and is only used in roaming scenarios, see Figure 10.8.5.1.1.

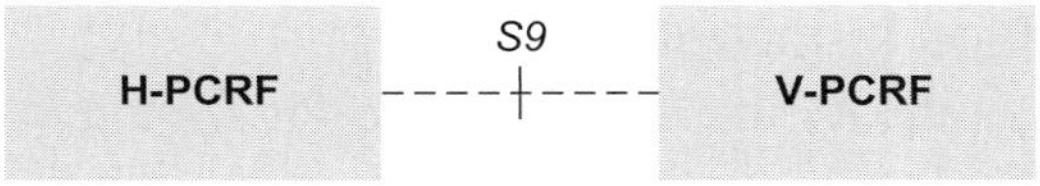

Figure 10.8.5.1.1 *S9 interface.*

10.8.5.2   Interface functionality

The main purpose of the S9 interface is to transfer policy decisions (i.e. PCC rules or QoS rules) generated in the home network into the visited network and transport the events that may occur in the visited network to the home network.

The S9 interface can also be used to transfer session information in specific roaming scenarios. The two main roaming scenarios are the home routed case (with PDN GW/PCEF in the home network) and the local breakout (with PDN GW/PCEF in the visited network). In the Local Breakout case, it is furthermore possible to use an AF either in the home network or in the visited network. When Local Breakout is used and the AF is in the visited network, the S9 interface carries service session information from the V-PCRF to the H-PCRF (see Section 8.2 for further details regarding PCC usage in roaming scenarios).

10.8.5.3   Protocol

The protocol over the S9 interfaces is based on Diameter. Two 3GPP vendor specific Diameter applications are used over the S9 interface: the S9 application and the Rx application.

The S9 Diameter application is a new vendor-specific application defined in 3GPP Rel-8. It is based on the Diameter base protocol and incorporating commands and AVPs from the DCCA defined in RFC 4006 [4006].

The Rx Diameter application is described above for the Rx interface. It is used over the S9 interface in case of Local Breakout with the AF in the visited network, as described above.

The protocol stack for S9 is illustrated in Figure 10.8.5.3.1.

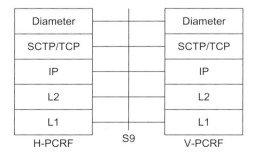

Figure 10.8.5.3.1 *Protocol stack for S9.*

The definition of the S9 interface and its functionality is given in 3GPP TS 23.203 [23.203]. The protocols over the S9 interface, including the S9 Diameter application, are defined in TS 29.215 [29.214]. The Rx Diameter application is defined in TS 29.214 [29.214].

10.8.6   SPR – PCRF (Sp)

10.8.6.1   General

The Sp interface is defined between the PCRF and the Subscriber Profile Repository (SPR), see Figure 10.8.6.1.1.

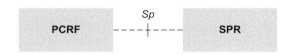

Figure 10.8.6.1.1   *Sp interface.*

10.8.6.2   Interface functionality

The Sp interface is used to transport subscription data from the SPR to the PCRF. The PCRF may request subscription data for a given user. The SPR may also notify the PCRF in case the subscription data has been modified.

The SPR may contain subscription data related to the transport level policies for the access network. The details regarding the subscription data contained in the SPR have not been further specified. A reason for not specifying the detailed subscription data is that the policies may depend significantly on the operator's business models and the types of subscriptions and services offered. It is thus reasonable to avoid detailed specification in order not to put unnecessary restrictions on the type of policies that can be kept in the SPR.

10.8.6.3   Protocol

The Sp interface and its functionality are specified in 3GPP TS 23.203 [23.203].

The protocol over the Sp interface is however not specified.

10.9 EIR-related interfaces

10.9.1 MME-EIR and SGSN-EIR interfaces (S13 and S13')

10.9.1.1 General

The interface S13 is defined between the Equipment Identity Register (EIR) and the MME and the interface S13' is defined between EIR and SGSN. The S13' interface applies to the S4-based SGSN only, see Figure 10.9.1.1.1.

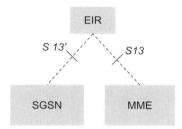

Figure 10.9.1.1.1 *S13 and S13' interfaces.*

10.9.1.2 Interface functionality

The S13 and S13' interfaces between the MME and the EIR and between the SGSN and the EIR respectively are used to check the status of the UE (e.g. if it has been reported stolen). The MME or SGSN checks the ME Identity by sending the Equipment Identity to an EIR and analysing the response.

10.9.1.3 Protocol

The same protocol is used on both S13 and S13'. This protocol is based on Diameter and is defined as a vendor specific Diameter application, where the vendor is 3GPP. The S13/S13' Diameter application is based on the Diameter base protocol but defines new Diameter commands and AVPs to implement the functions described above. Diameter messages over the S13 and S13' interfaces use the SCTP [2960] as a transport protocol. The protocol stack is illustrated in Figure 10.9.1.3.1.

Diameter			Diameter
SCTP			SCTP
IP			IP
L2			L2
L1			L1
MME/SGSN		S13/S13'	EIR

Figure 10.9.1.3.1 *Protocol stack for S13/S13'.*

The protocol over the S13/S13' interface, including the S13/S13' Diameter application is defined in 3GPP TS 29.272 [29.272].

10.10   I-WLAN-related interfaces

10.10.1   UE – ePDG (SWu)

10.10.1.1   General

The SWu interface is defined between the UE and the ePDG. The interface runs over an un-trusted non-3GPP IP Access, see Figure 10.10.1.1.1.

Figure 10.10.1.1.1   *SWu interface.*

10.10.1.2   Interface functionality

The SWu interface supports procedures for establishment or disconnection of an end-to-end tunnel between the UE and the ePDG. The tunnel establishment is always initiated by the UE, whereas the tunnel disconnection can be initiated by the UE or the ePDG. The SWu interface also supports tunnel modification procedures in order to update the tunnel in case the UE has acquired a new IP address from the un-trusted non-3GPP IP Access, for example in case the UE has moved to another un-trusted non-3GPP IP Access. For further details on tunnel management over un-trusted non-3GPP IP Access, see Section 7.3.

10.10.1.3   Protocol

The tunnel between UE and ePDG is an IPsec tunnel. The UE and ePDG use IKEv2 to establish the IPSec security association (SA) for the tunnel.

The UE uses standard DNS mechanisms in order to select a suitable ePDG. As input to the DNS query, the UE creates a FQDN based on the operator identity. As a reply from the DNS system, the UE receives one or more IP addresses of suitable ePDG(s). Once the ePDG has been selected, the UE initiates the IPsec tunnel establishment procedure using the IKEv2 protocol. Public key signature-based authentication with certificates, as specified for IKEv2, is used to authenticate the ePDG. EAP-AKA within IKEv2 is used to authenticate the UE. As part of the IKEv2 procedure, an IPSec SA is established. IPSec Encapsulated Security Payload (ESP) in tunnel mode shall be used for the IPSec tunnel between the UE and the ePDG.

SWu also supports the mobility extensions for IKEv2 defined by MOBIKE [4555]. This allows the IPSec SA to be updated in case the UE acquires a new IP address in the un-trusted non-3GPP IP Access.

For further details on IKEv2, MOBIKE and ESP, see Section 11.9.

The protocol stack for SWu is illustrated in Figure 10.10.1.3.1.

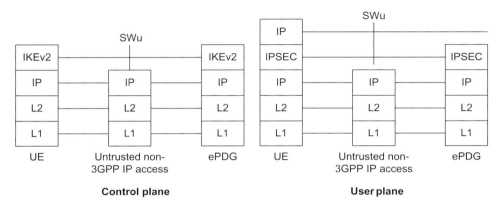

Figure 10.10.1.3.1 *SWu-PROTO. Protocol stack for SWu.*

The tunnel management procedures for SWu are defined in TS 33.402 [33.402].

10.11 ANDSF-related interfaces

The Access Network Discovery and Selection Function (ANDSF) is a mechanism that allows the UE to be provided with relevant parameters for intersystem mobility policy and access network discovery. This is done using the S14 interface, which utilizes Open Mobile Alliance (OMA) Device Management (DM). A brief outline of OMA DM is provided below in order to place the discussion regarding the ANDSF-UE interface in the correct context. A detailed description of OMA DM is beyond the scope of this book and the interested reader is referred to [Brenner, 2008]. A general overview of ANDSF can be found in Section 6.4.

The OMA DM v1.2 specifications are based on the OMA DM v1.1.2 specifications and make use of the OMA SyncML Common v1.2 specifications as specified in the OMA SyncML Common specifications Enabler Release Definition for SyncML (ERELDSC).

DM is the technology that allows the ANDSF to configure the UE on behalf of the operator and the end-user. Using DM, the ANDSF remotely sets parameters via the use of a Managed Object (MO). The MO is organized in nodes, interior nodes and leaf nodes. The leaf nodes contain the actual parameter values, see Figure 10.11.1. The ANDSF MO contains nodes related to Policy, Discovery Information and UE Location. There is also an additional node defined, Ext, for vendor-specific requirements. These are discussed briefly below:

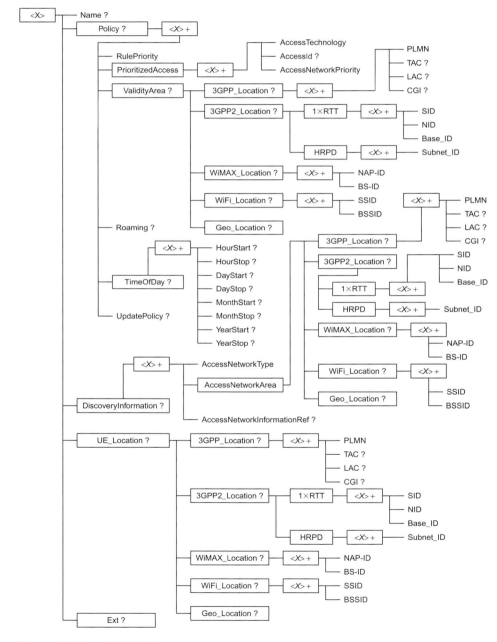

Figure 10.11.1   *ANDSF MO.*

10.11.1   Policy node

The policy node acts as a placeholder for policies related to intersystem mobility, including the rules and also the priority in which the policies should be applied. It also allows the ANDSF to set a particular access technology as the prioritized access, meaning that this is the one that should be searched for first, including the access network ID. Also included within the Policy node is the Validity area for a particular rule; this is used in the case where a particular rule is applicable to a particular location that the UE may find itself in.

With regard to Validity Areas, these can be different for different access networks, as a result, there are leaves that describe the validity areas and rules associated with them for each radio access technology types; 3GPP, 3GPP2, WiMAX and WiFi (WLAN).

In the case where a UE is roaming, the roaming leaf contains roaming conditions that should be applied in this case. The UE shall only apply such rules, however, if the UE's roaming state matches the one specified in the roaming value.

Particular rules may be applicable at a certain time of day and as a result there is also a leaf named TimeOfDay that handles this scenario.

In the case that the UE finds itself in a situation where it feels that a rule is no longer valid, it uses the UpdatePolicy leaf to determine whether it needs to request an update of its intersystem mobility policy or not.

10.11.2   Discovery Information node

Using the Discovery Information node, an Operator may provide information on the access networks that are available for a UE to connect to. The UE, meanwhile, may use this information in order to decide which access network to connect to.

The Discovery Information node, therefore, provides information regarding the type of access network, the area that the access network covers. The leaf describing the access network areas again covers all of the different network access types; 3GPP, 3GPP2, WiMAX, WiFi (WLAN). It also covers the use of Geo_Location, which acts as a placeholder for the geographic location of one or more access networks.

10.11.3   UE location node

The UE Location node acts as a placeholder for location descriptions; a UE therefore inserts information regarding all of the access networks that it can discover into this node. For 3GPP networks, this includes: PLMN, Tracking Area Code, Location Area Code and Cell Global Identity. For 3GPP2 networks, this includes SID, NID and Base ID. Whilst for WiMAX networks, UE Location includes: NAP-ID and BS-ID. For WLAN networks, SSID and BSSID are captured.

10.11.4   Ext node

Ext is the node where vendor-specific information about the ANDSF MO is placed. For the purposes of this node, vendor means application vendor, device vendor, etc. This is generally indicated by a vendor-specific name under the Ext node. As can be expected, since this is a vendor-specific node, the leaves under the Ext node are left undefined. If a vendor wishes to utilize extensions, they define the interior nodes and leaves themselves. These are therefore naturally not within in the scope of the standardization.

The MO for the transfer of intersystem mobility policy between the ANDSF and the UE is shown in Figure 10.11.1. For further details on the ANDSF MO see 3GPP TS 24.312 [24.312].

10.12   HRPD IW-related interfaces

10.12.1   Optimized handover and related interfaces (S101 and S103)

In order to support optimized handover between LTE and eHRPD networks, as explained in preceding chapter, one control plane (S101) and one user plane (S103) interface have been added to the architecture. Below we describe the functions and protocol supported over these two interfaces.

10.12.2   MME ↔ eHRPD access network (S101)

S101 is a tunnel between MME and eHRPD Access network where messages are carried over the serving access network towards a target access network (where the handover may occur) in order for preparation for handover via pre-registration, then maintain the resources via session maintenance and then the actual handover. These are messages tunneled over S101-AP where the GTPv2-C protocol functions are used with explicit utilization for S101 interface. The GTPv2-C message type used for S101 are as follows:

Message Type value (Decimal)	Message
0	Reserved
1	Echo Request
2	Echo Response
3	Version Not Supported Indication
4	Direct Transfer Request message
5	Direct Transfer Response message
6	Notification Request message
7	Notification Response message
8–24	For future S101 interface use
25–31	Reserved for Sv interface
32–255	Reserved for GTPv2-C spec

The protocol itself is segmented to provide the pre-configured tunnel via Path management general messages and then for specific messages used for information transfer over the control plane, see Figure 10.12.2.1.

Figure 10.12.2.1 *S101 interface.*

3GPP TS 29.276 [29.276] and TS 29.274 [29.274] describe the details for the messages. Procedures are covered in TS 23.402 [23.402].

The messages carried over this interface is not modified by MME but rather forwarded to/from the source and the target access network (in this case between eNodeB and HRPD AN). Each message must have unique identifier (also known as Session Identifier) to be able to identify uniquely in a global network the individual terminal the message is destined to or coming from.

The pre-configured tunnel carries messages from the MME or eHRPD access network towards its peer in the target network using Direct Transfer Message and Response. Depending on where the message is originating, the content is according to that specific access network (e.g. if an HRPD message is to be tunnelled then this message shall be carried over a transparent container), see Figure 10.12.2.2.

In case of Notification Request Message, the information mainly carries events such as completion of handover process.

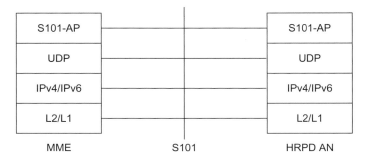

Figure 10.12.2.2   *Protocol stack for S101.*

The GTPv2-C Path management and Reliability procedures are used to manage the pre-configured tunnel S101.

The S101 message header takes the following form (Bits marked by * are spare and set to zero):

Octets	Bits 8	7	6	5	4	3	2	1
1	Version = 010			(*)	T = 0	E = 0	(*)	(*)
2	Message Type							
3	Length (1st Octet)							
4	Length (2nd Octet)							
5	Sequence Number (1st Octet)							
6	Sequence Number (2nd Octet)							
7	Spare							
8	Spare							

Examples of Direct Transfer Message information elements are described below. One can see the parameter usage during the optimized handover procedure described in Chapter 12 of the book.

Information element	Presence requirement
IMSI	Mandatory
HRPD Sector ID	Conditional
S101 Transparent Container	Mandatory
PDN GW PMIP GRE Tunnel Info	Conditional
S103 GRE Tunnel Info	Conditional
S103 HSGW IP Address	Conditional
Handover Indicator	Conditional
Tracking Area Identity	Conditional
Recovery	Conditional
Private Extension	Optional

Note that for the interested reader, the 3GPP TS 29.276 [29.276] is the specification for S101.

10.12.3 *Serving GW ↔ HSGW (S103)*

The S103 interface from the Serving GW to the HSGW in the CDMA HRPD network, see Figure 10.12.3.1.

Figure 10.12.3.1 *S103 interface.*

This interface provides support for forwarding of downlink data during hand-over from LTE to HRPD. The purpose of this interface is to reduce loss of user data during the handover procedure. Signalling procedures and parameters available on the S101 interface are used to set up GRE tunnel on the S103 interface. For details on GRE tunnel, see Section 11.6 for the overview and usage of GRE. The protocol stack for S103 is described in Figure 10.12.3.2.

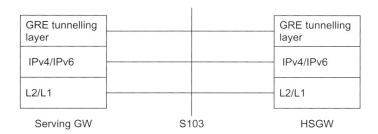

Figure 10.12.3.2 *Protocol stack for S103.*

S103 interface must be able to identify the user data traffic on a per terminal and per PDN connection basis.

10.13 Interface to external networks

10.13.1 *General*

The SGi interface is defined between the PDN GW and external IP networks (also called 'PDNs') as shown in Figure 10.13.1.1. The external networks may be Internet and/or Intranets and IPv4 and/or IPv6 may be used.

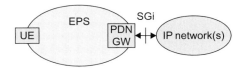

Figure 10.13.1.1   *IP network interworking.*

The PDN GW is the access point of the EPS and the EPS will look like any other IP network or subnetwork. From the external IP network's point of view, the PDN GW is seen as a normal IP router.

10.13.2   Functions

Access to Internet, Intranet or an ISP involves functions such as IPv4 address allocation, IPv6 address autoconfiguration, and may also involve specific functions such as authentication, authorization, secure tunnelling to Intranet/ISP.

An operator may offer direct transparent access to the Internet and operator services and will in that case offer at least basic ISP functions. An operator may also offer so called non-transparent access to an Intranet or ISP.

In both the transparent and non-transparent case the UE is given an IPv4 address and/or an IPv6 prefix. The difference is that in the transparent case the IP addresses belong to the operator while in the non-transparent case the addresses belong to the Intranet/ISP addressing space.

IPv4 address and/or IPv6 prefix are assigned either statically tied to the subscription or dynamically allocated at PDN connectivity establishment. This IPv4 address and/or IPv6 prefix is used for packet forwarding between the Internet and the PDN GW and within the packet domain. With IPv6, Stateless Address Autoconfiguration shall be used to assign an IPv6 address to the UE. The PDN GW may use a local IP address pool, or use DHCP or AAA protocols to retrieve UE IP addresses from the external IP network. For more information on IP address allocation for different cases see 3GPP TS 29.061 [29.061] and 23.401 [23.401].

The PDN GW prevents IP spoofing by verifying the source IP address of the IP packets issued by the UE and compare it against the allocated address.

To support IMS the PDN GW will also need to provide a list of P-CSCF addresses to the UE on request. The UE need this information in order to register for IMS services. In addition IMS requires a bearer for the IMS signalling and bearers for media when IMS sessions are established. This means that PDN GW needs to be configured to allow the IMS signalling and it need to support the Gx interface to the PCRF in order to allocated bearers for the IMS media flows.

11

Protocols

11.1   Introduction

This chapter provides an overview of the main protocols used in the EPS with the aim to give a basic overview of these protocols and their basic properties.

11.2   GPRS tunnelling protocol overview

The original version of the GTP protocol is what the GSM standards developed to cater to the specific needs such as mobility and bearer management and tunnelling of user data traffic for GPRS. Then 3GPP further enhanced GTP for usage in 3G UMTS. During the development of EPS, the GTP track of the architecture was enhanced considerably to improve the bearer handling and thus the GTP control plane protocol was upgraded to GTPv2-C.

The two main components of GTP are the control plane part of GTP (GTP-C) and the user plane part of GTP (GTP-U). GTP-C is used to control and manage tunnels for individual terminals connecting to the network in order to establish user data path. For the GTP-U uses a tunnel mechanism to carry the user data traffic. There also exists GTP', which is defined under the GTP protocol umbrella for the purpose of charging, but in this book we will not discuss this legacy protocol usage of GTP. There exists three versions of GTP-C: GTPv0, GTPv1 and GTPv2 and there exists two versions of GTP-U: GTPv0 and GTPv1. In this book we will provide some background on GTPv1-C for better understanding of GTPv2-C which is used exclusively for EPS. We will also discuss some details on GTPv1-U.

In order to understand the functions of the GTP protocol, it is useful to have a look at how GTP has been used in GPRS and 3G Packet Core. Figure 11.2.1 illustrates the interfaces that use GTP.

In case of GPRS and 3G packet core, the Gn interface between SGSNs and between SGSN and GGSN (when the entities are within an operator's PLMN) and Gp interface between SGSN and GGSN (inter-PLMN or inter-operator

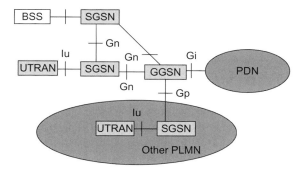

Figure 11.2.1 *GTP interfaces for GPRS.*

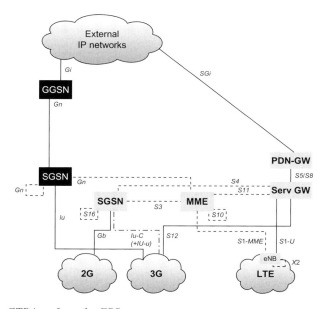

Figure 11.2.2 *GTP interfaces for EPS.*

may be more common term) support GTPv1-C and GTPv1-U protocols. For 3G Packet core using WCDMA/HSPA radio accesses, Iu supports GTPv1 user plane protocol (Figure 11.2.2).

In case of EPS, the interfaces between SGSN and MME, between MMEs, between SGSNs, between Serving GW and PDN GW and between Serving GWs use the GTPv2-C and the interface between HRPD Access Network and MME uses GTPv2-C tunnels to carry the tunnelled messages. The GTPv1-U is used between eNodeB and Serving GW, between RNC and Serving GW and between SGSN and Serving GW as well as between Serving GWs. Thus the GTPv2-C is used on S3, S4, S5, S8, S10, S11 and S16 interfaces and GTPv1-U is used on S1-U, Iu-U, S4, S5, S8, S12 and X2-U interfaces.

So what is this protocol suite and how does it work? As can be easily seen from the 3GPP architecture, the entities supporting the GTP protocol need to support one to many and many to many relationships with each other. A single SGSN must beable to connect to multiple RNCs, SGSNs as well as many GGSNs within and between different operators' networks. Similarly, a GGSN must be able to connect to multiple SGSNs from different operators' networks spanning significant geographical areas in order to support its own subscribers who may be in their home network or in a roaming partner's network. Development of GTP protocol caters to such diverse deployment requirements that are corner-stone for success of mobile systems worldwide.

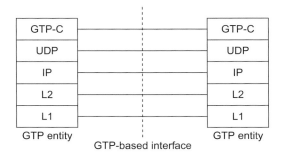

Figure 11.2.1.1   *GTP Control Plane Protocol Stack.*

If we look at the original GTP message structure (see Table 11.2.1), it becomes quite obvious that it serves to manage a cellular network by developing and grouping messages according to the functional needs of a cellular system. Then when we look at the GTPv2 messages developed for EPS (see Table 11.2.2) then we clearly see the evolution of the protocol in a more generic manner catering to more flexible bearer management as well as simplified/unified network elements interactions and support for mobility/common core functions for non-3GPP access networks and better error/failure/network management as well as restoration and recovery handling for network elements such as MME, Serving GW/PDN GW. The key functions that GTP was built upon and then additions from EPS can be categorised as below:

1. **Mobility Management**: The set of messages created as part of this function includes managing mobile device's identification and maintain presence/status among various network elements in a coordinated manner, handling data transfer between entities during/at handover/relocation of the mobile terminal.

2. **Tunnel Management**: Involves creation and deletion of the end user's session and creation, modification and deletion of bearers established during the period the user is connected and actively involved in services by the network. Simply stated, these messages keep the user's different service requirements maintained in the network as the user moves around within and between PLMNs.

3. **Service specific Functions**: For GTPv1 it includes mainly support of MBMS related functions. For GTPv2, during the writing of this book, MBMS service is being developed and impacts if any on GTPv2 is not yet addressed. GTPv2 provides messages in order to function CS Fallback, Optimized handover with 3GPP2, non-3GPP mobility.

4. **Mobile Terminal information transfer**: For GTPv2 this is incorporated within Mobility Management and is only supported for GERAN/UTRAN accesses.

5. **System maintenance(path management/error handling/restoration and recovery/trace)**: Supporting network level functions in order to handle overall robustness of the tunnels and recover from failure in network entities. These messages (such as Echo Request/Response) have been supported in GTPv1 and now in GTPv2, but wherever possible, for GTPv2 improvements have been made in error handling and recovery procedures.

Some messages have been removed from GTPv1 to GTPV2 since the functions associated with them are no longer supported in the system, one example is the messages related to the function Network Initiated PDP Context Set up.

11.2.1 Protocol structure

Let us now first take a look at the GTPv1 protocol structure. It can be shown as in Figure 11.2.1.1, where the GTP-C protocol provides the messages to carry out functions such as mobility management, bearer management (also referred to as tunnel management), location management as well as mobile terminal status reporting. GTPv2 follows similar structure but some groups of messages are not required for the systems operation and thus not supported, as discussed in Section 11.2 above. It should be quite clear that the GTP-C and GTP-U tunnels are associated with each other for any single specific user since their role is to establish connections throughout the network so that the terminal can send/receive data. The following Table 11.2.1 illustrates the key GTPv1 control plane messages for GPRS.

Table 11.2.1   *GTPv1 Control plane Messages (GPRS).*

Message Type value (Decimal)	Message	GTP-C
0	For future use. Shall not be sent. If received, shall be treated as an Unknown message	
1	Echo Request	X
2	Echo Response	X
3	Version Not Supported	X
4	Node Alive Request	
5	Node Alive Response	
6	Redirection Request	
7	Redirection Response	
8–15	For future use. Shall not be sent. If received, shall be treated as an Unknown message	
16	Create PDP Context Request	X
17	Create PDP Context Response	X
18	Update PDP Context Request	X
19	Update PDP Context Response	X
20	Delete PDP Context Request	X
21	Delete PDP Context Response	X
22	Initiate PDP Context Activation Request	X
23	Initiate PDP Context Activation Response	X
24–25	For future use. Shall not be sent. If received, shall be treated as an Unknown message	
26	Error Indication	
27	PDU Notification Request	X
28	PDU Notification Response	X
29	PDU Notification Reject Request	X
30	PDU Notification Reject Response	X
31	Supported Extension Headers Notification	X
32	Send Routeing Information for GPRS Request	X
33	Send Routeing Information for GPRS Response	X

(Continued)

Table 11.2.1   *(Continued)*

Message Type value (Decimal)	Message	GTP-C
34	Failure Report Request	X
35	Failure Report Response	X
36	Note MS GPRS Present Request	X
37	Note MS GPRS Present Response	X
38–47	For future use. Shall not be sent. If received, shall be treated as an Unknown message	
48	Identification Request	X
49	Identification Response	X
50	SGSN Context Request	X
51	SGSN Context Response	X
52	SGSN Context Acknowledge	X
53	Forward Relocation Request	X
54	Forward Relocation Response	X
55	Forward Relocation Complete	X
56	Relocation Cancel Request	X
57	Relocation Cancel Response	X
58	Forward SRNS Context	X
59	Forward Relocation Complete Acknowledge	X
60	Forward SRNS Context Acknowledge	X
61-69	For future use. Shall not be sent. If received, shall be treated as an Unknown message	
70	RAN Information Relay	X
71–95	For future use. Shall not be sent. If received, shall be treated as an Unknown message	
96–105	MBMS	X
106–111	For future use. Shall not be sent. If received, shall be treated as an Unknown message	
112–121	MBMS	X
122–127	For future use. Shall not be sent. If received, shall be treated as an Unknown message	

(Continued)

Table 11.2.1   *(Continued)*

Message Type value (Decimal)	Message	GTP-C
128	MS Info Change Notification Request	X
129	MS Info Change Notification Response	X
130–239	For future use. Shall not be sent. If received, shall be treated as an Unknown message	
240	Data Record Transfer Request	
241	Data Record Transfer Response	
242–254	For future use. Shall not be sent. If received, shall be treated as an Unknown message	
255	G-PDU	

For GTPv1-C, some example messages that carry out the functions mentioned above are provided here for the readers before delving into the protocol details.

For GTPv1-C some example message flows between SGSN and GGSN:

Functions	Message name	Entities	Interface
Mobility management	SGSN context Request	SGSN-SGSN	Gn
	Forward Relocation request	SGSN-SGSN	Gn
Tunnel management	Create PDP context	SGSN -> GGSN	Gn/Gp
	Update PDP context	SGSN -> GGSN	Gn/Gp
Path management	Echo request	SGSN-GGSN	Gn/Gp

Similar messages for EPS network are shown in detail under the interface details later on.

Example of GTPv1-U message can be described similarly, though note that the main purpose of these control messages is to ensure 'smooth' user data traffic handling for uplink and downlink direction for the end-user. These messages include Echo request/response for path management purposes and Error Indication messages for exception handling. A GTP entity may use the Echo request to find out if the other GTP entity is alive. The Error Indication messages can be used to inform the other GTP entity that there is no EPS bearer (or PDP context in case of GPRS) corresponding to a received user plane packet. The actual control signalling for GTP-U is performed over S1-AP (for MME and eNodeB) and GTPv2-C (for the core network entities) and over RANAP and GTPv1/v2-C for RNC and core network entities (Figure 11.2.1.2).

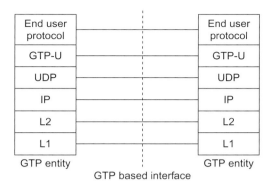

Figure 11.2.1.2   *GTP-U Protocol stack.*

Let us now get into a bit more detail into the GTP tunnels and their basic structure. For those readers interested in the details of the protocols such as all the messages, the coding of the parameters and the interworking of the formats themselves, we recommend the specifications where GTP-C protocols are defined in 3GPP TS 29.060 [29.060] (GTPv1) and TS 29.274 [29.274] (GTPv2-C) and GTP-U protocol is defined in 3GPP TS 29.060 [29.060] and TS 29.281 [29.281].

A few concepts must be described before one can understand the GTP protocol handling. GTP is a tunnelling protocol over UDP/IP (can be either IPv4 or IPv6). GTP is a tunnel with its specific tunnel definition and its tunnel identifiers.

GTP tunnels are used between two corresponding GTP nodes communicating over a GTP-based interface to separate traffic into different communication flows. A local Tunnel Endpoint Identifier (TEID), the IP address and the UDP port uniquely identifies a tunnel endpoint in each node, where the TEID assigned by the receiving entity must be used for the communication. Figure 11.2.2.1 illustrates an example of the GTP-C and GTP-U tunnel representation in EPS for terminals. Note that it is a simplified high level view for illustration purposes on how the GTP tunnels are represented in the system.

A GTP path is identified in each node with an IP address and a UDP port number. A path may be used to multiplex GTP tunnels and there may be multiple paths between two entities supporting GTP.

Another important feature of GTP protocol is its usage of Cause values in response messages. Cause values represent the actual status of the action requested (e.g. Accept/Reject) as well as additional useful information which would facilitate the receiving entity to make a more informed decision on the possible course of action. For EPS, a list of these Cause values can be found in the specification TS29.274 [29.274].

11.2.2   Control plane (GTPv2-C)

Through GTP-C messages tunnels are established, used, managed and released. A path may be maintained by keep-alive echo messages. The GTPv2-C protocol stack is shown in Figure 11.2.1.1.

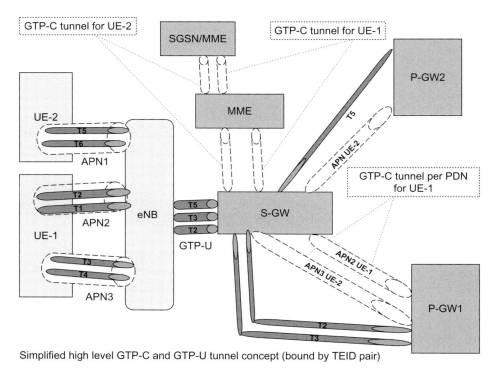

Simplified high level GTP-C and GTP-U tunnel concept (bound by TEID pair)

Figure 11.2.2.1   *GTP Tunnel representation.*

For the control plane, for each endpoint of a GTP-C tunnel there is a control plane TEID (TEID-C). The scope of the GTP tunnel and the TEID-C depends on the interface and its functions (such as if the interface is used on a per terminal connection basis such as the S3 interface or per PDN connection basis like for S5/S8 interface)

- The TEID-C is unique per PDN-Connection on GTP-based S5 and S8. The same tunnel is shared for the control messages related to all bearers associated to the PDN-Connection. A TEID-C on S5/S8 interface is released after all its associated EPS bearers are deleted.
- There is only one pair of TEID-Cs per UE on each of the S3 and the S10 interfaces. The same tunnel is shared for the control messages related to the

same UE operation. A TEID-C on S3/S10 interface is released after its associated UE context is removed or the UE is detached.

- There is only one pair of TEID-C per UE over the S11 and the S4 interfaces. The same tunnel is shared for the control messages related to the same UE operation. A TEID-C on S11/S4 interface is released after all its associated EPS bearers are deleted.

GTP defines a set of messages between two associated EPC entities. The messages are defined in 3GPP TS 29.274 [29.274] and shown here for illustration purposes. For most detailed and up to date information, the reader should look at the latest version of the specification.

Message types for GTPv2 are listed in Table 11.2.2 below.

Table 11.2.2 *GTPv2-Control plane Messages (EPS).*

Message Type value (Decimal)	Message	GTP-C
0	Reserved	
1	Echo Request	X
2	Echo Response	X
3	Version Not Supported Indication	X
4 to 24	Reserved for S101 interface	
25 to 31	Reserved for Sv interface	
	SGSN/MME to PGW (S4/S11, S5/S8)	
32	Create Session Request	X
33	Create Session Response	X
34	Modify Bearer Request	X
35	Modify Bearer Response	X
36	Delete Session Request	X
37	Delete Session Response	X
	SGSN to PGW (S4, S5/S8)	
38	Change Notification Request	X
39	Change Notification Response	X
40 to 63	For future use	

(Continued)

Table 11.2.2 *(Continued)*

Message Type value (Decimal)	Message	GTP-C
	Messages without explicit response	
64	Modify Bearer Command (MME/SGSN to PGW –S11/S4, S5/S8)	X
65	Modify Bearer Failure Indication (PGW to MME/SGSN –S5/S8, S11/S4)	X
66	Delete Bearer Command (MME/SGSN to PGW –S11/S4, S5/S8)	X
67	Delete Bearer Failure Indication (PGW to MME/SGSN –S5/S8, S11/S4))	X
68	Bearer Resource Command (MME/SGSN to PGW –S11/S4, S5/S8)	X
69	Bearer Resource Failure Indication (PGW to MME/SGSN –S5/S8, S11/S4)	X
70	Downlink Data Notification Failure Indication (SGSN/MME to SGW –S4/S11)	X
71	Trace Session Activation	X
72	Trace Session Deactivation	X
73	Stop Paging Indication	X
74 to 94	For future use	
	PGW to SGSN/MME (S5/S8, S4/S11)	
95	Create Bearer Request	X
96	Create Bearer Response	X
97	Update Bearer Request	X
98	Update Bearer Response	X
99	Delete Bearer Request	X
100	Delete Bearer Response	X
	PGW to MME, MME to PGW, SGW to PGW, SGW to MME (S5/S8, S11)	
101	Delete PDN Connection Set Request	X
102	Delete PDN Connection Set Response	X
103 to 127	For future use	

(Continued)

Table 11.2.2 *(Continued)*

Message Type value (Decimal)	Message	GTP-C
	MME to MME, SGSN to MME, MME to SGSN, SGSN to SGSN (S3/10/S16)	
128	Identification Request	X
129	Identification Response	X
130	Context Request	X
131	Context Response	X
132	Context Acknowledge	X
133	Forward Relocation Request	X
134	Forward Relocation Response	X
135	Forward Relocation Complete Notification	X
136	Forward Relocation Complete Acknowledge	X
137	Forward Access Context Notification	X
138	Forward Access Context Acknowledge	X
139	Relocation Cancel Request	X
140	Relocation Cancel Response	X
141	Configuration Transfer Tunnel	X
142 to 148	For future use	
	SGSN to MME, MME to SGSN (S3)	
149	Detach Notification	X
150	Detach Acknowledge	X
151	CS Paging Indication	X
152	RAN Information Relay	
153 to 159	For future use	
	MME to SGW (S11)	
160	Create Forwarding Tunnel Request	X
161	Create Forwarding Tunnel Response	X
162	Suspend Notification	X
163	Suspend Acknowledge	X
164	Resume Notification	X

(Continued)

Table 11.2.2   *(Continued)*

Message Type value (Decimal)	Message	GTP-C
165	Resume Acknowledge	X
166	Create Indirect Data Forwarding Tunnel Request	X
167	Create Indirect Data Forwarding Tunnel Response	X
168	Delete Indirect Data Forwarding Tunnel Request	X
169	Delete Indirect Data Forwarding Tunnel Response	X
170	Release Access Bearers Request	X
171	Release Access Bearers Response	X
172 to 175	For future use	
	SGW to SGSN/MME (S4/S11)	
176	Downlink Data Notification	X
177	Downlink Data Notification Acknowledge	X
	SGW to SGSN (S4)	
178	Update Bearer Complete	X
179 to 191	For future use	
	Other	
192 to 255	For future use	

11.2.3   User plane (GTPv1-U)

GTP-U tunnels are used to carry encapsulated payload (original Packet Data Unit to be tunnelled) and signalling messages between a given pair of GTP-U Tunnel Endpoints. The TEID-U which is present in the GTP header indicates which tunnel a particular payload belongs to. Thus packets are multiplexed and de-multiplexed by GTP-U between a given pair of Tunnel Endpoints.

In case of LTE/EPC, the GTP-U tunnels are established using S1-MME or GTP-C (e.g., EPS bearer establishment process) and in case of 3G packet core, it is established as mentioned before using RANAP and GTP-C (e.g., PDP context activation process). The protocol stack for GTP-U is shown in Figure 11.2.1.2.

As there exists different protocol versions, the version-not-supported indicator is used to determine what version the peer GTP endpoint supports.

11.2.4   Protocol format

The control plane GTP uses a variable length header. Control Plane GTP header length is of a multiple of four octets as shown in the example below according to TS 29.274 [29.274]:

| | **Bits** | | | | | | | |
Octets	8	7	6	5	4	3	2	1
1	Version			P	T	Spare	Spare	Spare
2	Message Type							
3	Message Length (1st Octet)							
4	Message Length (2nd Octet)							
m to k(m+3)	If T flag is set to 1, then TEID shall be placed into octets 5-8. Otherwise, TEID field is not present at all.							
n to (n+1)	Sequence Number							
(n+2) to (n+3)	Spare							

The GTP-C header may be followed by subsequent information elements dependent on the type of control plane message. The format of a GTPv2-C message is illustrated below.

| | **Bits** | | | | | | | |
Octets	8	7	6	5	4	3	2	1
1 to m	GTP-C header							
m+1 to n	Information Element(s)							

In GTPv2-C, the information elements are added for new parameters if needed in future instead of using extension headers that used to be in use for GTPv1-C.

For EPS, the GTPv2-C header takes the following form (EPC functional message specific header format which does not include messages such as Echo type etc.):

| | **Bits** | | | | | | | |
Octets	8	7	6	5	4	3	2	1
1	Version			Spare	T=1	Spare	Spare	Spare
2	Message Type							
3	Message Length (1st Octet)							
4	Message Length (2nd Octet)							
5	Tunnel Endpoint Identifier (1st Octet)							
6	Tunnel Endpoint Identifier (2nd Octet)							
7	Tunnel Endpoint Identifier (3rd Octet)							
8	Tunnel Endpoint Identifier (4th Octet)							
9	Sequence Number (1st Octet)							
10	Sequence Number (2nd Octet)							
11	Spare							
12	Spare							

Whereas a user plane GTP header would, for example, have the following format as specified in 3GPP TS 29.281 [29.281]:

Bits

Octets	8	7	6	5	4	3	2	1
1	Version			PT	(*)	E	S	PN
2	Message Type							
3	Length (1st Octet)							
4	Length (2nd Octet)							
5	Tunnel Endpoint Identifier (1st Octet)							
6	Tnnel Endpoint Identifier (2nd Octet)							
7	Tunnel Endpoint Identifier (3rd Octet)							
8	Tunnel Endpoint Identifier (4th Octet)							
9	Sequence Number (1st Octet)[1, 4]							
10	Sequence Number (2nd Octet)[1, 4]							
11	N-PDU Number[2, 4]							
12	Next Extension Header Type[3, 4]							

Notes * This bit is a spare bit. It shall be sent as '0'. The receiver shall not evaluate this bit.
1 This field shall only be evaluated when indicated by the S flag set to 1.
2 This field shall only be evaluated when indicated by the PN flag set to 1.
3 This field shall only be evaluated when indicated by the E flag set to 1.
4 This field shall be present if and only if any one or more of the S, PN and E flags are set.

11.3   Mobile IP
11.3.1   General

The basic IP stack does not provide support for mobility. If a UE has been allocated an IP address, this IP address is used not only to identify the UE in the sense that packets sent to this IP address is really destined to that UE. The IP address is also used to identify the network where the UE has attached. Each global IP address belongs to a certain IP sub-network. Routers connecting different sub-networks will, with the help of routing protocols, make sure that packets destined to this IP address will reach the sub-network to which this IP address 'belongs'. If the UE connects to another IP sub-network the IP packets destined to the old IP address will still be routed to the old sub-network. The UE will thus no longer be reachable using the old IP address. Furthermore, even packets sent by the UE in its new sub-network may be discarded. The reason is that routers or firewalls may perform egress filtering of traffic leaving the sub-network and discard packets sent with IP addresses not belonging to the network. This has been illustrated in Figure 11.3.1.1. The change of sub-network may, for example, occur if the UE moves and connects to another network using the same interface (e.g., a UE that move between WLAN hotspots) or connects to another network using another access technology (e.g., goes from using a 3GPP access to using WLAN).

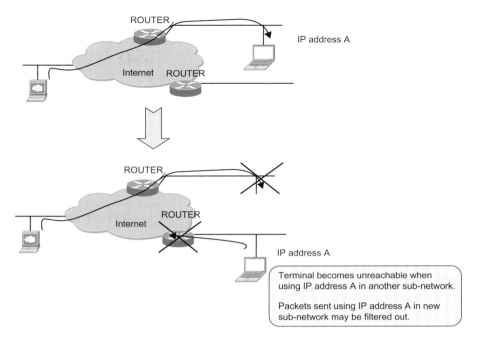

Figure 11.3.1.1 *A node becomes unreachable with its original IP address when moving to another IP sub-network.*

Packets destined to the UE's old IP address will continue to end up on the old sub-network corresponding to that IP address. The UE thus needs to change IP address and get an IP address from the IP address range corresponding to the new point-of-attachment. In this way the UE will be reachable at its new point-of-attachment using its new IP address. However, if the UE replaces its old IP address with a new one, ongoing IP sessions need to be terminated and then restarted with the new IP addresses.

Mobile IP (MIP) is intended to solve these problems by providing mobility support on the IP layer. Mobile IP allows the UE to change its point-of-attachment (i.e., sub-network) while continuing to use the same IP address and maintaining ongoing IP sessions. Since Mobile IP operates on the IP layer, it can provide mobility support for different kinds of lower layers. Mobile IP is thus suitable to provide mobility not just when moving across different IP networks for the same access technology, but also across heterogeneous access technologies. How this works is explained below.

EPS makes use of Mobile IP to provide IP level mobility when the UE moves between different access technologies, for example, from an access in the 3GPP family of accesses to a WLAN access.

Mobile IP is specified by the IETF. In fact, IETF has specified different variants of Mobile IP applicable to IPv4, IPv6 or both IPv4 and IPv6. The different variants are more or less related with one another. Mobile IPv4 [3344] is applicable to IPv4 and was specified first. The Mobile IP version for IPv6, Mobile IPv6 (MIPv6) [3775], reuses many of the basic concepts developed for Mobile IPv4, but is still a distinct protocol. Dual-stack MIPv6 [5555] is based on MIPv6 and contains the necessary enhancements for dual-stack IPv4/IPv6 operation. There is also a network-based version of MIPv6 called Proxy Mobile IPv6 (PMIPv6) [5213]. Figure 11.3.1.2 illustrates the different variants and also indicates their relation. In addition, there are numerous RFCs containing amendments, optimizations and enhancements, for example, to improve handover performance (not illustrated in the figure). There have also been proposals for Proxy Mobile IPv4 and dual-stack Mobile IPv4 variants, but these are not covered here.

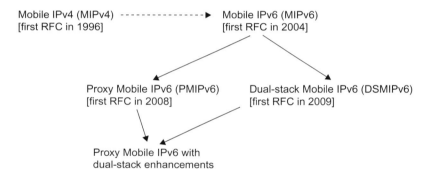

Figure 11.3.1.2   *Mobile IP family tree.*
The Proxy Mobile IPv6 with dual-stack enhancements was, when this book was prepared, still an Internet Draft but assumed to be soon be approved as RFCs.

It is not feasible in a book like this to describe all different variants of Mobile IP or even all aspects and details of a single Mobile IP variant. Instead we provide a high-level overview of primarily MIPv6 and dual-stack MIPv6. In Section 11.4, PMIPv6 is covered. These are the main Mobile IP-based protocols used in EPS. Also Mobile IPv4 is supported by EPS to some extent. However, we regard dual-stack MIPv6 and PMIPv6 as the more general and future proof Mobile IP protocols and also more relevant to EPS. Therefore Mobile IPv4 is only discussed briefly in this chapter, mainly pointing out the differences compared to MIPv6. The description also focuses on those aspects of MIPv6 that are most relevant for its use in EPS. A reader interested in a more complete description of all features and different options of Mobile IP in general and MIPv6 in particular should, for example, consult dedicated books on the topic or the relevant RFCs.

11.3.2 Host-based and network-based mobility mechanisms

Before going into the details on how Mobile IP works, it is useful to take a high-level view of different mobility concepts. As was described in Section 6.4, IP level mobility protocols could be roughly classified into two basic types; host-based mobility protocols and network-based mobility protocols. Mobile IP is a host-based mobility protocol where the UE has functionality to detect movement and to exchange Mobile IP signalling with the network in order to maintain IP-level session continuity. The other type of mobility protocol, or mobility scheme, is the network-based mobility management scheme. In this case the network can provide mobility services for a UE that is not explicitly exchanging mobility signalling with the network. It is in this case a task of the network to keep track of the UE's movements and ensure that the appropriate mobility signalling is executed in the core network in order for the UE to maintain its session while moving. PMIPv6 described in Section 10.4 is an example of a network-based mobility protocol. GTP is another example of a network-based protocol that is used to support mobility.

11.3.3 Basic principles of mobile IP

Before going into the actual mechanisms of Mobile IP it is necessary to describe the terms and concepts used. The description in this section is to most degree covering Mobile IP concepts in general, but is geared towards MIPv6 in specific matters, dual-stack MIPv6 is covered further below. Even though EPC supports both MIPv4 (in Foreign Agent mode) and DSMIPv6 for mobility between heterogeneous accesses, DSMIPv6 is the more general and future proof protocol of the two. Mobile IPv4 in FA mode is supported primarily for interworking with legacy CDMA and WiMAX systems.

As mentioned above, Mobile IP allows a UE to be always reachable using the same IP address even when the UE moves between different IP sub-networks. This IP address is called the *Home Address* (HoA) and is an IP address assigned from the address space of the *home network* (also referred to as the *home link*).

Note that 'home network' in Mobile IP terminology is not the same as the 'home network' (or 'Home PLMN') used when discussing roaming. A home network in Mobile IP sense is the IP network where the UE's HoA has been allocated and is thus a term related to IP topology and IP routing. The 'home network' in case of roaming is however a term denoting the network of the home operator or business entity where the subscriber has its subscription. The Mobile IP 'home network' may be located in the Home PLMN or the Visited PLMN depending on if the PDN GW is allocated from Home PLMN or Visited PLMN.

In Mobile IP terminology, the mobile UE is referred to as a *Mobile Node* (MN). However, in order to align with the terminology used in the rest of this book, we will continue to use the term 'UE' when referring to the Mobile Node also in this chapter.

When the UE is attached to its home network it can use the HoA in the usual way without any need for Mobile IP services. However, when the UE attaches to a different IP network where the HoA is not topologically located, this is no longer possible. In Mobile IP terminology, the UE is in this case attached to a 'foreign link' (or 'foreign network').

When the UE is attached in a foreign network it acquires a local IP address from that network. This IP address is in Mobile IP terminology called a *Care of Address* (CoA). The CoA is topologically located in the network the UE is currently accessing.

When the UE is at this foreign network, IP packets addressed to the CoA will reach the UE, while packets addressed to the HoA will reach the home network instead and not the UE. To solve this problem, Mobile IP introduces a network entity that maintains an association between the CoA and the HoA. This entity is called a *Home Agent* (HA) and is a router that is located on the UE's home network. (For EPS, the HA functionality is located in the PDN GW). The association between the two IP addresses is called a *binding*. When the UE has attached to a foreign link it informs the HA about its current point-of-attachment (i.e., its current local IP address, the CoA). The HA then intercepts packets that are routed to the home network addressed to the HoA, and forwards them in a tunnel to the UE's current location, that is, its CoA.

This behaviour, at least for the down-link, resembles the mail forwarding that can be used if a family moves from one city to another. The post office in the old city can be informed about the family's new address, and will 'intercept' and forward all mails addressed to the old address by placing the mails in a new envelope addressed to the family's new address. As we will see below, this comparison does however not really work when looking at up-link packets. In the Mobile IP case also up-link packets are typically sent via the HA in the home network, while in the example with the post office, the family can send letters from its new address without having to send them via the post office in its old home town. An exception to this principle is MIPv6 Route Optimization (RO) where traffic is not sent via the HA. However, since RO is not supported in EPS, this is only briefly discussed below.

Before describing in more detail how MIPv6 works, we should also introduce the third entity in the MIPv6 architecture, the *Correspondent Node* (CN). The CN is an IP node with which the UE is communicating. It could, for example, be a server of some kind or another UE with which the Mobile IP UE is communicating. The CN does not need any Mobile IP functionality.

The basic Mobile IP operation will be described below by going through an example use case where a binding is created and updated.

11.3.3.1 Bootstrapping

When the UE is powered on, it connects to a network and acquires an IP address from the local network. This IP address becomes the CoA. In order to utilize Mobile IP, the UE needs to have the IP address of the HA, a security association with the HA and a HoA. The process for establishing this information is called *bootstrapping*. Even though this information may be statically pre-configured in the UE and HA, it is in many cases beneficial to establish this information dynamically. In particular, in an EPS deployment with a large number of subscribers the option to pre-configure the UE does not scale very well and would be difficult for operators to manage. Therefore dynamic bootstrapping mechanisms are used.

The MIPv6 capable UE also needs to determine whether it needs to invoke Mobile IP or not. The UE does this by performing *home link detection* to determine whether it is attached to its home link or to a foreign link.

Several different methods have been defined for how the UE discovers an IP address of a suitable HA. Also EPS supports different procedures for how the HA IP address is provided to the UE. It may be discovered using DNS or be provided to the UE using other means depending on what access technology the UE is using. For more details, see Section 9.2.6.

Once the UE knows the HA IP address, it can contact the HA to set up a security association. MIPv6 uses IPSec to protect the Mobile IP signalling and IKEv2 to establish the IPSec SA. During the IKEv2 procedure, the UE and HA perform mutual authentication and the HA can also deliver the HoA to the UE. See Section 11.3.4 for further aspects related to MIPv6 security.

When the UE has acquired its HoA it performs home link detection by checking whether the HoA is 'on-link' or not, that is, whether or not the HoA belongs to the local network where the UE is currently attached. If the UE is attached to its home network no Mobile IP services are needed. The UE can use its HoA in the usual way.

11.3.3.2   Registration

If the UE is attached to a foreign network, the UE needs to inform the HA about the current CoA. The UE does this by sending a Mobile IP Binding Update (BU) message to the HA. The BU message contains the HoA and the CoA and is protected using the IPSec SA previously established. The HA maintains a *Binding Cache* containing the HoAs and CoAs for each UE that have registered with the HA. When receiving the BU for a new UE, the HA creates a new entry in the Binding Cache and replies to the UE with a Binding Acknowledgement (BA). The MIPv6 registration is illustrated in Figure 11.3.3.2.1. For a more detailed call for initial attach using DSMIPv6, see Section 12.2.4.

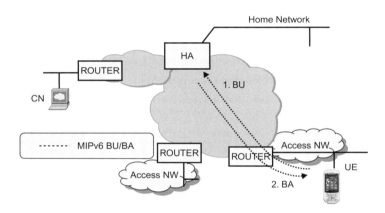

Figure 11.3.3.2.1   *A node registers with the Home Agent by sending a Binding Update.*

11.3.3.3   Routing of packets

When the UE is attached to a foreign network and a binding cache entry has been created in the HA, the HA intercepts all packets routed to the home network and destined to the UE's HoA. The HA then encapsulates the packets in a new IP header and forwards the packet to the UE's CoA. When receiving the packet, the UE de-capsulates it and processes it in the normal way. When the UE sends packets, the UE tunnels the packets to the HA which de-capsulates the packets and forwards the packets towards the final destination. This bi-directional tunnelling of packets between the UE and the HA is illustrated in Figure 11.3.3.3.1.

An alternative to bi-directional tunnelling would have been for the UE to sent the up-link packets directly to the destination, without tunnelling them to the HA first. This would have created a 'triangular routing' where down-link packets are routed to the home network and passes through the HA while up-link packets

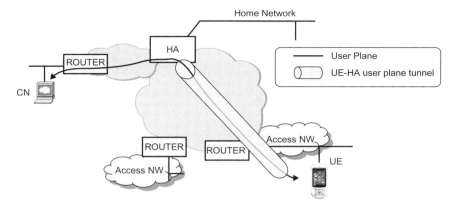

Figure 11.3.3.3.1 *User plane tunnelled bi-directionally between UE and Home Agent.*

are routed directly to the final destination, bypassing the HA. MIPv6 always uses bi-directional tunnelling while Mobile IPv4 allows both triangular routing and bi-directional tunnelling (called reverse tunnelling for Mobile IPv4). It can also be noted that MIPv6 allows for a feature referred to as Route Optimization where both up-link and down-link user plane is sent directly between the UE and the CN, without passing the HA. Route optimization is not used in EPS, but is briefly described in Section 11.3.7.

11.3.3.4 Binding lifetime extension

A binding in the HA has a certain lifetime. Unless the binding is renewed before the lifetime expires, the HA will remove the binding. This is used, for example, to clean up bindings belonging to terminals that are no longer attached to the network and that did not cancel the binding properly when they were disconnected. In order for the terminal to refresh the binding, the terminal sends a new BU well before the expiry of the binding lifetime.

11.3.3.5 Movement and update of the binding

In case the UE moves to different point-of-attachment and receives a new local IP address the UE again performs home link detection to determine whether it is now connected to the home link. If the UE determines that it has moved to another network different from its home network, the UE needs to inform the HA about the new CoA acquired in the new network. If not, the HA would continue forwarding the IP packets to the old foreign network. The UE thus sends a new BU to the HA containing its HoA and the new CoA. When receiving the BU, the HA updates the binding cache entry for the HoA with the new CoA and starts forwarding traffic to the new CoA. The movement, MIPv6 BU/BA signalling and the new user plane tunnel are illustrated in Figure 11.3.3.5.1.

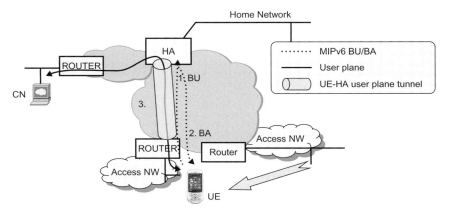

Figure 11.3.3.5.1 *Movement, update of binding and switch of tunnel towards new point-of-attachment.*

11.3.3.6 Movement and de-registration

If the UE moves to its home link it does not need the Mobile IP service anymore since it can use the HoA in the usual way. The UE therefore sends a BU to inform the HA that it is now on its home network and that the HA no longer needs to intercept and forward packets on behalf of the UE. The user plane tunnel between UE and HA is also removed. In EPS the UE is always considered to be on its home link when using a 3GPP access. Therefore deregistration occurs, for example, when the UE moves from a non-3GPP access where S2c is used to a 3GPP access (Figure 11.3.3.6.1).

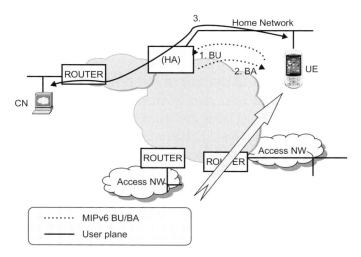

Figure 11.3.3.6.1 *Return home procedure.*

11.3.3.7 Binding revocation

If the UE is located on a foreign link with a binding registration in the HA, the HA may in some cases want to terminate the Mobile IP session. This may, for example, happen if the user is no longer authorized to use Mobile IP. In this case the HA can send a Binding Revocation Indication (BRI) to the UE. The UE then replies with a Binding Revocation Acknowledgement (BRA) and the Mobile IP session is terminated. The BRI and BRA messages are defined in [draft-ietf-mext-binding-revocation].

11.3.4 Mobile IPv6 security

The Mobile IP signalling extends between the UE and the HA. It is therefore important to ensure that this signalling is properly protected. Mobile IPv4 and MIPv6 use different security solutions, and in line with the rest of this chapter we will focus on MIPv6.

Even for MIPv6 there exist different security solutions. The MIPv6 RFC requires that IPSec is used to protect the BU and BA between the UE and the HA. Originally, MIPv6 security was based on the old IPSec architecture, either using manual configuration or using IKEv1 to establish the IPSec security association (see Section 11.9 for more information on IPSec). This is described in RFC 3775 [3775] and RFC 3776 [3776]. Recently, the MIPv6 specification has been updated to support also the revised IPSec architecture and IKEv2. The usage of MIPv6 with the revised IPSec architecture is described in RFC 4877 [4877].

The UE and the HA must support the use of ESP in transport mode to protect the BU and BA messages. Integrity protection is mandatory while ciphering is optional.

In addition to the IPSec-based solutions, an alternative security mechanism has been documented in RFC 4285 [4285]. Instead of using IPSec, this security solution provides integrity protection by adding message authentication mobility options to the MIPv6 signalling. This solution was developed for use in networks based on the legacy 3GPP2 standard. The motivation was that it would be more lightweight than the IPSec-based solutions and that it would provide sufficient security in the specific 3GPP2 deployments. However, for EPS-based systems, only the IKEv2-based security solution is supported.

For more details on Mobile IP security in EPS, see the description of the S2c interface in Section 10.5.

11.3.5   Packet format

11.3.5.1   Mobile IP signalling (control plane)

In order to understand the MIPv6 packet format, it is useful to recapture a few basic aspects about the IPv6 header. IPv6 defines a number of extension headers that can be used to carry the 'options' of the IP packet. The extension headers, if they are present, follow after the 'main' IPv6 header and before the upper layer header (e.g., TCP or UDP header). One of the extension headers, the hop-by-hop header, contains information intended for each router on the path. This header therefore has to be examined by each router on the path. In general however, the extension headers contain information only intended for the final destination of the packet. This means that these extension headers do not need to be examined by every router on the path. Examples of extension headers containing information for the final destination of the packet are the ESP header (for IPSec) and the fragmentation header (in case the packet is fragmented). The ESP and fragmentation headers are extension headers defined for explicit purposes. Another way to provide options to the final destination is to use the Destination Options extension header. This header can contain a variable number of options. Figure 11.3.5.1.1 illustrates an IPv6 packet containing 'main' header, extension headers and payload.

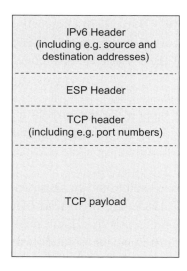

Figure 11.3.5.1.1   *Example IPv6 packet containing main IPv6 header, an extension header as well as an upper layer header and payload.*

MIPv6 defines a new extension header, called the *Mobility Header* (MH), to carry the MIPv6 messages. All messages used in MIPv6, including the BU and BA, are defined as MH types. The MH format is shown in Figure 11.3.5.1.2.

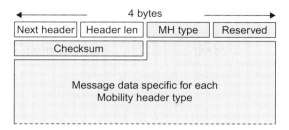

Figure 11.3.5.1.2 *Mobility Header format.*

The Next Header and Header Length fields are not specific to the MH but are present in all extension headers. The Next Header field indicates what type of header (e.g., extension header or upper layer header) follows this header. The Header Length field contains the length of the header. The MH type field indicates what particular MIPv6 message this is, for example, a BU, BA, BRI, Binding Error, etc. The Checksum field contains a checksum of the MH. The Message Data part contains information specific for each message (see below).

This means that MIPv6 messages are carried as part of the IPv6 header information and not as payload of the IPv6 packets. This is different from Mobile IPv4 that is carried as UDP encapsulated payload in an IPv4 packet.

MIPv6 also defines other IPv6 header fields. A new option for the Destination options header is used to carry the HoA. MIPv6 also defines a new routing header variant (Routing Header type 2) as well as a number of new ICMPv6 messages. Below we describe the BU and BA messages. It is however not the intent of this book to go through all MIPv6 messages and headers. An interested reader is referred to the MIPv6 RFC 3775 [3775] for more details.

Figure 11.3.5.1.3 illustrates the Binding Update message. It contains the main IPv6 header, the ESP header (for protecting the message), a Destinations options header carrying the Home Address as well as the Binding Update Mobility Header. The Binding Update Mobility Header is further detailed in Figure 11.3.5.1.4.

The A, H, L, K, M, R, P and F fields in the BU MH contain flags for different purposes. As we will see in Section 11.4, the P flag is used by PMIPv6. The Sequence number is used by the receiver to determine the order in which the BUs were sent by a UE. This is, for example, useful in case the UE rapidly moves between different accesses and sends multiple BUs within a short interval. The Lifetime is the time that remains before the binding expires. The Mobility options field may contain additional options. One example is the Alternate CoA mobility option. The CoA is used as source address of the BU, but including it in the CoA

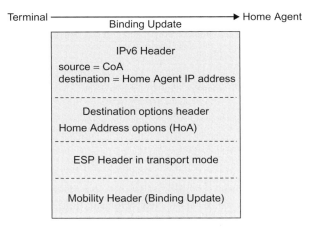

Figure 11.3.5.1.3   *Binding Update message.*

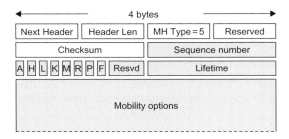

Figure 11.3.5.1.4   *Mobility Header for the Binding Update message.*

mobility option as well allows it to be protected by ESP (ESP in transport mode does not protect the IP header). In response to a BU, the HA sends a BA. Figure 11.3.5.1.5 illustrates the BA message. It contains the main IPv6 header, the ESP header (for protecting the message), a type 2 Routing Header carrying the HoA as well as the BA MH.

The MH for the BA message is illustrated in Figure 11.3.5.1.6. The status field indicates the result of the BU. The sequence number sent in the BA is the same as that received in the BU. This allows the Mobile IP client to match Updates with Acknowledgements. The Lifetime includes the time granted by the HA until the binding expires. To maintain the binding in the HA, the UE must refresh the binding before it expires by sending a new BU message to the HA.

11.3.5.2   User plane

When the MIPv6 session has been established, all user plane packets for the HoA are tunnelled between the UE and the HA (except in case where Route

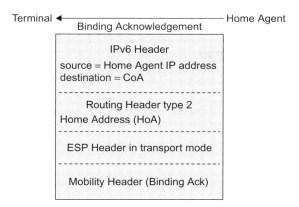

Figure 11.3.5.1.5 *Binding Acknowledgement message.*

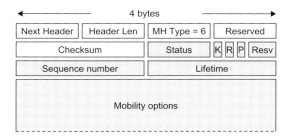

Figure 11.3.5.1.6 *Mobility Header for the Binding Acknowledgement message.*

Optimization is used, see Section 11.3.7. This tunnelling is performed using IPv6 encapsulation defined in RFC 2473 [2473]. Note however that additional encapsulation protocols are defined for the dual-stack version of MIPv6.

11.3.6 Dual-stack operation

The text above has described the basics of MIPv6. MIPv6 was designed for IPv6 only and thus supports only IPv6 traffic and IPv6 capable networks. Mobile IPv4 on the other hand was designed for IPv4 and supports only IPv4 traffic and IPv4 capable networks. An IPv4-only node can thus use Mobile IPv4 to maintain connectivity while moving between IPv4 networks and an IPv6-only node can use MIPv6 to maintain connectivity while moving between IPv6 networks. However, this situation is not optimal for a dual-stack UE supporting both IPv4 and IPv6. Such a UE would need to use Mobile IPv4 for its IPv4 stack and MIPv6 for its IPv6 stack so that it can move between IPv4 and IPv6 subnets. There are a few drawbacks with this solution for dual-stack UEs. First of all it requires that the dual-stack UE needs to support two sets of mobility management protocols, which increases the complexity of the UEs. Also, it needs to send two sets of

Mobile IP signalling messages on every handover, to inform both the Mobile IPv4 HA and the MIPv6 HA about the move. Furthermore, since Mobile IPv4 requires an IPv4 CoA and MIPv6 requires an IPv6 CoA, all access networks need to be dual-stack in order to provide mobility for both the IPv4 and IPv6 sessions. For example, a dual-stack UE attempting to connect via an IPv4-only network would not be able to maintain connectivity of its IPv6 applications and vice versa. Also to the operator this scenario has drawbacks since the operator needs to run and maintain two sets of mobility management systems on the same network.

The dual-stack extensions of MIPv6 avoid these drawbacks, by enhancing the protocol to support also IPv4 access network (i.e., IPv4 CoA) and IPv4 user plane traffic (i.e., using an IPv4 HoA). The dual-stack version of MIPv6 is usually referred to as DSMIPv6 and is specified in RFC 5555. The solution defines extensions for carrying the mobile node's IPv4 CoA, IPv4 HoA and IPv4 address of the HA in the MIPv6 signalling messages. It should be noted that DSMIPv6 requires that the terminal is always assigned an IPv6 HoA.

As indicated above, DSMIPv6 supports more network scenarios than basic MIPv6. Scenarios supported by DSMIPv6 are illustrated in Figure 11.3.6.1. A requirement to support both IPv4 and IPv6 traffic is that the HA supports both IPv4 and IPv6. Even though only single-stack foreign networks are shown in the figure, the foreign network may of course support both IPv4 and IPv6. In the latter case, the terminal should prefer using an IPv6 CoA.

As explained in the previous section, all MIPv6 messages are defined as native IPv6 packets (using IPv6 extensions headers, etc.). In an IPv4-only foreign network the UE can however only acquire an IPv4 CoA and send IPv4 packets. In order to send a MIPv6 message to the HA, the MIPv6 packets must be encapsulated in IPv4 and sent to the IPv4 address of the HA. An example of a BU message for an IPv4-only foreign network is shown in Figure 11.3.6.2. In order to support private IPv4 addresses and NAT traversal, UDP encapsulation may be used.

Also additional user plane tunnelling formats are needed to support IPv4 user data and IPv4-only foreign networks. The IPv4 or IPv6 user plane data is encapsulated in either IPv4 or IPv6 depending on the IP version of the foreign network. Furthermore, in order to support private IPv4 addresses and NAT traversal, also UDP encapsulation is supported. The user data tunnelling formats for the different scenarios are shown in Figure 11.3.6.1.

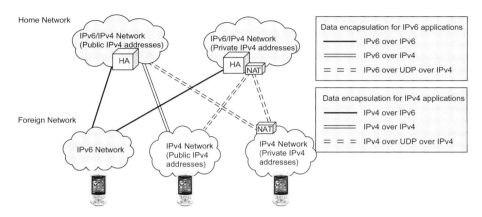

Figure 11.3.6.1 *Network scenarios for dual-stack MIPv6.*

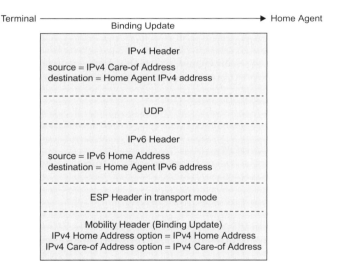

Figure 11.3.6.2 *DSMIPv6 Binding Update message for an IPv4 CoA.*

11.3.7 Additional MIPv6 features – route optimization

MIPv6 is a quite extensive protocol and so far we have only provided a very high level description of a few main functions. One feature that was only briefly mentioned above is Route Optimization (RO). RO is supported for MIPv6 but not available in Mobile IPv4. It is an alternative to the bi-directional tunnelling between UE and HA. With RO, the user plane traffic is sent directly between the UE and the Correspondent Node (CN) without passing the HA.

RO is not supported by EPS and there are different reasons for that. In EPS the HA is located in the PDN GW and it is always assumed that the user plane traffic goes through one PDN GW where charging, policy enforcement and lawful intercept can take place. Furthermore, MIPv6 RO is limited to IPv6 traffic and IPv6 foreign networks. RO is thus not supported for IPv4 traffic even when DSMIPv6 is used. EPS provides other solutions that can enable efficient routing. In roaming situations it is, for example, possible to assign a PDN GW located in the visited PLMN, thus avoiding the transport of all user plane traffic to the home PLMN. Also PDN GW selection functions in EPS have impact on the routing, for example, by selecting a PDN GW that is geographically close to the UE.

MIPv6 RO allows a UE to inform a CN about its current CoA. The UE basically sends a BU to the CN and the CN in turn creates a binding in between the HoA and the CoA. When the CN sends a packet to a specific IP address, it checks its bindings for an entry (i.e. a HoA) that matches the IP address. If a match is found, the CN can communicate with the UE using the CoA. Traffic sent by the CN will thus be routed to the foreign network directly without passing the home network. MIPv6 defines special messages as well as security mechanisms to set up the route binding in the CN. Considering that RO is not used in EPS, and this is a book on EPS, we will not go into further details on this topic. The interested reader is instead referred to the MIPv6 RFC 3775 [3775].

11.4 Proxy Mobile IPv6

11.4.1 General

As explained in Section 11.3, mobility schemes can often be classified as either being host-based or network-based. MIPv6, described in the previous section, is a host-based mobility management solution where the UE has functionality to detect movement and to exchange IP mobility signalling with the network in order to maintain IP-level session continuity.

PMIPv6, defined in RFC 5213 [5213], on the other hand is network-based mobility management protocol that has a similar purpose as MIPv6, that is, to facilitate IP-level session continuity. PMIPv6 reuses much of the concepts and packet formats that have been defined for MIPv6. A key difference between MIPv6 and PMIPv6 is however that with PMIPv6 the UE does not have Mobile IP software and does not participate in the IP mobility signalling. A key intent of PMIPv6 is in fact to enable IP-level mobility also for those UEs that do not have Mobile IP client functionality. Instead it is mobility agents in the network that track the UE's movement and perform IP mobility signalling on behalf of the UE. A mobility agent in the network acts as a proxy for the UE when it comes to IP mobility signalling, hence the name Proxy Mobile IPv6.

Since PMIPv6 reuses many parts of MIPv6 such as packet format, the description of PMIPv6 in this section will to a large extent build on the description of Mobile IP in Section 11.3.

PMIPv6 is used on the S2a, S2b interfaces and as a protocol alternative on the S5/S8 interface. Specific aspects related to PMIPv6 usages in EPS are described in previous chapters, for example, regarding EPS bearers [Section 6.3], mobility [Section 6.4], PCC [Section 8.2], and so on. For more details on the PMIP-based interfaces, see Section 10.4. Below we describe the PMIPv6 protocol as such.

11.4.2 Basic principles

PMIPv6 introduces two new network entities; the Mobile Access Gateway (MAG) and the Local Mobility Anchor (LMA). The MAG is the mobility agent that acts essentially as the Mobile IP client on behalf of the UE. The LMA is the mobility anchor point and its role is similar to that of the HA for MIPv6, that is, to maintain a binding between the HoA of the UE and its current point-of-attachment. The MAG is located in the access network while the LMA is located in the network where the HoA is topologically located. The PMIPv6 architecture is illustrated in Figure 11.4.2.1.

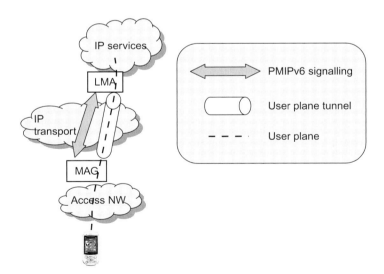

Figure 11.4.2.1 *Proxy Mobile IP.*

The responsibility of the MAG is to detect the movement of a UE and to initiate the appropriate IP mobility signalling. A key function of the MAG is also to emulate the UE's home network, that is, to make sure that the UE does not detect any change in its layer-3 attachment even after the UE has changed its point-of-attachment. The UE shall be allocated the same IP address and other IP configuration parameters after the move as it had before the move. Furthermore, the target MAG is also updated with other parameters such as IPv6 link-local address to be used by the MAG. This is to ensure that, in a handover, the target MAG uses the same link-local address as the source MAG. This gives the UE the impression that it is still on the same local network even after the handover. How this works we will see in an example scenario below.

Below we provide an example for how PMIPv6 works and can be used in a network.

When the UE connects to an access network it typically performs access authentication and authorization for that access. During the access authentication, the UE also provides the user identity (the IMSI in case of EPS) and the security (e.g., encryption) may be set up. When the UE has successfully attached to the access network and provided its identity, it may, for example, request an IP address using DHCP. It should be noted that the details regarding the signalling between UE and the MAG depends very much on the type of access used. For example, access authentication and IP address allocation may be done in different ways in different access systems. Please see Chapter 7 and Section 6.3 for further details on access-specific aspects.

The MAG in the access network now initiates PMIPv6 signalling towards the LMA to inform the LMA of this user's current point-of-attachment. To do this the MAG first has to select a suitable LMA (this is similar to how the UE must discover a suitable HA in case of Mobile IP). When PMIPv6 is used in EPS, the MAG performs LMA discover by resolving the APN using the DNS functions (for more details, see Chapter 9). The PMIPv6 RFC 5213 [5213] also describes other means for the MAG to find a suitable LMA.

Once an LMA has been selected, the MAG sends a Proxy Binding Update (PBU) message to the LMA. The PBU contains the Proxy CoA which is the IP address of the MAG. This allows the LMA to create an association (binding) between the Proxy CoA and the UE's HoA in a much similar way as the MIPv6 HA creates a binding between the CoA and the HoA. The difference here is that with PMIPv6 the UE is not aware of the Proxy CoA.

The LMA then replies with a Proxy Binding Acknowledgement (PBA). This message contains the IPv6 prefix allocated to the terminal. (With the dual-stack amendments described further below, the PBA may also contain an IPv4 address allocated to the terminal). The PBA also carries other IP parameters associated with the home network such as the MAG IPv6 link-local address. Once the MAG receives the PBA it can provide the allocated IP address/prefix to the UE (e.g., using DHCP).

When this is done, the MAG and LMA establish a tunnel where the user plane for the UE is forwarded. All user plane data sent by the UE is intercepted by the MAG and forwarded to the LMA via the MAG-LMA tunnel. The LMA in turn de-capsulates the packets and forwards them towards their final destination. In the other direction, all traffic addressed to the HoA is intercepted by the LMA in the home network and forwarded to the MAG (via the tunnel) which in turn sends it to the UE (Figure 11.4.2.2).

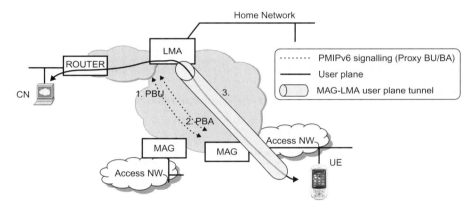

Figure 11.4.2.2   *UE connected to a network using PMIPv6.*

To the UE it looks like it was really connected to its home network since it is allocated the same IP address (the HoA) and other IP parameters associated with the home network. The MAG is emulating the home link towards the UE and the UE can continue its IP sessions as if it was still connected to the home network. See Section 12.2.2 for a more detailed call flow for an attach procedure using PMIP.

If the UE now moves and connects to a different access network, either using the same access technology or a different access technology, the access attach procedure will again be performed. The MAG in the new access network will detect that the UE has attached. In order to provide session continuity, the new MAG has to send a PBU to update the binding in the LMA. The new MAG therefore sends the PBU to the same LMA informing about the Proxy CoA of the new MAG. The LMA updates its binding and replies with a PBA containing, for example, the HoA and other parameters. The new MAG now assigns the same IP address (i.e., the HoA) as in the old access. The user plane tunnel is moved to the new MAG. The

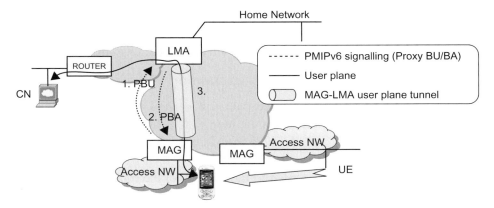

Figure 11.4.2.3   *PMIP-based handover.*

UE again thinks it is connected to the home network and can thus continue to use the HoA as before the change in point-of-attachment (Figure 11.4.2.3).

For a more detailed call flow of a PMIP-based handover procedure in EPS, please see Section 12.4.3.

If the LMA for some reason wants to disconnect the terminal, it sends a Binding Revocation Indication (BRI) to the MAG. The MAG in turn removes its mobility context, disconnects the terminal, and sends a Binding Revocation Acknowledgement (BRA) to the LMA. The format of the BRI and BRA messages are the same for both MIPv6 and PMIPv6 [draft-ietf-mext-binding-revocation].

One important aspect to note in the example above is that the signalling between UE and MAG is access specific. There is no IP level mobility signalling between the UE and the network as was the case for MIPv6. PMIPv6 is used inside the network in order to establish, to modify or to remove the mobility session and to provide the MAG in the access network with the home network information (e.g., HoA). In this way the MAG can emulate the home network by assigning the same IP-layer parameters as they would have been assigned in the home network. The UE is thus unaware of the topology of the network. The user plane tunnel between MAG and LMA enables the UE to use the HoA from any access link where PMIPv6 is used.

11.4.3   PMIPv6 security

Since PMIPv6 is a network-based mobility protocol it has different security requirements than MIPv6 which is a host-based mobility protocol.

PMIPv6 performs mobility signalling on behalf of a UE that has attached to its network. PMIPv6 therefore requires that proper access authentication and authorization have been performed so that there is a trusted connection between the UE and the MAG before the MAG initiates PMIPv6 signalling. If this trusted connection is not required, a malicious UE might, for example, trigger a MAG to perform mobility signalling on another user's behalf. See Chapter 7 for further details on access security.

Also the PMIPv6 signalling itself needs to be properly protected. The PMIPv6 RFC supports protection of the PMIPv6 signalling between the MAG and the LMA using IPSec (ESP in transport mode with integrity protection). It is also possible to use other security mechanisms depending on deployment. In EPS, the Network Domain Security (NDS/IP) is the general framework for protecting signalling messages between network nodes. It is used also for PMIPv6 signalling in EPS. For further details, see the brief description of NDS/IP in Section 7.4.

11.4.4   PMIPv6 packet format

As already mentioned, PMIPv6 inherits many concepts from MIPv6, including the packet format. The format of the PBU and PBA messages are the same as the MIPv6 BU and BA messages with the difference that the P flag has been introduced to indicate that the BU/BA refers to a proxy registration. For details regarding the format of the BU and BA messages, see Section 11.3.5. PMIPv6 also introduces new Mobility Options for use with the PBU and PBA messages. See RFC 5213 [5213] for further details.

The user plane is sent between MAG and LMA in a bi-directional tunnel. The tunnels may be unique per UE or shared by multiple UEs depending on deployment. The tunnels may be statically configured at the MAGs and LMAs in the network or be dynamically established and torn down. The MAG and LMA may use IPv6 encapsulation [RFC 2473] for this tunnel but also GRE tunnelling is supported [draft-ietf-netlmm-grekey-option]. EPS requires that GRE tunnelling is used where the GRE key field is used to uniquely identify a specific PDN connection, see Section 11.6.

In some environments there is a need to include specific information elements not defined as part of the main PMIPv6 specifications in the PMIPv6 messages. In this case it is possible to use vendor-specific Mobility Options that can be included with the Mobility Header. The general format of these vendor-specific options is defined in RFC 5094 [5094]. 3GPP EPS makes use of such vendor-specific options to transport, for example, the Protocol Configuration Options (PCO) fields, Charging ID and 3GPP-specific error codes, as defined in TS 29.275 [29.275].

11.4.5   Dual-stack operation

The dual-stack enhancements for PMIPv6 reuses the dual-stack features defined for DSMIPv6. This means that the PBU and PBA may contain IPv4 CoA options and IPv4 HoA options. This allows PMIPv6 to support also IPv4-only access networks as well as IPv4 HoAs. A key difference compared to DSMIPv6 is however that with PMIPv6 the IPv6 HoA is not mandatory. It is allowed to assign only an IPv4 HoA to the terminal. The dual-stack extensions for PMIPv6 are defined in [draft-ietf-netlmm-pmip6-ipv4-support]. (Note that when this book was prepared, the dual-stack extensions for PMIPv6 was still an Internet Draft but soon to reach RFC status. Please refer to the references section for further details.) In a similar way as for DSMIPv6, when PMIPv6 is used over an IPv4 transport network, the PMIPv6 signalling messages are encapsulated in IPv4 packets. The user plane tunnel MAG and LMA are either encapsulated in IPv6, IPv4, UDP-over-IPv4 or GRE-over-IPv4/IPv6, depending on network scenario. As mentioned above, EPS uses GRE tunnelling over IPv4 or IPv6 transport. Dual-stack scenarios for PMIPv6 are illustrated in Figure 11.4.5.1. Note

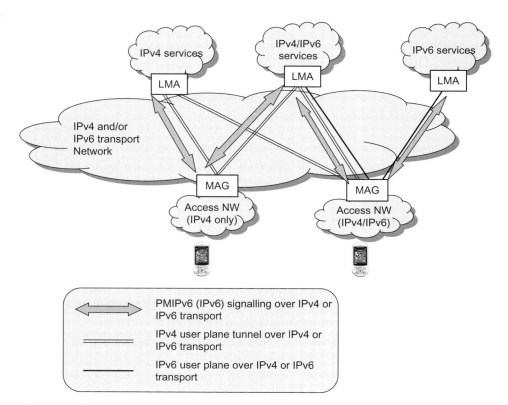

Figure 11.4.5.1   *Example scenarios for dual-stack PMIPv6.*

that the access network may be supporting IPv6 only (for simplicity this is not shown in the figure).

11.5   Diameter

11.5.1   Background

Diameter is a protocol originally designed for Authentication, Authorization and Accounting (AAA) purposes. It is an evolution of its predecessor, the RADIUS protocol. The Diameter protocol name is 'derived' from RADIUS in the sense that the diameter is twice the radius. The RADIUS protocol has been commonly and successfully deployed to provide AAA services for fixed dial-up accesses as well as for cellular CDMA systems. It is also used in GPRS networks on the Gi interface.

The Diameter protocol was designed to overcome several shortcomings of RADIUS. For example, Diameter supports improved failure handling, more reliable message delivery, bigger size information elements, improved security, better possibilities for extensibility, more flexible discovery of other Diameter nodes, etc. Furthermore, in contrast to RADIUS, Diameter provides a full specification of intermediate entities such as proxies. At the same time Diameter was constructed to provide an easy migration and compatibility with RADIUS. For example, a Diameter message, like a RADIUS message, conveys a collection of Attribute Value Pairs (AVP).

3GPP makes extensive use of Diameter on numerous interfaces. It should however be noted that 3GPP basically only uses RADIUS on the Gi/SGi interface and does, therefore, not have a significant legacy of RADIUS usage. Instead Diameter is used exclusively on several interfaces, without a RADIUS legacy. Nevertheless, comparisons between Diameter and RADIUS may be useful for readers more familiar with the RADIUS protocol. Such comparisons have therefore been included below where applicable.

11.5.2   Protocol structure

The Diameter protocol is constructed as a base standard and additional extensions called applications. The core of the Diameter protocol is defined in the Diameter base standard, RFC 3588 [3588]. This RFC specifies the minimum requirements for a Diameter implementation and includes a few general Diameter messages (called Commands in Diameter) as well as AVPs that can be carried by the commands. Extensions (called Applications in Diameter) are then created on top of the Diameter base protocol to support specific requirements. The applications may

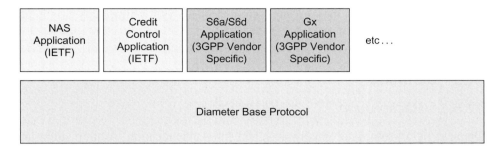

Figure 11.5.2.1   *Structure of the Diameter protocol, consisting of a base protocol and extensions called applications.*

define new commands as well as new AVPs as needed. A Diameter application is thus not a program or application in the usual sense, but a protocol based on Diameter. The applications benefit from the general capabilities of the Diameter base protocol. Applications may also be based on existing, already defined, applications. In this case they inherit Diameter commands and mandatory AVPs from the application(s) they are based on, but they use new application identifiers, add new AVPs and modify the protocol state machines according to their own procedures. Figure 11.5.2.1 illustrates the protocol structure of Diameter including a few example Diameter applications.

Several Diameter applications have been standardized by IETF but it is also possible to define vendor-specific applications. A 'vendor' in this context it not necessarily a vendor making products, it is instead someone (e.g., an organization or company) that has requested a Diameter application identifier from the Internet Assigned Numbers Authority (IANA). As will be seen in the interfaces section, 3GPP has defined several vendor-specific Diameter applications that are used over the Diameter-based interfaces such as S6a, S6b, SWa, SWx and so on. In many cases the 3GPP vendor-specific applications are based on existing Diameter applications defined by IETF. Several of these applications were discussed together with the corresponding interfaces in Chapter 10.

11.5.3   Diameter nodes

The network entities implementing Diameter act in a certain role in the network architecture. Diameter defines three types of Diameter nodes depending on the role the node plays: client, server and agent. The role a certain Diameter node plays depends on the network architecture.

The client is typically the entity requesting a service from a Diameter server and thus originates the request to initiate a Diameter session with a server.

Diameter agents are Diameter nodes that can bring flexibility into the network architecture. They can be used to support a system where different parts of the network are operated by different administrations, such as in a roaming scenario. They are also used in the routing of Diameter messages to aggregate Diameter requests destined to a specific realm. The agents examine the received requests and determine the right target. This can provide load balancing features and simplified network configurations. Certain agents can also perform additional message processing.

There are four types of agents: relay, proxy, redirect and translation agents. This can be compared to RADIUS where basically only a single type of intermediary node, the RADIUS proxy, exists.

A relay agent is used to forward a message to the appropriate destination, depending on the information contained in the message. More information on Diameter message routing is provided below. The relay agent needs to understand the Diameter base protocol but need not understand the Diameter application used.

A proxy agent is similar to a relay agent, with the difference that it can perform additional processing of the Diameter messages, for example, to implement certain policy rules. Since the proxy agent can modify messages it needs to understand the service being offered and thus also understand the Diameter application being used.

A redirect agent also provides a routing function, for example, to perform realm to server resolution. The redirect agent does, however, not forward a received message towards the destination, but rather replies with another message to the node that sent the request. This reply contains information that allows the node to send the request again, but now directly to the server. The redirect agent is thus not on the routing path of the Diameter messages.

A translation agent may perform translation between Diameter and other protocols. A typical example that could be used to support migration scenarios is a translation agent translating between Diameter and RADIUS.

11.5.4 Diameter sessions, connections and transport

Diameter uses either the Transmission Control Protocol (TCP) or Stream Control Transport Protocol (SCTP) to transport the messages between two Diameter peers. Since TCP and SCTP are both connection oriented protocols, a connection between the two peers has to be established before any Diameter command can be sent. Both TCP and SCTP provide a reliable transport. This can be compared to RADIUS which uses UDP as transport protocol, which provides a connection-less and unreliable transport. For more details on SCTP and the differences compared to TCP, please see Section 11.11.

The Diameter *connection* between two peers should be distinguished from the Diameter *session* being established between a client and a server. While a connection is a transport level connection between two peers, the Diameter session is a logical concept describing the application level association between a client and server (possible spanning Diameter agents) identified by a session identifier. The Diameter peer connection and Diameter session are illustrated in Figure 11.5.4.1.

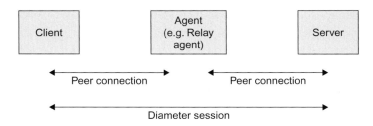

Figure 11.5.4.1   *Diameter connections and Diameter sessions.*

Diameter messages are protected using Transport Layer Security (TLS) or IPSec. The Diameter base specification mandates that all Diameter nodes must support IPSec while TLS is optional to support in the Diameter client. This protection is provided hop-by-hop between the Diameter peers.

In a 3GPP environment however, the general framework of NDS/IP is used for all IP-based control signalling, including Diameter. There is thus no need to provide a specific security association between the Diameter nodes in EPS. For more details on NDS/IP, please see Chapter 7.

In order to handle the flexibility of Diameter in terms of applications, security features, etc, in a dynamic fashion, two Diameter peers establishing a connection also performs a capability exchange. This exchange allows each peer to learn about the other peer's identity and its capabilities (protocol version number, supported Diameter applications, supported vendor-specific attributes, security mechanisms, etc.).

11.5.5   Diameter request routing

As mentioned above, Diameter agents may assist in the routing of a Diameter command towards its final destination, the Diameter server.

A Diameter agent forwarding a command typically performs the routing based on the destination realm as well as the application used. It may thus use a different destination based on the application identification. The Diameter node maintains a list of supported realms and known Diameter peers as well as the peer's capabilities (e.g., supported applications).

An agent can perform realm to server resolution and can thus be used to aggregate requests from different sources destined to a specific destination realm. This allows the agent to act as a centralized routing entity.

This feature can also be utilized in EPS in case a network deploys multiple Diameter servers, for example, multiple HSS or PCRF nodes. For the HSS, the Diameter client may not know which HSS node that handles the subscription record for a specific user. In this case a Diameter redirect agent or proxy agent can perform the resolution from a realm and user name into a server name for the HSS holding the subscription record. Diameter agents also have a specific usage for PCRF selection as is further described in Chapter 9.

11.5.6   Peer discovery

Each Diameter entity must be able to find the next hop Diameter node. With RADIUS, each RADIUS client/proxy has to be statically configured with information about the RADIUS servers/proxies with which it may need to communicate. This could cause a high burden on the network management system to keep these configurations up to date. Diameter still supports the option to statically configure Diameter peers, but in addition it is possible to use more dynamic peer discovery mechanisms, for example, by utilizing the DNS.

Diameter clients can then depend on the realm info together with the desired Diameter application and security level to look up suitable first-hop Diameter nodes to which they can forward Diameter messages. The discovered peer location as well as routing configuration will be stored locally and used when making routing decisions.

The Diameter base protocol also includes mechanisms to support transport failure handling between peers, for example, using watchdog messages to detect transport failures as well as peer failover/fallback mechanisms.

11.5.7   Diameter message format

The Diameter messages are called commands. The content of the Diameter commands consists of a Diameter header followed by a number of AVPs. The Diameter header contains a unique command code that identifies the command and consequently the intention of the message. The actual data is carried by the AVPs contained in the message. The Diameter base protocol defines a set of commands and AVPs but a Diameter application can define new commands and/or new AVPs. The base protocol, for example, defines a set of base AVP formats that can be reused, essentially in an object oriented fashion, when defining new AVPs. In an application it is thus possible not only to define new AVPs but also new commands which makes Diameter very extensible and allows the construction of applications to suit the needs of 3GPP. The Diameter message format is shown in Figure 11.5.7.1.

The Application ID identifies for which Diameter application the message is for. The hop-by-hop identifier is used for matching requests with responses.

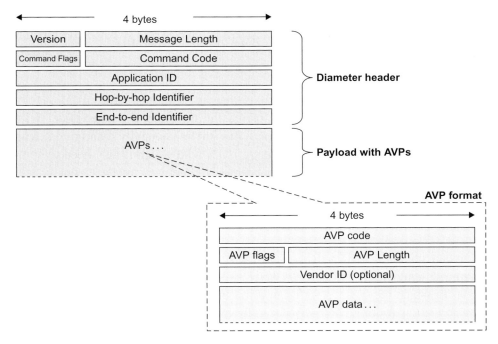

Figure 11.5.7.1   *Diameter message format.*

The end-to-end identifier is used to detect duplicate messages. Each AVP is identified using a unique AVP code. If the AVP is vendor specific, also the Vendor ID is used to uniquely identify the AVP. For more detailed information on the Diameter header and basic AVP formats, please see RFC 3588 [3588].

11.6 Generic routing encapsulation

11.6.1 Background

The GRE is a protocol designed for performing tunnelling of a network layer protocol over another network layer protocol. It is generic in the sense that it provides encapsulation of one arbitrary network layer protocol (e.g., IP or MPLS) over another arbitrary network layer protocol. This is different from many other tunnelling mechanisms where one or both of the protocols are specific, such as IPv4-in-IPv4 [2003] or Generic Packet Tunnelling over IPv6 [2473].

GRE is used for many different applications and in many different network deployments also outside the telecommunications area. It is not the intent of this book to discuss aspects for all those scenarios. Instead we focus on the properties of GRE that are most relevant to EPS.

11.6.2 Basic protocol aspects

The basic operation of a tunnelling protocol is that one network protocol, which we call the payload protocol, is encapsulated in another delivery protocol. It can be noted that encapsulation is a key component of any protocol stack where an upper layer protocol is encapsulated in a lower layer protocol. This aspect of encapsulation should however not be considered as tunnelling. When tunnelling is used, it is often the case that a layer-3 protocol such as IP is encapsulated in a different layer-3 protocol or another instance of the same protocol. The resulting protocol stack may look like in Figure 11.6.2.1.

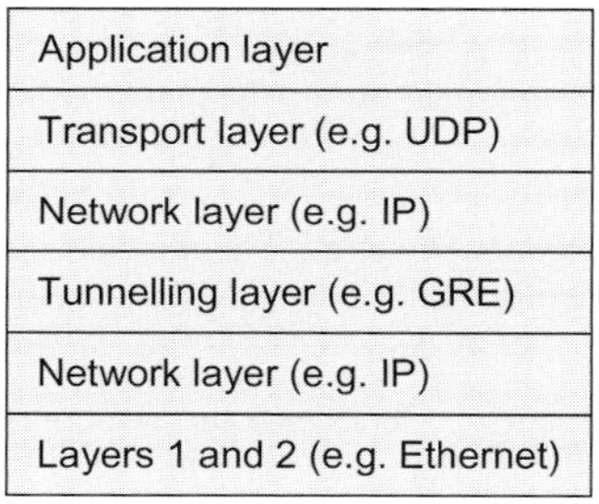

| Application layer |
| Transport layer (e.g. UDP) |
| Network layer (e.g. IP) |
| Tunnelling layer (e.g. GRE) |
| Network layer (e.g. IP) |
| Layers 1 and 2 (e.g. Ethernet) |

Figure 11.6.2.1 *Example of protocol stack when GRE tunnelling is used.*

We use the following terminology:

- Payload packet and payload protocol: The packet and protocol that needs to be encapsulated. (The three top most boxes in the protocol stack in Figure 11.6.2.1.)
- Encapsulation (or tunnel) protocol: The protocol used to encapsulate the payload packet, that is, GRE. (The third box from below in Figure 11.6.2.1.)
- Delivery protocol: the protocol used to deliver the encapsulated packet to the tunnel endpoint. (The second box from below in Figure 11.6.2.1.)

The basic operation of GRE is that a packet of protocol A (the payload protocol) that is to be tunnelled to a destination is first encapsulated in a GRE packet (the tunnelling protocol). The GRE packet is then encapsulated in another protocol B (the delivery protocol) and sent to the destination over a transport network of the delivery protocol. The receiver then de-capsulates the packet and restores the original payload packet of protocol type A. Figure 11.6.2.2 shows an example of tunnelling an IPv6 packet in a GRE tunnel over an IPv4 delivery protocol.

GRE is specified in RFC 2784 [2784]. There are also additional RFCs that describe how GRE is used in particular environments or with specific payload and/or delivery protocols. One extension to the basic GRE specification which is of particular importance for EPS is the GRE Key field extension specified by RFC 2890 [2890]. The Key field extension is further described as part of the packet format below.

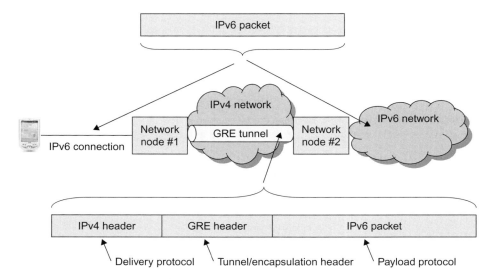

Figure 11.6.2.2   *Example of GRE tunnel between two network nodes with IPv4 delivery protocol and IPv6 payload protocol.*

11.6.3 GRE packet format

The GRE header format is illustrated in Figure 11.6.3.1.

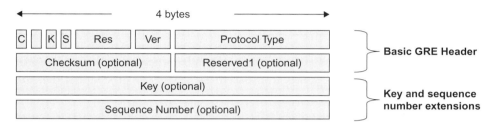

Figure 11.6.3.1 *GRE header format including the basic header (RFC 2784) as well as the key and sequence number extensions (RFC 2890).*

The C flag indicates whether or not the Checksum and Reserved1 fields are present. If the C flag is set, the Checksum and Reserved1 fields are present. In this case the Checksum contains a checksum of the GRE header as well as the payload packet. The Reserved1 field, if present, is set to all zeroes. If the C flag is not set, the Checksum and Reserved1 fields are not present in the header.

The K and S flags, respectively, indicate whether or not the Key and/or Sequence number is present.

The Protocol Type field contains the protocol type of the payload packet. This allows the receiving endpoint to identify protocol type of the de-capsulated packet.

The intention of the Key field is to identify an individual traffic flow within a GRE tunnel. GRE does in itself not specify how the two endpoints establish which Key field(s) to use. This is left to implementations or is specified by other standards making use of GRE. The Key field could, for example, be statically configured in the two endpoints, or be dynamically established using some signalling protocol between the endpoints. One specific usage of the Key field in EPS occurs when GRE is used as tunnel format for the PMIP-based interfaces. The Key field is here dynamically established as part of the PMIP signalling and used to identify a particular PDN connection between the MAG and LMA (see also description of PMIP in Section 11.4).

The Sequence number field is used to maintain the sequence of packets within the GRE tunnel. The node that performs the encapsulation inserts the Sequence number and the receiver uses it to determine the order in which the packets were sent.

11.7 S1-AP

In contrast to many of the other protocols in EPS, the S1-AP protocol is designed for a single interface, namely the MME to eNodeB interface. The protocol is named after the interface name (S1) and the addition of AP (Application Part) which is the 3GPP term for signalling protocol between two nodes.

S1-AP supports all mechanisms necessary for the procedures between MME and eNodeB and it also supports transparent transport for procedures that are executed between the MME and the UE.

S1-AP consists of Elementary Procedures. An Elementary Procedure is a unit of interaction between the eNodeBs and the MME. These Elementary Procedures are defined separately and are intended to be used to build up complete sequences in a flexible manner. The Elementary Procedures may be invoked independently of each other as stand alone procedures, which can be active in parallel. In some cases the independence between some Elementary Procedures is restricted, in this case the particular restriction is specified in the S1-AP protocol specification TS 36.413 [36.413]. An Elementary Procedure consists of an initiating message and possibly a response message.

The S1-AP protocol supports the following functions:

- Setup, modification and release of E-RABs.
- Establishment of an initial S1 UE context in the eNodeB (to setup the default IP connectivity, to setup one or more E-RAB(s) if requested by the MME, and to transfer NAS signalling-related information to the eNodeB if needed.)
- Provide the UE capability info to MME (when received from the UE).
- Mobility functions for UEs in order to enable intra and inter RAT HO.
- Paging: This functionality provides the EPC the capability to page the UE.
- S1 interface management functions, for example, Reset, Error Indication, Overload indication, Load balancing and S1 Setup functionality for initial S1 interface setup.
- NAS signalling transport function between the UE and the MME:
  - S1 UE context Release
  - UE context modification function.
- Status transfer functionality (transfers PDCP SN status information from source eNodeB to target eNodeB in support of in-sequence delivery and duplication avoidance for intra LTE handover).
- Trace of active UE's.
- Location reporting.

- S1 CDMA tunnelling (to carry CDMA signalling between UE and CDMA RAT over the S1 interface).
- Warning message transmission.

There is no version negotiation in S1-AP and the forwards and backwards compatibility of the protocol is instead ensured by a mechanism where all current and future messages, and IEs or groups of related IEs, include ID and criticality fields that are coded in a standard format that will not be changed in the future. These parts can always be decoded regardless of the standard version.

S1-AP relies on a reliable transport mechanism and is designed to run on top of SCTP.

11.8   Non-access stratum (NAS)

NAS denotes the protocols between the UE and the MME, which implements the mobility management and session management procedures. The NAS protocols, EPS Mobility management (EMM) and EPS Session Management (ESM), are tailor-made for E-UTRAN access and defined by 3GPP in TS 24.301 [24.301].

The EPS NAS protocols are defined as new protocols but they have many similarities with the NAS protocols used for 2G/3G. The EMM procedures are used to support UE mobility, security and signalling connection management services for ESM. The ESM procedures are used to activate, deactivate or modify EPS bearers.

11.8.1   EPS mobility management

The EMM procedures are used to keep track of the UE, to authenticate the UE, to provide security keys and to control integrity protection and ciphering. The network can assign new temporary identities and also request the identity information from the UE. In addition the EMM procedures provide the UE capability information to the network and the network may also inform the UE of information regarding specific services in the network. The EMM procedures are:

- Attach
- Detach
- Tracking area update
- GUTI reallocation
- Authentication
- Security mode control
- Identification

- MM information
- NAS message transport (used for SMS to CS fallback enabled UE).

An optimization compared to 2G/3G is that the attach procedure is always combined with an ESM procedure that activates a default bearer. This means that the UE will receive at least one bearer to a PDN by the completion of the combined procedure.

EMM procedures can only be performed if a NAS signalling connection has been established between the UE and the network. If there is no active signalling connection the EMM layer has to initiate the establishment of a NAS signalling connection. The NAS signalling connection is established by a service request procedure from the UE. For down-link NAS signalling the MME first initiates a paging procedure which triggers the UE to execute the service request procedure. The connection management procedures rely on services from the underlying S1-AP protocol on the S1-MME interface and RRC on the E-UTRAN-Uu interface to establish connectivity.

11.8.2 EPS session management

The ESM procedures are used to manage the bearers and PDN connections for a UE. This includes procedures for default and dedicated bearer establishment, bearer modification and deactivation. As noted above the default bearer establishment is always combined with the attach procedure but it can be used as stand alone procedures to establish additional default or dedicated bearers. The ESM procedures used for E-UTRAN are in principle network-initiated but the UE may also request the network to modify the bearer resource or ask the network to execute the EPS bearer activation and deactivation procedures.

ESM procedures are:

- Default EPS bearer context activation procedure
- Dedicated EPS bearer context activation procedure
- EPS bearer context modification procedure
- EPS bearer context deactivation procedure
- UE requested PDN connectivity procedure
- UE requested PDN disconnect procedure
- UE requested bearer resource modification procedure.

The NAS protocol is implemented as standard 3GPP L3 messages according to 3GPP TS 24.007 [24.007]. Standard L3 according to 24.007 and its predecessors have been used for NAS signalling messages in GSM and WCDMA/HSPA. The encoding rules have been developed to optimize the message size over the air interface and to allow extensibility and backwards compatibility without the need for version negotiation.

11.8.3 Message structure

Each NAS message contains a Protocol Discriminator and a Message Identity. The Protocol Discriminator is a 4-bit value that indicates the protocol being used, that is, for EPS NAS messages either EMM or ESM. The Message Identity indicates the specific message that is sent.

EMM messages also contain a security header that indicates if message is integrity protected and/or ciphered. ESM messages on the other hand contain an EPS Bearer Identity and a Procedure Transaction Identity. The EPS Bearer Identity indicates the assigned bearer identity and the Protocol Transaction Identifier indicates a particular NAS message exchange between the UE and the MME.

The rest of the information elements in the EMM and ESM NAS messages are tailored for each specific NAS message.

The organization of a normal EMM NAS message is shown below. Each row corresponds to one octet and bit 8 is the most significant bit in the octet.

General message organization example for a normal EMM NAS message

8	7	6	5	4	3	2	1	
Security header				Protocol discriminator				octet 1
Message type								octet 2
Other information elements as required								octet 3
								octet n

The EMM Service Request message is an exception that breaks the normal rules since it has been tweaked to fit into a single initial RRC message and hence optimizing the performance of the system. The structure of an EMM service request message is shown below.

General message organization for an EMM Service Request message

8	7	6	5	4	3	2	1	
Security header				Protocol discriminator				octet 1
Security parameters and message authentication information								octet 2
								octet 3
								octet 4

The structure of ESM messages is shown below.

General message organization example for an ESM NAS message

8	7	6	5	4	3	2	1	
EPS bearer identity				Protocol discriminator				octet 1
Procedure transaction identity								octet 2
Message type								octet 3
Other information elements as required								octet 4 ... octet n

11.8.4   Security-protected NAS messages

When a NAS message is security protected the normal EMM and ESM messages above can be ciphered and/or integrity protected and encapsulated as shown below.

General message organization example for a security-protected NAS message

8	7	6	5	4	3	2	1	
Security header type				Protocol discriminator				octet 1
Message authentication code								octet 2 ... octet 5
Sequence number								octet 6
NAS message								octet 7 ... octet n

Note that the service request message is handled as a special case and is never sent as a security-protected NAS message.

Further details on the EPS NAS messages and the information elements are available in 3GPP TS 24.301 [24.301] and 24.007 [24.007].

11.8.5   Message transport

The NAS messages are transported by S1-AP between MME and eNodeB and by RRC between eNodeB and UE. The underlying protocols S1-AP (MME-eNodeB) and RRC (eNodeB-UE) provide a reliable transport for NAS messages as long as the UE remains within a cell. The NAS protocol includes reliability mechanisms to cater for event like mobility and loss of coverage.

11.8.6   Future extensions and backward compatibility

The UE and network are in principle specified to ignore information elements that they do not understand. It is hence possible for a later release of the system to add new information elements in the NAS signalling without impacting the UEs and network that implement earlier releases of the specifications.

11.9 IP security

11.9.1 Introduction

IPSec is a very wide topic and many books have been written on this subject. It is not the intent and ambition of this chapter to provide a complete overview and tutorial on IPSec. Instead we will give a high-level introduction to the basic concepts of IPSec focusing on the parts of IPSec being used in EPS.

IPSec provides security services for both IPv4 and IPv6. It operates at the IP layer and offers protection of traffic running above the IP layer and it can also be used to protect the IP header information on the IP layer. EPS uses IPSec to secure communication on several interfaces, in some cases between nodes in the core network and in other cases between the UE and the core network. For example, IPSec is used to protect traffic in the core network as part of the NDS/IP framework, see Section 7.4. IPSec is also used on the SWu interface to protect user plane traffic between the UE and the ePDG, as well on the S2c interface to protect DSMIPv6 signalling between the UE and the PDN GW. For more details on SWu and S2c, see Sections 10.5 and 10.6.

In the next section we give an overview of basic IPSec concepts. We then discuss the IPSec protocols for protecting user data; the ESP and the AH. After that we discuss the Internet Key Exchange (IKE) protocol used for authentication and establishing IPSec Security Associations (SA). Finally we briefly discuss the IKEv2 Mobility and Multi-homing Protocol (MOBIKE).

11.9.1.1 IPSec overview

The IPSec security architecture is defined in RFC 4301 [4301]. RFC 4301 is an update of the previous IPSec security architecture specification found in RFC 2401 [2401]. The set of security services provided by IPSec include:

- Access control
- Data origin authentication
- Connectionless integrity
- Detection and rejection of replays
- Confidentiality
- Limited traffic flow confidentiality.

By access control we mean the service to prevent unauthorized use of a resource such as a particular server or a particular network. The data origin authentication service allows the receiver of the data to verify the identity of the claimed sender of the data. Connection-less integrity is the service that ensures that a receiver can detect if the received data has been modified on the path from the sender. It does, however, not detect if the packets have been duplicated (re-played) or re-ordered. Data origin authentication and connection-less integrity are typically used together. Detection and rejection of replays is a form of partial sequence integrity, where the receiver can detect if a packet has been duplicated. Confidentiality is the service that protects the traffic to be read by unauthor-ized parties. The mechanism to achieve confidentiality with IPSec is encryp-tion, where the content of the IP packets are transformed using an encryption algorithm so that it becomes unintelligible. Limited traffic flow confidentiality is a service whereby IPSec can be used to protect some information about the characteristics of the traffic flow, for example, source and destination addresses, message length or frequency of packet lengths.

In order to use the IPSec services between two nodes, the nodes use certain security parameters that define the communication such as keys, the encryption algorithms and so on. In order to manage these parameters, IPSec uses Security Associations (SAs). An SA is the relation between the two entities defining how they are going to communicate using IPSec. An SA is uni-directional, so to pro-vide IPSec protection of bi-directional traffic a pair of SAs is needed, one in each direction. Each IPSec SA is uniquely identified by a Security Parameter Index (SPI) together with the destination IP address and security protocol (AH or ESP, see below). The SPI can be seen as an index to a Security Associations database maintained by the IPSec nodes and containing all SAs. As will be seen below, the IKE protocol can be used to establish and maintain IPSec SAs.

IPSec also defines a nominal Security Policy Database (SPD) which contains the policy for what kind of IPSec service is provided to IP traffic entering and leaving the node. The SPD contains entries that define a subset of IP traffic, for example, using packet filters, and points to an SA (if any) for that traffic.

11.9.2 Encapsulated security payload and authentication header

IPSec defines two protocols to protect data, the Encapsulated Security Payload (ESP) and the Authentication Header (AH). The ESP protocol is defined in RFC 4303 [4303] and AH in RFC 4302 [4302], both from 2005. Previous versions of ESP and AH are defined in RFC 2406 [2406] and 2402 [2402] respectively.

ESP can provide integrity and confidentiality while AH only provides integrity. Another difference is that ESP only protects the content of the IP packet (including the ESP header and part of the ESP trailer) while AH protects the complete IP

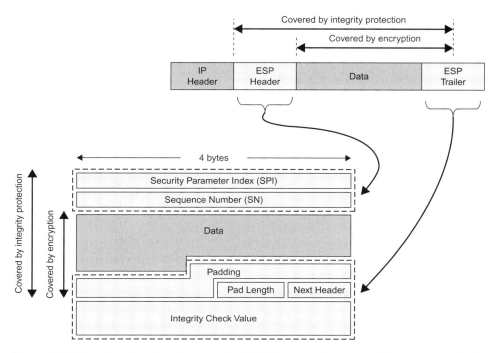

Figure 11.9.2.1 *IP packet (data) protected by ESP. (Note that the same packet is shown twice.)*

packet, including the IP header and AH header. See Figure 11.9.2.1 and Figure 11.9.2.2 for illustrations of ESP- and AH-protected packets. The fields in the ESP and AH headers are briefly described below. ESP and AH are typically used separately but it is possible, although not common, to use them together. If used together, ESP is typically used for confidentiality and AH for integrity protection.

The SPI is present in both ESP and AH headers and is a number that, together with the destination IP address and the security protocol type (ESP or AH), allows the receiver to identify the SA to which the incoming packet is bound. The Sequence number contains a counter that increases for each packet sent. It is used to assist in replay protection. The Integrity Check Value (ICV) in the AH header and ESP trailer contains the cryptographically computed integrity check value. The receiver computes the integrity check value for the received packet and compares it with the one received in the ESP or AH packet.

ESP and AH can be used in two modes: transport mode and tunnel mode. In transport mode ESP is used to protect the payload of an IP packet. The Data field as depicted in Figure 11.9.2.1 would then contain, for example, a UDP or TCP header as well as the application data carried by UDP or TCP. See Figure 11.9.2.3 for an illustration of a UDP packet that is protected using ESP in transport mode. In tunnel mode on the other hand, ESP and AH are used to protect

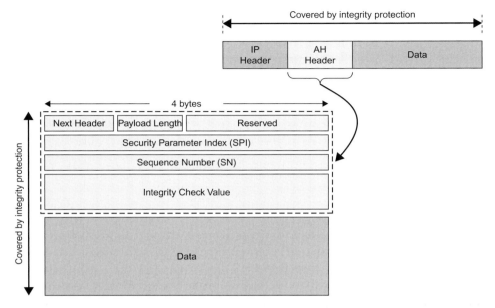

Figure 11.9.2.2 *IP packet (data) protected by AH.*

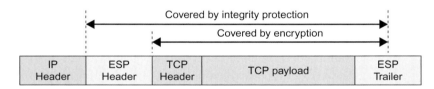

Figure 11.9.2.3 *Example of IP packet protected using ESP in transport mode.*

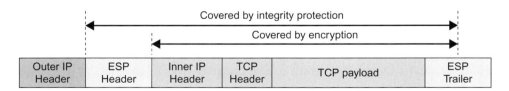

Figure 11.9.2.4 *Example of IP packet protected using ESP in tunnel mode.*

a complete IP packet. The Data part of the ESP packet in Figure 11.9.2.1 now corresponds to a complete IP packet, including the IP header. See Figure 11.9.2.4 for an illustration of a UDP packet that is protected using ESP in tunnel mode.

Transport mode is often used between two endpoints to protect the traffic corresponding to a certain application. Tunnel mode is typically used to protect all IP traffic between security gateways or in VPN connections where a UE connects to a secure network via an un-secure access.

11.9.3 Internet key exchange

In order to communicate using IPSec, the two parties need to establish the required IPSec SAs. This can be done manually by simply configuring both parties with the required parameters. However, in many scenarios a dynamic mechanism for authentication, key generation and IPSec SA generation is needed. This is where the Internet Key Exchange (IKE) comes into the picture. IKE is used for authenticating the two parties and for dynamically negotiating, establishing and maintaining SAs. (One could view IKE as the creator of SAs and IPSec as the user of SAs.) There are in fact two versions of IKE, IKE version 1 (IKEv1) and IKE version 2 (IKEv2).

IKEv1 is based on the Internet Security Association and Key Management Protocol (ISAKMP) framework. ISAKMP, IKEv1 and their use with IPSec are defined in RFC 2407, RFC 2408 [2408] and RFC 2409 [2409]. ISAKMP is a framework for negotiating, establishing and maintaining SAs. It defines the procedures and packet formats for authentication and SA management. ISAKMP is however distinct from the actual key exchange protocols in order to cleanly separate the details of security association management (and key management) from the details of key exchange. ISAKMP typically uses IKEv1 for key exchange, but could be used with other key exchange protocols. IKEv1 has later been replaced by IKEv2 which is an evolution of IKEv1/ISAKMP. IKEv2 is documented in a single document, RFC 4306 [4306], which thus replaces the three RFCs used for documenting IKEv1 and ISAKMP. Improvements have, for example, been made in terms of reduced complexity of the protocol, simplification of the documentation (one RFC instead of three), reduced latency in common scenarios, support for Extensible Authentication Protocol (EAP) and mobility extensions (MOBIKE). Even though IKEv1 is replaced by IKEv2, IKEv1 is still in operational use.

The establishment of an SA using IKEv1 or IKEv2 occurs in two phases. (On this high level, the procedure is similar for IKEv1 and IKEv2). In phase 1 an IKE SA is generated which is used to protect the key exchange traffic. Also mutual authentication of the two parties takes place during phase 1. When IKEv1 is used, authentication can be based on either shared secrets or certificates by using a public key infrastructure (PKI). IKEv2 in addition also supports the use of the EAP and therefore allows a more wide range of credentials to be used, such as SIM cards. See Section 11.10 for more information on EAP. In phase 2, another SA is created which is called the IPSec SA in IKEv1 and child SA in IKEv2. (We will for simplicity use the term IPSec SA for both versions.) This phase is protected by the IKE SA established in phase 1. The IPSec SAs are used for the IPSec protection of the data using ESP or AH. After phase 2 is completed, the two parties can start to exchange traffic using EPS or AH.

EPS makes use of both IKEv1 and IKEv2. The NDS/IP standard allows both IKEv1 and IKEv2 to be used, see Section 7.4. On other interfaces in EPS, it is however primarily IKEv2 that is used. For example, on the SWu interface between UE and ePDG, and on the S2c interface between UE and PDN GW, IKEv2 is used.

11.9.4   IKEv2 mobility and multi-homing

In the IKEv2 protocol, the IKE SAs and IPSec SAs are created between the IP addresses that are used when the IKE SA is established. In the base IKEv2 protocol, it is not possible to change these IP addresses after the IKE SA has been created. There are however scenarios where the IP addresses may change. One example is a multi-homing node with multiple interfaces and IP addresses. The node may want to use a different interface in case the currently used inter-face suddenly stops working. Another example is a scenario where a mobile UE changes its point-of-attachment to a network and is assigned a different IP address in the new access. In this case the UE would have to negotiate a new IKE SA and IPSec SA which may take a long time and result in service interruption.

In EPS, this may occur if a user is using WLAN to connect to an ePDG. The user traffic between the UE and the ePDG (i.e., on the SWu interface) is pro-tected using ESP in tunnel mode. The IPSec SA for ESP has been set up using IKEv2. (See Section 10.10 for more details). If the user now moves to a different network (e.g., to a different WLAN hotspot) and receives a new IP address from the new network it would not be possible to continue using the old IPSec SA. A new IKEv2 authentication and IPSec SA establishment have to be performed.

The MOBIKE protocol extends IKEv2 with possibilities to dynamically update the IP address of the IKE SAs and IPSec SAs. MOBIKE is defined in RFC 4555 [4555].

MOBIKE is used on the SWu interface to support scenarios where the UE moves between different un-trusted non-3GPP accesses.

11.10   Extensible authentication protocol

11.10.1   Overview

The Extensible Authentication Protocol (EAP) is a protocol framework for performing authentication, typically between a UE and a network. It was first introduced for the Point to Point Protocol (PPP) in order to allow additional authentication methods to be used over PPP. Since then it has been introduced also in many other scenarios, for example, as an authentication protocol for IKEv2 as well as for authentication in Wireless LANs using the IEEE 802.11i and 802.1x extensions.

EAP is extensible in a sense that it supports multiple authentication protocols and allows for new authentication protocols to be defined within the EAP framework. EAP is not an authentication method in itself, but rather a common authentication framework that can be used to implement specific authentication methods. These authentication methods are typically referred to as EAP methods.

The base EAP protocol is specified in RFC 3748 [3748]. It describes the EAP packet format as well as basic functions such as the negotiation of the desired authentication mechanism. It also specifies a few simple authentication methods, for example, based on one time passwords as well as a challenge-response authentication similar to CHAP. In addition to the EAP methods defined in RFC 3748, it is possible to define additional EAP methods. Such EAP methods may implement other authentication mechanisms and/or utilize other credentials such as public key certificates or (U)SIM cards. A few of the EAP methods standardized by IETF are briefly described below:

- EAP-TLS is based on TLS and defines an EAP method for authentication and key derivation based on public key certificates. EAP-TLS is specified in RFC 5216 [5216].
- EAP-SIM is defined for authentication and key derivation using the GSM SIM card. EAP-SIM also enhances the basic GSM SIM authentication procedure by adding support for mutual authentication. EAP-SIM is specified in RFC 4186 [4186].
- EAP-AKA is defined for authentication and key derivation using the UMTS SIM card and is based on the UMTS AKA procedure. EAP-AKA is specified in RFC 4187 [4187].
- EAP-AKA' is a small revision of EAP-AKA that provides for improved key separation between keys generated for different access networks. EAP-AKA' is work in progress in IETF and is defined in RFC 5448 [5448].

In addition to the standardized methods, there are also proprietary EAP methods that have been deployed in corporate WLAN networks.

EPS makes extensive use of EAP-AKA and EAP-AKA' on various interfaces. EAP-AKA is supported for access authentication in un-trusted non-3GPP accesses interworking with the EPC (SWa interface), for tunnel authentication towards the ePDG (SWu and SWm interfaces) as well as for establishing the IPSec SA to be used for DSMIPv6 (S2c and S6b interfaces). EAP-AKA' is supported for access authentication in trusted and un-trusted non-3GPP accesses interworking with the EPC (STa and SWa interfaces). When EAP-based access authentication over STa/SWa is performed it occurs prior to invoking the mobility protocol (PMIPv6, DSMIPv6 or MIPv4). For more details, see the corresponding interface descriptions as well as Chapter 7.

11.10.2   Protocol

The architecture for the EAP protocol distinguishes three different entities:

1. The EAP peer: This is the entity requesting access to the network, typically a UE. For EAP usage in WLAN (802.1x), this entity is also known as the supplicant.
2. The Authenticator: This is the entity performing access control, such as a WLAN access point or an ePDG.
3. The EAP Server: This is the backend authentication server providing authentication service to the authenticator. In EPS, this is the 3GPP AAA Server.

The EAP protocol architecture is illustrated in Figure 11.10.2.1.

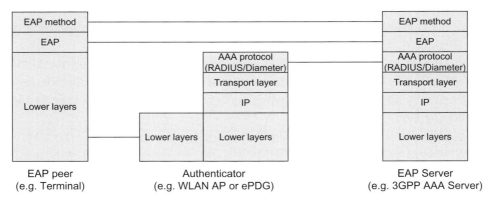

Figure 11.10.2.1   *EAP protocol architecture.*

EAP is often used for network access control and thus takes place before the UE is allowed access and before the UE is provided with IP connectivity. Between the UE (EAP peer) and the authenticator, EAP messages are typically transported directly over data link layers such as PPP or WLAN (IEEE 802.11), without requiring IP transport. Instead the EAP messages are encapsulated directly in the underlying link layer protocol. There are different specifications for how this transport is provided. For example, RFC 3748 [3748] describes EAP usage for PPP, while IEEE 802.1x describes EAP usage over IEEE 802 links such as WLAN. EAP can also be used for authentication with IKEv2, and in this case is transported over IKEv2 and IP (the IKEv2 and IPSec layers are not illustrated in Figure 11.10.2.1).

Between the authenticator and the EAP server, the EAP messages are typically carried in an AAA protocol such as RADIUS or Diameter.

The EAP communication between the peer and the server is basically transparent to the authenticator. The authenticator therefore need not support the specific EAP method used but needs only to forward the EAP messages between the peer and the server.

An EAP authentication typically begins by negotiating the EAP method to be used. After the EAP method has been chosen by the peers, there is an exchange of EAP messages between the UE and the EAP server where the actual authentication is performed. The amount of round trips needed and the types of EAP messages exchanged depend on the particular EAP method used. When the authentication is complete, the EAP server sends an EAP message to the UE to indicate whether the authentication was successful or not. The authenticator is informed about the outcome of the authentication using the AAA protocol. Based on this information from the EAP server, the authenticator can provide the UE with access to the network, or continue blocking access.

Depending on EAP method, the EAP authentication is also used to derive keying material in the EAP peer and the EAP server. This keying material can be transported from the EAP server to authenticator via the AAA protocol. The keying material can then be used in the UE and in the authenticator for deriving access-specific key needed to protect the access link.

Figure 11.10.2.2 provides an example of an authentication using EAP-AKA. Although not explicitly illustrated in the figure, the EAP messages between the

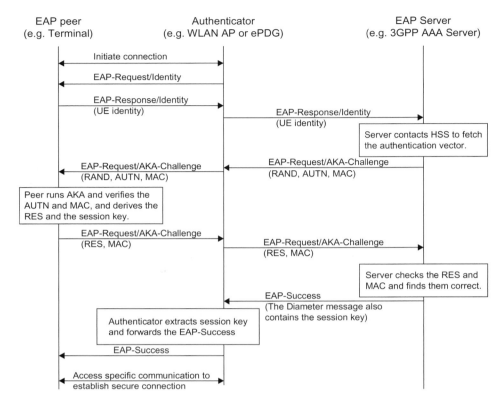

Figure 11.10.2.2   *Example message exchange for authentication using EAP-AKA.*

EAP peer and the authenticator are carried over an underlying protocol specific to the type of access. The EAP messages between the authenticator and the EAP server are carried in an AAA protocol such as Diameter. A reader not interested in the particulars of authentication using AKA can disregard the details of the call flow. The interested reader may however want to compare EAP-AKA message exchange in Figure 11.10.2.2 with the EPS-AKA message exchange over E-UTRAN described in Section 7.3.2. The EPS-AKA over E-UTRAN and the EAP-AKA for accesses supporting EAP are two ways to perform AKA-based authentication.

11.11   Stream control transmission protocol

11.11.1   Background

The SCTP is a transport protocol, operating at an equivalent level in the stack as UDP (User Datagram Protocol) and TCP. Compared to TCP and UDP, SCTP is richer in functionality and also more tolerant against network failures. Even though both TCP and UDP are used as transport protocols in EPS, we will not describe them in any detail in this book since we assume that most readers have a basic understanding of those protocols. The SCTP on the other hand, also used as transport protocol at several interfaces in EPS, is a less known transport protocol and therefore briefly presented in this section.

SCTP is used on several interfaces in the EPC architecture. In particular the mandated Diameter transport protocol on the S6a/S6d interfaces. SCTP is also used for the transport of S1-AP on the S1-MME interface. More details can be found together with the description of the interfaces in Chapter 10.

Compared to UDP [768] from 1980 and TCP [793] from 1981, SCTP is a rather new protocol originally specified in RFC 2960 [2960] from year 2000. The SCTP specification has since then been updated in RFC 4960 [4960] from 2007. The motivation for designing SCTP was to overcome a number of limitations and issues with TCP that are of particular relevance in telecommunication environments. These limitations as well as similarities and differences between UDP/TCP and SCTP are discussed below.

11.11.2   Basic protocol features

SCTP shares many basic features with UDP or TCP. SCTP provides (similar to TCP and in contrast to UDP) a reliable transport ensuring that data reaches the destination without error. Also similar to TCP and in contrast to UDP, SCTP is a connection-oriented protocol meaning that all data between two SCTP endpoints are transferred as part of a session (or association as it is called by SCTP).

The SCTP association must be established between the endpoints before any data transfer can take place. With TCP, the session is set up using a three-way message exchange between the two endpoints. One issue with TCP session setup is that it is vulnerable to so called SYN flooding attacks that may cause the TCP server to overload. SCTP has solved this problem by using a four-way message exchange for the association setup, including the use of a special 'cookie' that identifies the association. This makes the SCTP association setup somewhat more complex but brings additional robustness against these types of attacks. An SCTP association as well as the position of SCTP in the protocol stack is illustrated in Figure 11.11.2.1. As is also indicated in the figure, an

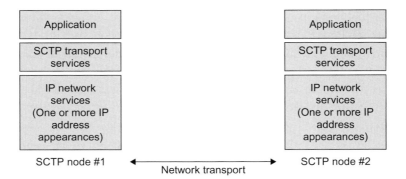

Figure 11.11.2.1   *SCTP Association.*

SCTP association may be utilizing multiple IP addresses at each endpoint (this aspect is further elaborated below).

Similar to TCP, SCTP is rate adaptive. This means that it will decrease or increase the data transfer rate dynamically, for example, depending on the congestion conditions in the network. The mechanisms for rate adaptation of a SCTP session are designed to behave cooperatively with TCP sessions attempting to use the same bandwidth.

SCTP is, similar to UDP, message-oriented which means that SCTP maintains message boundaries and delivers complete messages (called chunks by SCTP). TCP on the other hand is byte-oriented in the sense that it provides the transport of a byte stream without any notion of separate messages within that byte stream. This is desirable to deliver, for example, a data file or a web page, but may not be optimal to transfer separate messages. If an application sends a message of X bytes and another message of Y bytes over a TCP session, the messages would be received as a single stream of X + Y bytes at the receiving end. Applications using TCP must therefore add their own record marking to separate their messages. Special handling is also needed to ensure that messages are 'flushed out' from the send buffer to ensure that a complete message is transferred in a reasonable time. The reason is that TCP normally waits for the send buffer to exceed a certain size before sending any data. This can create considerable delays if the two sides are exchanging short messages and must wait for the response before continuing.

A comparison between SCTP, TCP and UDP is provided in Table 11.11.2.1. More details on the multi-streaming and multi-homing are provided below.

Table 11.11.2.1 *Comparison between SCTP, TCP and UDP.*

	SCTP	TCP	UDP
Connection oriented	Yes	Yes	No
Reliable transport	Yes	Yes	No
Preserves message boundary	Yes	No	Yes
In-order delivery	Yes	Yes	No
Un-ordered deliver	Yes	No	Yes
Data checksum	Yes (32-bit)	Yes (16-bit)	Yes (16-bit)
Flow and congestion control	Yes	Yes	No
Multiple streams within a session	Yes	No	No
Multi-homing support	Yes	No	No
Protection against SYN flooding attacks	Yes	No	N/A

11.11.3 Multi-streaming

TCP provides both reliable data transfer and strict order-of-transmission delivery of data while UDP does provide neither reliable transport nor strict order-of-transmission delivery. Some applications need reliable transfer but are satisfied with only partial ordering of the data and other applications would want reliable transfer but do not need any sequence maintenance. For example, in telephony signalling it is only necessary to maintain the ordering of messages that affect the same resource (e.g., the same call). Other messages are only loosely correlated and can be delivered without having to maintain a full sequence ordering for the whole session. In these cases, the so called head-of-line blocking caused by TCP may result in unnecessary delay. Head-of-line blocking occurs, for example, when the first message or segment was lost for some reason. In this case the subsequent packets may have been successfully delivered at the destination but the TCP layer on the receiving side will not deliver the packets to the upper layers until the sequence order has been restored.

SCTP solves this by implementing a multi-streaming feature (the name Stream Control Transmission Protocol comes from this feature). This feature allows data to be divided into multiple streams that can be delivered with independent message sequence control. A message loss in one stream will then only impact the stream where the message loss occurred (at least initially) while all other streams could continue to flow. The streams are delivered within the same SCTP association and are thus subject to the same rate and congestion control. The overhead caused by SCTP control signalling is thus reduced.

Multi-streaming is implemented in SCTP by decoupling the reliable transfer of data from the strict-order-of-transmission of the data (Figure 11.11.3.1). This is

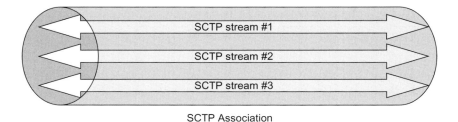

SCTP Association

Figure 11.11.3.1   *Multi-streaming with SCTP.*

different from TCP where the two concepts are coupled. In SCTP, two types of sequence numbers are used. The Transport Sequence Number is used to detect packet loss and to control the retransmissions. Within each stream, SCTP then allocates an additional sequence number, the Stream Sequence Numbers. The Stream Sequence Numbers determine the sequence of data delivery within each independent stream and is used by the receiver to deliver the packets in sequence order for each stream.

SCTP also makes it possible to bypass the sequenced delivery service completely, so that messages are delivered to the user of SCTP in the same order they successfully arrive. This is useful for applications that require reliable transport but do not need in-order delivery, or have their own means to handle sequencing of received packets.

11.11.4   Multi-homing

Another key aspect of SCTP that is an enhancement compared to TCP is the multi-homing features. In telecommunications network it is very important to maintain reliable communications paths to avoid service outage and other problems due to core network transmission problems. Even though the IP routing protocols would be able to find alternative paths in case of a network failure, the time delays until the routing protocol converge and the connectivity is recovered are typically unacceptable in a telecommunications network. Also, if a network node is single homed, that is, has only a single network connection, the failure of that particular connection would make the node unreachable. Redundant network paths and network connections are thus two components in high-available telecommunications systems.

A TCP session involves a single IP address at each endpoint and if one of those IP addresses becomes unreachable, the session fails. It is therefore complicated to use TCP to provide highly-available data transfer capability using multi-homed hosts, that is, where the endpoints are reachable over multiple

IP addresses. SCTP on the other hand is designed to handle multi-homed host and each endpoint of an SCTP association can be represented by multiple IP addresses. These IP addresses may also lead to different communication paths between the SCTP endpoints. For example, the IP addresses may belong to different local networks or to different backbone carrier networks.

During the establishment of a SCTP association, the endpoints exchange lists of IP addresses. Each endpoint can be reached on any of the announced IP addresses. One of the IP addresses at each endpoint is established as the primary and the rest become secondary. If the primary should fail for whatever reason, the SCTP packets can be sent to the secondary IP address without the application knowing about it. When the primary IP address becomes available again, the communications can be transferred back. The primary and secondary interfaces are checked and monitored using a heartbeat process that tests the connectivity of the paths (Figure 11.11.4.1).

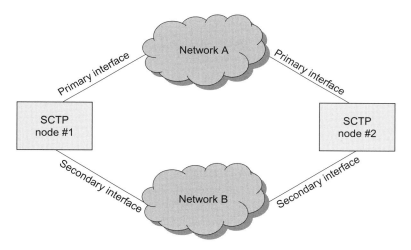

Figure 11.11.4.1   *Multi-homing with SCTP.*

11.11.5   *Packet structure*

The SCTP packet is composed of a Common Header and chunks. A chunk contains either user data or control information (Figure 11.11.5.1).

The first 12 bytes make up the Common Header. This header contains the source and destination ports (SCTP uses the same port concept as for UDP and TCP). When an SCTP association is established, each endpoint assigns a Verification Tag. The Verification Tag is then used in the packets to identify the association. The last field of the Common Header is a 32-bit checksum that allows the

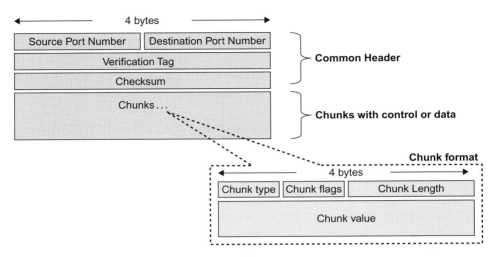

Figure 11.11.5.1   *SCTP header and chunk format.*

receiver to detect transmission errors. This checksum is more robust than the 16-bit checksum used in TCP and UDP.

The chunks containing control information or user data follow the Common Header. The chunk type field is used to distinguish between different types of chunks, that is, whether it is a chunk containing user data or control information, and also what type of control information it is. The chunk flags are specific to each chunk type. The chunk value field contains the actual payload of the chunk. RFC 4960 [4960] defines 13 different chunk type values and the detailed format of each chunk type.

12

Procedures

This chapter provides a brief introduction to some of the procedures used in EPS. It should be noted that it is not feasible within this book to include a complete description of all procedures that exist in EPS. Instead, we have chosen a few key procedures that should give a good overview for some of the most important use cases. An interested reader can consult the 3GPP technical specifications 23.060 [23.060], 23.401 [23.401] and 23.402 [23.402] for complete descriptions.

12.1   Attach and detach for E-UTRAN

12.1.1   Attach procedure for E-UTRAN

Attach is the first procedure the UE executes after being switched on. The procedure is performed to make it possible to receive services from the network. An optimization in the SAE system is that the attach procedure also includes the establishment of a default EPS bearer ensuring that always-on IP connectivity for UE/users of the EPS is enabled. An example of the attach procedure is outlined in Figure 12.1.1.1.

The procedure is briefly described in the following steps.

A. The UE sends an Attach Request message to the eNodeB. The eNodeB checks the MME ID transferred in the Radio Resource Control (RRC) layer. If the eNodeB has a link to the identified MME, it forwards the Attach Request to that MME. If not, the eNodeB selects a new MME and forwards the Attach Request.
B. The MME has changed and the MME uses the old MME ID in the GUTI to find the old MME and retrieves the UE context.
C. *Authentication and security procedures*. The ME identity is also retrieved in conjunction with this step.
D. If the MME has changed, the MME informs the HSS that the UE has moved. The HSS stores the MME address and it instructs the old MME to cancel the UE context.
E. The default bearer is authorized by the PCRF and established between Serving and PDN GW.

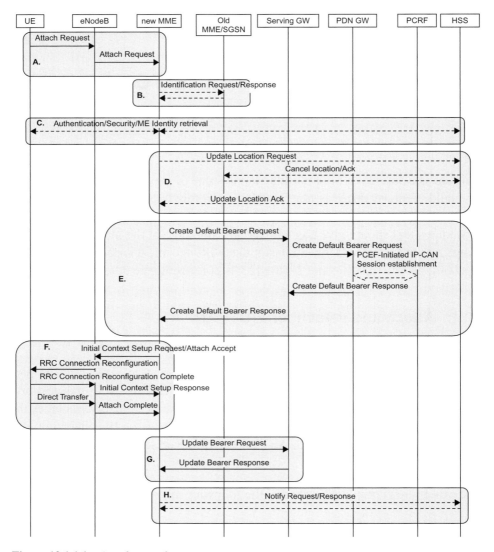

Figure 12.1.1.1 *Attach procedure.*

F. The default bearer is established over the radio interface and the Attach Accept is sent to the UE.

G. MME informs the Serving GW of the eNodeB Tunnel Endpoint Identifier (TEID) which completes the setup of the default bearer as it can now be used in both uplink and downlink.

H. If the MME has selected a PDN GW that is not the same as the one in the received subscription information, it will send a notification of the new PDN GW identity to the HSS.

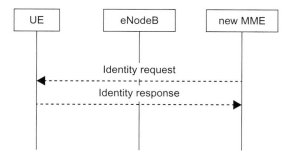

Figure 12.1.1.2 *Identity Request.*

In addition there are some additional steps that may be executed together with the attach procedure. For example if the UEs temporary ID (GUTI) is unknown in both the old MME and new MME (after step A and B), the new MME will request the UE to send its permanent subscription identity (IMSI) as shown in Figure 12.1.1.2.

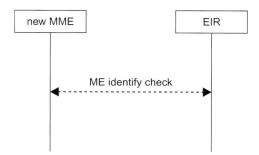

Figure 12.1.1.3 *ME identity check.*

The MME may check the ME identity with an Equipment Identity Register (after step C). The EIR can be used to blacklist, for example, stolen UEs. Depending on the response from the EIR, the MME may continue the attach procedure or reject the UE (see Figure 12.1.1.3).

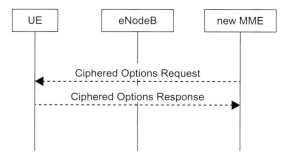

Figure 12.1.1.4 *Ciphered options request.*

If the UE would like to send an APN or PCO, it sets a flag in the initial Attach Request message. The MME will then request the information from the UE after the ciphering has started in step C. This way there is no need to send the APN or PCO unencrypted over the radio interface. The ciphered options request procedure is used to transfer the APN and/or PCO to the MME (see Figure 12.1.1.4).

12.1.2   Detach procedure for E-UTRAN

The detach procedure is used to remove bearers and clear states in the network when, for example, the UE is turned off. The detach procedure can also be used by the network to remove the bearers and states for a UE that has missed to perform TA update because it is out of coverage. There may also be subscription or maintenance reasons to detach the UE, for example, if an MME is taken out of service.

Note that in the normal case, the MME address is not removed from the HSS and the MME can retain the UE context. This saves signalling with HSS since it is rather likely that the UE will re-attach in the same MME and then there is no need to inform the HSS and download the subscription data.

12.1.2.1   UE-initiated detach procedure
The procedure is shown in Figure 12.1.2.1.1 and is briefly described in the following steps.

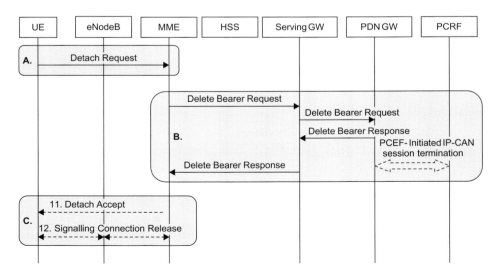

Figure 12.1.2.1.1   *UE-initiated detach procedure.*

A. The UE sends a Detach Request to the MME because it is turned off.
B. The MME instructs the Serving and PDN GW to delete any bearers for the UE and the PCEF in the PDN GW informs the PCRF that the bearers are being removed.
C. The MME may confirm the detach with a Detach Accept message and remove the signalling connection.

If the UE has not communicated with the network for a long time (longer than the TA update timer) the MME may initiate the detach procedure. In that case, the MME may try to inform the UE with a Detach Request message and then it deletes the bearers as in the UE-initiated detach flow step B (Figure 12.1.2.1.1).

In some special cases, the HSS may also use the detach procedure by sending a Cancel location with cause value subscription withdrawn. This will trigger the MME to remove the UE context, send a Detach Request to the UE and delete the bearers as in step B (Figure 12.1.2.1.1).

12.2 Attach and detach for non-3GPP accesses

12.2.1 General

When the UE is powered on and if a non-3GPP access is available, it may be decided, either automatically by the UE based on policies or by manual choice, to make the initial attach in the non-3GPP access. The UE may, for example, use policies received from the ANDSF. Furthermore, the IPMS mechanisms are used to decide on which mobility protocol is used (see Section 6.4). Finally, it has to be determined whether to treat the non-3GPP access as a trusted or untrusted non-3GPP access.

In this section, we show two examples of attach/detach procedures. First, we describe attach and detach procedures for untrusted non-3GPP access using PMIPv6 on S2b. We then describe the attach and detach procedure for a trusted non-3GPP access when using DSMIPv6 (S2c). Note that the other combinations, that is, trusted access with PMIPv6 and untrusted access with DSMIPv6, are of course also possible but not explicitly showed in the book. For a complete description of the available attach and detach procedures, see TS 23.402 [23.402].

The descriptions are for non-roaming scenarios. It can be noted that in a roaming scenario, the call flows would also include a 3GPP AAA proxy as well as a visited PCRF.

The attach procedure is performed to establish the IP connection and to make it possible to receive services from the network, for example, when the UE is turned on. The detach procedure is used to terminate the IP connections and to clear states in the network when, for example, the UE is turned off.

12.2.2   Attach procedure in untrusted non-3GPP access using PMIPv6 (S2b)

The procedure is shown in Figure 12.2.2.1 and is briefly described in the following steps.

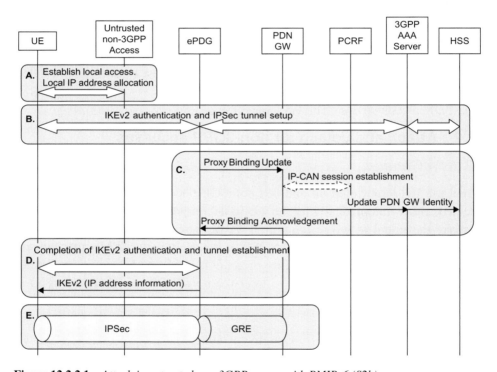

Figure 12.2.2.1   *Attach in untrusted non-3GPP access with PMIPv6 (S2b).*

A. The UE establishes a connection to the untrusted non-3GPP access (e.g. WLAN) and receives a local IP address from the access network. There may be access authentication performed (e.g. based on EAP-AKA) but this is optional and not shown in the call flow for simplicity.

B. The UE discovers the IP address of an ePDG using DNS and then initiates the IKEv2 procedure towards the ePDG. EAP-AKA is used for authentication as described in Section 7.3.5. Diameter is used between ePDG and the AAA server, and between the AAA server and the HSS.

C. When the IKEv2 procedure has progressed, the ePDG sends a PBU to the PDN GW. The PDN GW notifies the PCRF about the new connection. The PDN GW also sends the PDG GW Identity to the HSS (via the AAA server). The PDN GW then responds with a PBA to the ePDG. The PBA contains the IP address for the PDN connection.

D. The final steps of the IKEv2 procedure are executed. The ePDG includes the IP address of the PDN connection in an IKEv2 message to the UE.

E. When the attach procedure is complete, an IPSec tunnel has been established between UE and ePDG and a GRE tunnel between ePDG and PDN GW.

12.2.3 Detach procedure in untrusted non-3GPP access using PMIPv6 (S2b)

There is no specific detach procedure in untrusted non-3GPP access. Instead, each PDN connection is terminated separately. In order to detach a UE, all active PDN connections have to be closed. The UE and the network may trigger a PDN disconnection. The HSS may, for example, trigger a termination of all active PDN connections, in case the subscription has been cancelled. For a description of the available procedures, see TS 23.402 [23.402].

The UE-initiated PDN disconnection procedure is shown in Figure 12.2.3.1

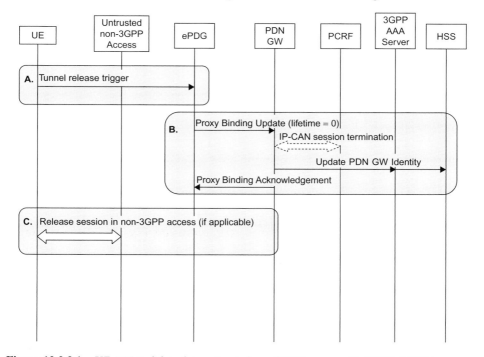

Figure 12.2.3.1 *UE-initiated detach in untrusted non-3GPP access with PMIPv6 (S2b).*

The procedure is briefly described in the following steps.

A. The IPSec tunnel is released.
B. Since the IPSec tunnel is released, the ePDG sends a PBU to the PDN GW with the lifetime parameter set to zero to indicate that the PDN connection shall be terminated. The PDN GW informs the PCRF that the IP-CAN session is closed. The PDN GW informs the HSS (via the 3GPP AAA Server) to remove the PDN GW identity information that is stored for this PDN connection. The PDN GW also replies with a PBA to the ePDG.
C. The UE may release any resource it has in the non-3GPP access. It may, for example, release the local IP address.

12.2.4   Attach procedure in trusted non-3GPP access using DSMIPv6 (S2c)

The procedure is shown in Figure 12.2.4.1 and is briefly described in the following steps.

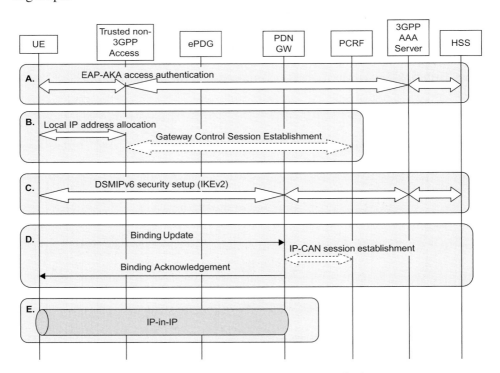

Figure 12.2.4.1   *Attach in trusted non-3GPP access with DSMIPv6 (S2c).*

A. The UE performs access authentication in the trusted non-3GPP access.
B. The local IP connection in trusted non-3GPP access is stabled and the UE receives a local IP address from the non-3GPP network. The trusted non-3GPP access also initiates a gateway control session establishment with the PCRF.

C. If the UE has not bootstrapped DSMIPv6 security before, it initiates the IKEv2 procedure at this point. Authentication and IPSec SA establishment based on IKEv2 and EAP-AKA is performed. The IPSec SA will be used for protecting the DSMIPv6 signalling.

D. The UE sends a Binding Update to the PDN GW. The PDN GW initiates an IP-CAN session establishment towards the PCRF. The PDN GW replies with a Binding Acknowledgement to the UE.

E. When the attach procedure is complete, there is an IP-in-IP tunnel between UE and the PDN GW. For more details on the operation of DSMIPv6, see Section 11.3.

12.2.5   Detach procedure in trusted non-3GPP access using DSMIPv6 (S2c)

There is no specific detach procedure when DSMIPv6 is used. Instead, each PDN connection is terminated separately. In order to detach a UE, all active PDN connections have to be closed. The UE and the network may trigger a PDN disconnection. The HSS may, for example, trigger a termination of all PDN connections in case the subscription has been cancelled. For a description of the available procedures, see TS 23.402 [23.402].

The UE-initiated PDN disconnection procedure is shown in Figure 12.2.5.1.

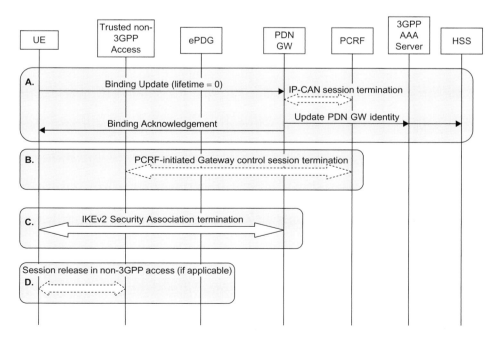

Figure 12.2.5.1   *UE-initiated detach in trusted non-3GPP access with DSMIPv6 (S2c).*

The procedure is briefly described in the following steps.

A. The UE sends a Binding Update with lifetime zero to indicate that the PDN connection shall be closed. The PDN GW informs the PCRF that the IP-CAN session is closed. The PDN GW informs the HSS (via the 3GPP AAA Server) to remove the PDN GW identity information that is stored for this PDN connection. The PDN GW also replies with a Binding Acknowledgement to the UE.
B. The PCRF sends a message to the BBERF in the trusted non-3GPP access to remove the QoS rules associated with the PDN connection that has been closed.
C. The UE terminates the IKEv2 security association for the given PDN connection.
D. The UE may release any resource it has in the non-3GPP access. It may, for example, release the local IP address.

There are also procedures for allowing the HSS to detach a UE, for example, in case the subscription has been cancelled. For a description of the available procedures, see TS 23.402 [23.402].

12.3 Tracking Area update

The TA update procedure is one of the mobility management procedures that are used to ensure that the MME knows in which TA, or set of TAs, the UE is currently located in. TA update is performed when the UE moves to a TA outside the list of TAs it was assigned in the previous TA update or attach procedure. TA update is also performed periodically even if the UE remains in the assigned TAs.

12.3.1 Tracking Area update procedure

In its simplest form, the TA update is just a couple of messages between the UE and the MME. The procedure in the simple form is used when the UE has

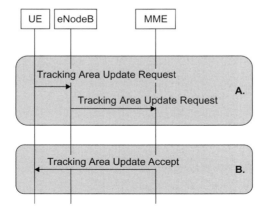

Figure 12.3.1.1 *Tracking area update.*

previously performed an attach or a TA update towards the same MME. The trigger for the TA update may be expiry of the periodic timer or that the UE has moved outside the assigned set of TAs.

The procedure is shown in Figure 12.3.1.1 and is briefly described in the following steps.

A. The UE decides to perform a TA update; in this case, the trigger can be either expiry of the periodic update timer or that the UE has moved outside its assigned set of TAs. When the eNodeB receives the TA update request, it routes the message to the correct MME based on the GUMMEI that the UE has placed in the RRC message that transports the TA Update message.

B. The MME recognizes the UE and resets the periodic update timer and sends a TA Update Accept message to the UE. The TA Update Accept message may contain a new list of TAs for the UE. The UE stores the new list of TAs.

12.3.2   TA update with MME change

When moving to LTE from 2G/3G or when the MME is changed, the TA update procedure needs to cater for the change of nodes and update all related nodes accordingly.

The procedure is shown in Figure 12.3.2.1 and is briefly described in the following steps.

A. The UE decides to perform a TA update; in this case, the trigger could be that the UE has moved outside its assigned set of TAs or that the UE moves from 2G/3G into LTE. When the eNodeB receives the TA update request, it detects that the indicated MME is not associated with the eNodeB and hence a new MME needs to be selected. The eNodeB performs an MME selection and forwards the TA update to the selected MME.

B. The new MME uses the GUTI (the MME Identity inside the GUTI) received from the UE to determine where the UE context resides. The MME requests the UE context from the old SGSN or old MME. The request also include the TA Update message, which allows the old MME/old SGSN to validate the integrity of the message. If the message passes the integrity check, the old MME or old SGSN sends the UE context to the new MME. If the integrity check fails, an error message is sent to the new MME which then authenticates

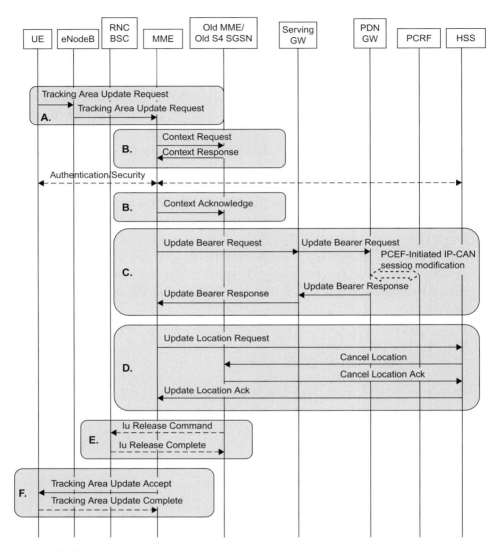

Figure 12.3.2.1　*Tracking area update with MME change.*

the UE. The new MME acknowledges the reception of the UE context message back to the old MME or old SGSN.

C. The new MME informs the Serving GW of the change of MME with an Update Bearer Request. The Serving GW in turn updates the PDN GW with the current RAT type and possibly the location of the UE. PDN GW may also inform the PCRF of the change in RAT type and location.

D. The MME informs the HSS that the UE has moved. The HSS stores the MME address and it instructs the old SGSN/MME to cancel the UE context. The HSS then acknowledges the Location Update.

E. If the old SGSN has an active Iu connection for the UE, the connection is released.

F. The MME completes the TAU procedure with a TA Update Accept message to the UE. The TA Accept message may contain a new list of TAs for the UE and a new GUTI. The UE confirms the new GUTI by sending a TA update complete.

It is also possible to change the Serving GW during a TA update procedure. The MME may decide to change the serving GW, for example, due to non-optimal routing of the user plane after mobility. If this is the case, the MME will inform the old MME/SGSN that the Serving GW is changed in the context transfer procedure. The old MME/SGSN will delete the bearers in the old Serving GW. A change of serving GW is likely a very rare event in most networks, and in networks with collocated Serving and PDN GW, there is no reason at all to relocate the Serving GW (since the PDN GW will anyway remain fixed).

A change of MME due to mobility within LTE will likely be a rare event in most networks, since MME pools can be used to allow several MMEs to share the UEs in a larger area. As long as the UE remain in the pool coverage area, the UE can stay connected to the same MME.

12.3.2.1 Triggers for TA update

TA update occurs when UE experiences any of the following conditions:

- UE detects it has entered a new TA that is not in the list of TAIs that the UE registered with the network;
- the periodic TA update timer has expired;
- UE was in UTRAN PMM_Connected state (e.g. URA_PCH) when it reselects to E-UTRAN;
- UE was in GPRS READY state when it reselects to E-UTRAN;
- the UE reselects to E-UTRAN when it has modified the bearer configuration modifications when on GERAN/UTRAN;
- the radio connection was released with release cause 'load re-balancing TAU required';
- the UE core network capability and/or UE-specific DRX parameters information is changed.

12.4   Handover procedure

12.4.1   Basic handover

If we consider radio access and packet core network level handover without worrying about the service continuity and session continuity aspects, the following possible handover combinations can be found:

1. Intra and Inter 3GPP access handover
   a. Intra E-UTRAN.
   b. E-UTRAN to/from UTRAN (GTP or PMIP).
   c. E-UTRAN to/from GERAN (GTP or PMIP).
   d. Intra GERAN, Intra UTRAN and GERAN to/from GERAN. (These handover cases are not covered in this book.)
2. 3GPP and non-3GPP handover
   a. Optimized handover E-UTRAN to/from HRPD (for GTP and PMIP).
   b. Basic non-optimized handover: trusted non-3GPP access (including HRPD) to/from GERAN/UTRAN/E-UTRAN (with GTP/PMIP on 3GPP access and PMIP/MIPv4FA/DSMIPv6 on non-3GPP access).
   c. Basic non-optimized handover: untrusted non-3GPP access (including HRPD) to/from GERAN/UTRAN/E-UTRAN (with GTP/PMIP on 3GPP access and PMIP/DSMIPv6 on non-3GPP access).

Note that all these scenarios can be Intra or Inter PLMN handover as described below in more detail. In case of handover, we have chosen to focus on trusted non-3GPP access via S2a interface, which implies a network-based mobility using PMIP protocol for non-3GPP accesses and either GTP or PMIP protocol for 3GPP accesses.

12.4.1.1   3GPP radio access

Within 3GPP (as well as in most other cellular technologies) a handover is defined as follows in a very narrow term (as per TS 22.129 [22.129]):

> *Handover is the process in which the radio access network changes the radio transmitters or radio access mode or radio system used to provide the bearer services, while maintaining a defined bearer service QoS.*

Handover is a key mobility mechanism for any cellular systems, whether moving within an access technology or between different access technologies. Core networks play a crucial role in the handover mobility process but in a majority of the cases, the decision to handover is based on the radio conditions.

The UE assists the network by radio measurements about the serving as well as neighbouring cells in the same or different access technologies that may be candidates for handover. The details of how and when the UE and E-UTRAN decides to trigger a handover is far beyond the scope of this book but the interested reader can find more details in Dahlman (2008).

Handover can be of many different types and forms and if we exclude the process of selecting the target access technology by the UE being handed over to, handover may cause the core network to be involved in the most simplest form of getting the target access network information the UE is connecting to and to a more complex form where one or more core network entities needs to be relocated to better serve the user. In addition to the process of actually changing the radio and/or core network entities, the handover process also needs to ensure the service continuity, that is maintaining the bearer characteristics for active services as far as possible. A system may use handover/cell-reselection mechanism to achieve service continuity for a UE actively involved in a session (transmitting and receiving data). Note that other mechanisms, such as SRVCC, also enable certain specific service continuity for specific type of service(s) which is addressed in a separate section of the book (see Section 12.6).

So what are the different kinds of handovers possible from an overall network perspective? Let us take the example diagram illustrated in Figure 12.4.1.1.1, where it depicts a simple scenario where Operator X and Operator Y have some of their radio networks connected to each other and their core network

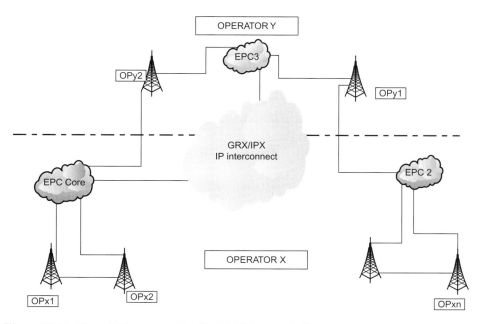

Figure 12.4.1.1.1 *Multi-operator simplified 3GPP network diagram.*

(here EPC) connected via a GRX/IPX interconnect. Operator X has two EPC networks connected to the RANs OPx1, OPx2, ..., OPxn, whereas Operator X has one EPC network connected to RANs OPy1 and OPy2. The first level of handover we can define may be whether the user is moving within Operator X networks thus causing *Intra PLMN handover*. If the user moves between RANs OPx1 and OPy2, then we have just encountered *Inter PLMN handover*. During Inter PLMN handover, when the user crosses radio access technology also, such as E-UTRAN to/from UTRAN, the network has also performed *Inter System Handover* by changing the radio access technology. Note that a UE will only be instructed to measure on neighbour cells to which handover is allowed.

In case of Inter PLMN handover, a number of aspects need to be understood/established before handover may be accomplished. First of all, in Inter PLMN handover, a session may not only cross an operator's boundary, but it may also cross country boundary. A call in the United States may, for example, arise in the upstate New York bordering Canada and continue inside Canada if Inter PLMN handover is supported between the two involved operators. As such, Inter PLMN handover is very much up to individual operator's discretion to support or not. An operator may choose to drop the session and then require the UE to register in the new PLMN instead and in this case, the service continuity is not maintained. Before proceeding with Inter PLMN handover, certain criteria need to be met as specified in 3GPP TS22.129 [22.129]:

- The ability to check with the home network whether the user is permitted to handover from the visited network to a target network.
- Invocation of the handover procedure only occurs if the target network provides the radio channel type required for the respective call.
- The avoidance of 'network hopping', that is, successive handover procedures between neighbouring networks for the same call.
- The possibility of user notification of Inter PLMN HO (e.g. possible tariff change) when it occurs.

During handover procedures, a network may operate in one of the three roles: Home PLMN, Serving PLMN and Visited PLMN. Home PLMN is where the subscriber has his/her subscription. Visited PLMN is the visited network for a roaming user where the user has performed successful registration process (i.e. HSS is aware of where the user is located and has performed all the procedures needed for updating the user's location). Serving PLMN/network is where the user may have been handed over to (e.g. the cell the user is being served by belongs to the Serving network operator) and has not yet performed the registration process. Most likely scenario is that the Serving PLMN becomes the Visited PLMN after registration unless the user moves out of the serving area/cell.

Shared networks have also support handover of all types, mentioned above, between shared and non-shared networks but some additional aspects, such as the core network it is connected to and the home network where the user has subscription, are also important to take into account.

Even though the most frequent cause of handover is the movement of the user (i.e. the UE), there are other additional triggers that can cause a handover to occur. Some examples of these reasons may be:

- Triggered due to a service requirement that may be met by a different RAT than the one the UE is being served by and core network may trigger such process.
- Various radio conditions such as change of radio access mode, capacity of a cell to be able to serve the user's current services.

Even though, in principle, handover shall not cause any significant loss/change of service or interruption of service, when multiple radio bearers are being handed over from one type of radio access to another, there may be need to drop certain bearers and maintain others based on, for example, priority and relevant QoS information, data rate, delay constraints, error rate, etc. Also, sometimes, certain QoS may be degraded to accommodate the handover of all PS bearers. Overall, instead of failing to handover at all, it would be preferable to be able to handover at least one bearer that is suited for the target radio access. Usually when moving from a higher bit rate to a lower one (e.g. UTRAN to GERAN), then decisions have to be made that suits the serving network operator (HPLMN when not roaming) and an operator may choose to drop all active bearers or based on certain pre-defined criteria choose to handover specific bearers.

So what has changed for EPS compared to what existing handover concepts did not already cover? For example, there is no longer a central entity control-ling the RAN like an RNC for UTRAN and a BSC for GERAN. Instead, the eNodeBs in E-UTRAN connects directly to the EPC core network entities, MME for the control plane for signalling and Serving GW for user plane traf-fic transferring data to/from a terminal for the user. In addition to EUTRAN, the EPS must also support handover to/from non-3GPP accesses such as HRPD network, WiMAX and Interworking WLAN (or I-WLAN as referred to in 3GPP). Thus it is expected that EPS must be able to support handover between heterogeneous access systems, where the non-3GPP access networks are not developed in 3GPP. These handovers also need to support service continuity, thus the ability to maintain IP connectivity within the EPC when moving between these heterogeneous access networks has taken a lot of efforts to accomplish. Even though Inter system handover is usually used in conjunction with 3GPP accesses, this term can easily be expanded to also cover Intra PLMN handover between 3GPP and non-3GPP access. For 3GPP accesses, both Inter PLMN and Intra PLMN handovers as well as Inter System handovers are supported with special emphasis on service continuity between UTRAN and E-UTRAN radio access. In addition, EPS also supports handover from E-UTRAN to pre-Release 8 3GPP networks, but note that the opposite direction is not supported. In this case, the source network (i.e. EPS) has to adopt the target network requirements due to the fact that the target network can not understand/interpret the EPS information due to that pre-Release 8 networks are not upgraded.

Another special aspect of E-UTRAN radio access is that it is a packet-only system and thus there is no support for circuit-switched bearers and CS Domain in the evolved system. So a handover from IMS voice on E-UTRAN EPS to CS voice on 2G/3G has been developed in 3GPP and it is known as Single Radio Voice Call Continuity (SRVCC). It is also possible to provide dual radio-based service continuity between 3GPP and other non-3GPP accesses with IMS, when the non-3GPP access is connected via EPS.

HRPD and 3GPP access

For HRPD, the UE can provide its capabilities regarding other radio accesses and systems it can support and relevant details such as single/dual radio, dual receiver, frequency, etc. via same mechanism as it would do for E-UTRAN access. E-UTRAN is able to configure which other access technology information the UE may be able to provide to the EPS system. Also for handover to/from non-3GPP access, the access technologies must be connected via EPS.

In case of HRPD networks, the requirements on E-UTRAN are very similar to existing 2G/3G 3GPP access networks. The following need to be supported by E-UTRAN and HRPD access network in order to facilitate smoother performance compared to other non-3GPP accesses. As stated previously, the operator requirements are directly linked to the existing networks and subscriber base for CDMA systems. E-UTRAN controls the trigger for UE to measure the HRPD information for preparation of handover, when handover is performed from E-UTRAN to HRPD direction. The HRPD system information block (SIB) has to be sent on the E-UTRAN broadcast channel. The UE monitors the broadcast channel in order to retrieve the HRPD system information for the preparation of cell reselection or handover from the E-UTRAN to HRPD system. HRPD system information may also be provided to the UE by means of dedicated signalling. The HRPD system information contains HRPD neighbouring cell information, CDMA timing information, as well as information controlling the HRPD pre-registration. Note that pre-registration is used only when optimized handover is supported in the EPS, for more details see later sections below.

General non-3GPP and 3GPP access

In case of general non-optimized handover, the UE needs to perform access attach procedure with target access. For handing over to a non-3GPP access, both network-based and host-based IP mobility management solutions are supported (see Section 11.3). For handovers from a non-3GPP access to a 3GPP access, only network-based mobility is supported in target non-3GPP access.

For handover using network-based mobility in target access, the terminal performs access attach with an indication that the attach is of type 'handover' in order for the target network (i.e. radio access, MME, Serving GW) to establish the necessary resources for the handover and also where possible, maintain the IP connectivity by maintaining the PDN GW and the terminal's IP address(es). For host-based mobility in target non-3GPP access, the terminal establishes a local IP connectivity in target access and the handover and IP address preservation is then executed using the host-based IP mobility mechanism. Depending on the number of PDN the UE was connected to before handover, and depending on what the target system can support, the PDN connectivity may be re-established in the target access by the network or by the UE itself or some of them are dropped during handover. So the handover, in this case, has very little interactions with access networks and is performed by the UE and the core network. For more details on these handover procedures, see sections below.

So what we can conclude is that the basic handover requirements remain the same as we move towards E-UTRAN and EPS, but we also continue to enhance the system to accommodate non-3GPP accesses as well as IMS service continuity as we evolve the systems. In the sections below, we will elaborate how the EPS level handovers are supported by the radio and evolved core networks.

12.4.2   Phases and details of handover procedure

12.4.2.1   Overall description

PS handover procedures for GSM were developed with the basic principles of two main phases: Handover preparation phase and handover execution phase. The same principles continue in the handover procedures for EPS as well.

We will briefly go over some of the principles of handover in the existing 2G/3G 3GPP Packet Core (GPRS PS Domain) and then go into the details of the EPS handovers. As in any handover case, there is a source cell in the RAN and a target cell in the RAN (within the same radio access in case of Intra RAT and in a different RAN in case of Inter RAT handover) where the terminal is planned to be moved to. For example, for a 2G system, Intra BSS handover may be performed maintaining the same SGSN (known as Intra SGSN handover) or also changing the SGSN (known as Inter SGSN handover). Or handover may be performed between different radio access type such as between BSS and UTRAN which is known as Inter RAT, where an Inter RAT HO can also be Intra SGSN or Inter SGSN handover. In case of 3G radio access with PS domain GPRS, there can be Inter RNC HO (where the RNC functions including SRNS function is moved) with Intra or Inter SGSN HO, SRNS relocation procedures with Intra or Inter SGSN HO, and Inter System HO with BSS to/from RNC handover is performed with Intra or Inter SGSN HO. In all these handover scenarios, the packet core network is involved and updated during the handover process. Note that the GGSN is not relocated or changed at any handover procedures.

The overall handover process may be described as the source access handling the procedures, such as UE and radio network measurements, to determine handover needs to be initiated, preparing the resources in the target radio and core network side, directing the UE to the new radio resources and releasing the resources in the source radio and core network where applicable, as well as handling gracefully any failure conditions that may occur to get back to a stable status, as well as ensuring that all control and user plane entities are properly connected in the new network. During this process, the uplink and downlink data may be buffered and then forwarded according to the most appropriate path determined by the specific process/handover type itself, thus minimizing any possible loss of user's data.

Let us analyse the high level view of the two phases (Preparation and Execution) for handover in brief. Figure 12.4.2.1.1 and Figure 12.4.2.1.2 outline the preparation phase and the execution phase for handover respectively.

Handover may fail at any time during the handover procedure and handover may also be rejected by the target RAN and source RAN may also cancel the handover due to conditions that could deem that the handover will not succeed or that the process has failed somewhere. In case the handover is rejected or cancelled by the target radio or source radio respectively, all acquired resources would be released and cleared of handover process both in radio and core networks. In case of handover failure, depending on whether the failure occurred

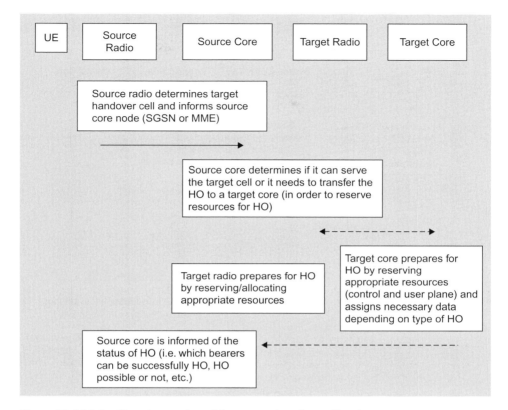

Figure 12.4.2.1.1 *Generic overview of the preparation phase of handover.*

during preparation or execution phase, the resources affected and actions needed would be different of course. In case UE fails to connect to the target cell during execution phase, it returns to the source radio and sends appropriate message of failure. If the UE has lost radio contact, then it is source radio's responsibility to inform source core and then the rest of the path in the target network is cleared/ released. If failure is caused due to core network procedures, then appropriate cause code would be sent to the UE and next actions determined (e.g. re-negotiate bearers, etc.).

Handover in EPS for 3GPP accesses
Now let us get started on the main focus of this book, which is to help understand the process of handover involving E-UTRAN and EPS.

In case of EPS and handover involving E-UTRAN, the MME and Serving GW are the involved entities for Intra RAT (LTE) handover. In case of PMIP protocol use, the BBERF located in the Serving GW may also be involved in order to update the PCRF with the right BBERF information. Compared to 2G/3G

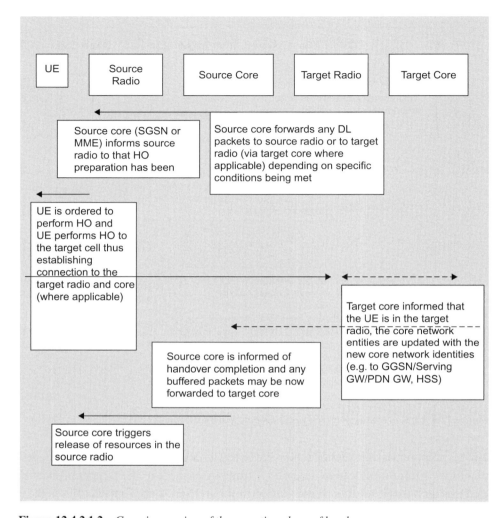

Figure 12.4.2.1.2   *Generic overview of the execution phase of handover.*

handover using GPRS, the EPS provides change/relocation during handover for the following possible combinations:

1. Intra E-UTRAN (between eNodeBs) only (Intra MME).
2. Inter E-UTRAN and MME (Inter MME).
3. Inter E-UTRAN and MME and Serving GW (Inter MME and Inter Serving GW).
4. Inter E-UTRAN and Serving GW (Inter Serving GW).
5. Inter RAT (E-UTRAN and GERAN/UTRAN) with core network entities combination of relocation (i.e. MME to/from SGSN relocation and then Serving GW may be relocated).

Inter 3GPP non-3GPP handover is covered in the subsequent sections.

Handover within E-UTRAN access

In certain extraneous conditions such as MME overload, an MME may trigger a relocation of users from the affected to a new MME. Note that compared to 2G/3G, the EPC is clearly separated as control and user plane entities in relation to eNodeB and MME is the control plane entity and Serving GW is the user plane entity. Thus depending on the type of handover performed, multiple entities may be relocated and updated with each other's information before the handover is completed. As users profile may restrict handover via roaming/area restrictions, MME provides this information to eNodeB via S1 interface and then during handover in active mode of the UE, eNodeB is responsible for verifying whether the handover is allowed or not.

As described in case of 2G/3G, LTE handover is also performed in two main phases, preparation and execution. Though due to the nature of X2-based handover, the execution phase has been further divided into execution and handover completion phase. In the handover completion phase, core network entities like MME and Serving GW then becomes aware of the handover completed and completes the necessary update to the control and user plane paths throughout the bearer connection path. One of the simplest cases is the Intra EUTRAN handover case, described and illustrated in Figure 12.4.2.1.3.

In this procedure, during the preparation phase, the core network entities are not involved. The source eNodeB makes the decision based on UE and radio level information as well as restriction data provided by the core network (i.e. MME) that handover needs to be performed and selects appropriate target eNodeB. eNodeBs are connected with each other via X2 reference point over an IP infrastructure. The source eNodeB is responsible for the decision for which of the EPS bearers are subject to forwarding of packets from the source eNodeB to the target eNodeB, EPC maintains the decision taken, that is no changes are performed. Both source eNodeB and target eNodeB may need to buffer data and during the handover phase, the source eNodeB establishes uplink and downlink data forwarding path for user plane traffic towards target eNodeB.

During the execution phase, the source eNodeB forwards any data received from the downlink path from Serving GW towards target eNodeB as long as the data is arriving and the source eNodeB is able to handle the data. Once the handover has been executed successfully (i.e. the UE is now connected to the target eNodeB), target eNodeB informs MME to switch path and MME then also informs Serving GW to switch the user plane path for the downlink data traffic (i.e. towards the UE) and informs source eNodeB via an end marker about the end of data transfer. This is the simplest type of handover from EPC perspective as can be readily seen.

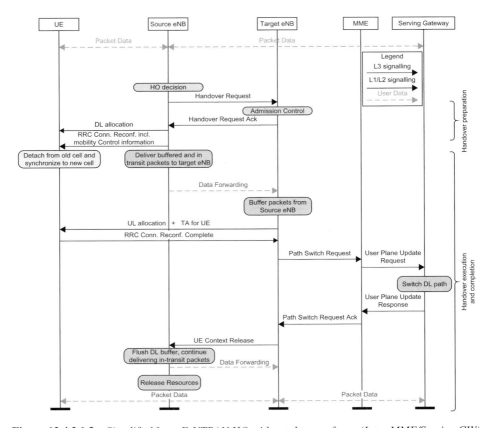

Figure 12.4.2.1.3   *Simplified Inter E-UTRAN HO without change of core (Intra MME/Serving GW).*

The procedure is a bit more complex if MME determines that Serving GW relocation may be needed in this case for reasons such as appropriate user plane connectivity, etc. But in order to be able to do a Serving GW relocation, there must be full IP connectivity between source eNodeB and source Serving GW, between source Serving GW and target eNodeB and between target Serving GW and target eNodeB.

At the reception of the Path Switch request from target eNodeB, MME requests target Serving GW to create new bearers as required after the handover completion and target Serving GW updates PDN GW with its address and updated user information, thus completing the new switched path between the UE and the target eNodeB and between the target Serving GW and the PDN GW. When the PDN GW gets the updated information, it starts sending downlink data via this updated path. MME informs Serving GW to release its resources for that UE and target eNodeB releases necessary resources in the source eNodeB, once path switch completion is indicated by MME.

Even though it seems natural to have X2-based handover, it is not reasonable to assume full IP connectivity between all eNodeBs directly from day 1 of E-UTRAN/EPS deployment and there may be other reasons where X2-based handover can not be performed. In such cases, S1-based handover via the core network is the mechanism available for all listed handover cases above.

In case of S1-based handover, either MME or both MME and Serving GW can be relocated even though MME change should not be done unless the target eNodeB belongs to a different MME pool area. If a source MME has selected a target MME for the handover, then it is the target MME's responsibility to decide if source Serving GW needs to be relocated or not, otherwise it is the source MME's responsibility to make that decision. The sequence diagram below shows the necessary steps for the handover; note that it is source eNodeB decision whether direct data forwarding will be performed via X2 interface or indirect data forwarding will be performed via source or target Serving GW, depending on whether the Serving GW is relocated or not. Source MME releases source Serving GW resources only when it has confirmation that target Serving GW path switch has been successfully completed.

In case the S1 handover is rejected for any reason, the UE then remains in the Source eNodeB/MME/Serving GW.

Note that we have focused on a GTP-based S5/S8 interface when describing the handover flows, in case PMIP is used, then the interactions between the Serving GW and PDN GW would be slightly different and there is also additional interaction due to the PCC using off path signalling causing additional interactions with PCRF. So when Serving GW change is required, the target Serving GW triggers the GW Session Control procedure to perform policy controlled functions, such as bearer binding, and also informs of RAT change if applicable and then updates the PDN GW. When source Serving GW is instructed to delete the bearers due to Serving GW relocation, it also ensures that PCRF association is removed between the BBERF in source Serving GW and PCRF. In case Serving GW is not relocated, the serving GW then updates PCRF with the updated QoS rules and session binding information which then triggers PDN GW updates for PCC as well. In case of GTP, since the policies are handled on path, these interactions are not needed between Serving GW and PDN GW, though PDN GW may trigger PCC procedures based on, for example, change in RAT type. Note that since the UE, the RANs as well as MME do not differentiate whether GTP or PMIP is used over S5/S8 reference points, there are no impacts on the main handover process due to the core network protocol selection process.

One aspect that is not supported during handover is dynamic protocol change (between GTP and PMIP) over S5/S8 reference points when handover may cause relocation of the Serving GW. The complexity of changing protocol 'on the fly' seems unnecessarily complex and the need for such dynamic change in the initial releases of LTE/EPS is not apparent yet. The situation can be made worse when a UE moves from one PLMN to another PLMN during handover and require both MME and Serving GW to be relocated, thus if at the same time a protocol change is also required which affects the PDN GW as well as PCRF.

Handover between E-UTRAN and other 3GPP access (GERAN, UTRAN) with S4-SGSN

When one has to cope with such wide deployment of GSM/UMTS and huge subscriber base (which implies number of terminals in the consumers hands) as well as radio network equipment (GERAN, UTRAN), minimizing impacts on the system when performing handover (and in overall functions) is a key aspect to overall success of EPS. Within 3GPP this type of intersystem handover was also developed during the specification of the 3G system. So the 'know how' and commitment were there to achieve the goal of an efficient intersystem handover. But the changes compared to 2G to 3G and 2G/3G to LTE/EPS is much more substantial both from radio and core network perspective, especially in the core network. Since the differences for the handover procedures compared to existing handovers are minimal, we will emphasize on the differences and any additional aspects that would benefit the readers' understanding of the process.

In case of IRAT handover where E-UTRAN is one radio access technology and EPC is the core network, the same principles apply as before, perform the preparation directed by the source radio network and in case of Inter RAT handover, it is always performed via the core network. The source RAN adapts the content and information flow that suits the target access network (this is crucial in case of handover from E-UTRAN to pre-Release 8 core networks), and as before the source decides to start the preparation of the handover. It is also source RAN that makes the final decision on the execution of the handover procedure.

Inter RAT HO is considered to be backwards handover where the radio resources are prepared in the target 3GPP radio access before the UE is commanded by the source 3GPP radio access to change to the target radio. The target access system is responsible for giving exact guidance for the UE on how to make the radio access there (such as radio resource configuration, target cell system information, etc.) during the preparation phase. Since the RAN connections are not established yet, the signalling and information transfer is done via the core network through the source radio access transparently to the terminal during the preparation phase of the handover.

In Inter RAT HO involving E-UTRAN access, core network entities involved during the preparation and transfer of data are the S4 SGSN and corresponding MME/Serving GW. Whether indirect forwarding is to be applied or not is configured in the MME and S4 SGSN as part of the operator specific data, the configuration indicated whether to perform indirect forwarding always, whether to perform indirect forwarding only for Inter PLMN Inter RAT handover or it is never performed.

The main aspect for this handover is that the MME to/from S4 SGSN relocation must always be performed in case of E-UTRAN to 2G/3G handover.

Handover between E-UTRAN and UTRAN and handover between E-UTRAN and GERAN are very similar; we have chosen to illustrate the procedure by using handover from E-UTRAN to UTRAN as the example. The following steps outline how the generic procedures above apply to handover between E-UTRAN and UTRAN.

- Once source eNodeB triggers the handover request to the MME, based on the information received, the MME determines that it is UTRAN Iu mode handover. It then selects the appropriate SGSN for the target RNC and initiates appropriate resource reservation process in the target system. MME is also responsible for mapping the EPS bearers that the source eNodeB has selected for handover to the PDP context bearers as applied to 2G/3G. This mapping is specified in 3GPP in order to provide consistent outcome during handover.
- In the target core network, the SGSN prioritizes the PDP context bearers and decides whether Serving GW needs to be relocated. If Serving GW is to be relocated, then SGSN selects the appropriate target Serving GW and triggers the appropriate resources allocation.
- SGSN also provides the target RNC with all relevant information provided by the source network in order to establish the radio resources required for handover. Once the target RNC has completed all necessary radio resource allocation towards the UE and user plane resources have been established for the Iu, target SGSN is informed of the completion of the process. At this point, the target RNC is prepared to receive user plane data from either the SGSN or Serving GW, depending on if direct tunnel is applied or not (in case of no direct tunnel, SGSN remains in the user plane path).
- Source MME and source eNodeB are then informed of successful handover resources reserved and preparation phase is completed.

- In case of indirect forwarding, the path is established by the SGSN in the target Serving GW and by the source MME in the source network.
- Once source MME informs source eNodeB about handover preparation completion, source eNodeB starts forwarding data and commands the UE to perform handover to the target RNC and provides the UE all necessary information provided by the target RNC. At this time, UE no longer receives/ sends any data via source eNodeB.
- The UE tunes into the target cell in which the radio bearers are established and the rest of the process continues in the core network where the target SGSN informs source MME of the handover completion when target RNC confirms it to the target SGSN. After that, the source MME and target SGSN needs to release any forwarding resources. The source MME also needs to release source Serving GW resources in case of Serving GW relocation and Serving GW updates the PDN GW with appropriate information to establish the user plane path between the RNC, SGSN, Serving GW and PDN GW.

Note that any EPS bearers that were not successfully transferred are deactivated by the SGSN and Serving GW and any downlink data are dropped in Serving GW. Also note that UE only re-establishes the bearers that were accepted by the target network for handover during preparation phase. Further information flows and much more detailed description of the handover procedure between E-UTRAN and GERAN/UTRAN are available in 3GPP TS 23.401 [23.401].

Handover for Gn/Gp-based SGSN

Interoperation scenarios for operating E-UTRAN with a PLMN maintaining Gn/ Gp SGSNs are supported for GTP-based S5/S8 interfaces. Thus, the PDN GW then acts as a GGSN supporting Gn/Gp interfaces towards the SGSN, the MME supports Gn interface towards SGSN. Note that HSS must also be able to work with Gr interface or supported via interworking function to enable interworking between S6a and Gr functions.

The main principles for handover between E-UTRAN and UTRAN/GERAN connected to Gn/Gp core networks are as follows:

- The handover procedures within E-UTRAN involving MME and Serving GW remain as in the case of using EPC core for GERAN/UTRAN.
- The handover procedures within GERAN/UTRAN involving existing GPRS procedures remain the same as currently specified without EPC.
- A handover from E-UTRAN to UTRAN/GERAN would imply executing Inter SGSN handover with the source SGSN represented by the MME for EUTRAN.

- A handover from UTRAN/GERAN to E-UTRAN would imply executing Inter SGSN handover with the source SGSN represented by the Gn/Gp SGSN for UTRAN/GERAN and target SGSN represented by MME for EUTRAN. In addition, the MME has to select the Serving GW and the Serving GW must also update the PDN GW with the appropriate S5/S8 related information.
- Serving GW establishes appropriate set up towards eNodeB.
- If indirect forwarding is to be performed, then MME performs the same procedure as for Inter EUTRAN handover described above.
- The PDN GW represents the GGSN function and the GGSN selected for the UTRAN/GERAN must be a PDN GW acting as a GGSN.
- Mapping between PDP context and EPS bearers as well as other parameters also handled in the appropriate entities (such as MME takes care of handling the security and QoS parameters).

Note that in order to allow for early implementation of UE and early deployment of networks there is an additional possibility to support mobility between E-UTRAN and UTRAN/GERAN. This mobility solution is based on RRC connection release with redirection information. The RRC connection release with redirection information is implemented in the Source eNodeB and does not require any additional support in the network. The redirection information points the UE towards GERAN or UTRAN where the UE will use existing routing area update procedure to recover the connection. This procedure has worse performance than the other handover solutions and may create a significant break in the connection. It may however have acceptable performance for early deployments with primarily less data only users.

Handover between GERAN and UTRAN access using S4 SGSN and GTP/PMIP protocol

For GERAN and UTRAN access, handovers have been developed and deployed for a long time, now using GPRS and GTP as its core protocol. During the SAE development process, there were no significant interest from any operators to develop PMIP or even for GTP-based 2G and 3G to access handover support when S4 SGSN is in use. But during the course of the standardization work, the interest in completing an EPC-only architecture for all 3GPP accesses including the ability to handover between 2G and 3G radio networks using EPC grew. The work did not require too much extra efforts from technical point of view, since it can be easily extended from the EPC support for 2G/3G. Specifications were developed such that S4 SGSN and Serving GW and PDN GW-based EPC architecture supports handover between GERAN and UTRAN as well as Intra GERAN and Intra UTRAN mobility with and without SGSN relocation and Serving GW relocation. The extra signalling required is to interact with Serving

GW in the appropriate time of the handover for bearer establishment; as well as in case of PMIP, the additional PCRF interactions are also required. 3GPP TS23.060 has described the procedural differences in a nut shell that includes appropriate message name used between the SGSNs following EPC messages; the EPS bearer is used as parameter and Serving GW and PDN GW are updated with appropriate bearer information as in the E-UTRAN to/from 2G/3G case.

12.4.3   Handover in EPS with non-3GPP accesses

12.4.3.1   Optimized handover for eHRPD access

For HRPD networks, there is a significant subscriber base already out there with major North American and Asian operators operating their networks. Even though the two technologies (one developed in 3GPP and the other in 3GPP2) have been competing over the last 20 years, the two bodies have also cooperated in many areas in order to develop common standards that are strategically important to operators overall, examples of these are IMS and PCC development. For CDMA a number of companies cooperated extensively inside and outside the standards forum in order to develop special optimized handover procedures between E-UTRAN and HRPD access. The resulting handover procedure has efficient performance and reduced service interruption. This work was brought into mainstream 3GPP standards under the SAE umbrella and further enhanced and aligned with the mainstream 3GPP work ongoing for SAE and thus produced the so-called Optimized Handover between E-UTRAN and HRPD. HRPD networks then became known as evolved HRPD (eHRPD) to highlight the changes required for interoperability and connectivity with EPC and E-UTRAN, though there are no changes to the actual radio network and its functions.

The optimized handover has been defined to work in two modes of operation: idle mode where the UE is idle (i.e. do not have any active radio connectivity in the system, ECM-IDLE in E-UTRAN and Dormant in HRPD) and active mode where the UE is active (i.e. active data transmission ongoing between the UE and the network). The actual handover is performed in two phases: preregistration phase where the target access and specific core network entity for the specific access (MME for E-UTRAN and HRPD Serving GW or HSGW for eHRPD access) is prepared ahead of time anticipating a possible handover (but there is no time association of how long a UE may be pre-registered in the system); and handover preparation and execution phase where the actual access network change occurs.

Note that in the early deployment of E-UTRAN in a CDMA network, it is considered more prevalent and thus important to support E-UTRAN to HRPD handover than the reverse direction, since it is assumed that the HRPD networks would have sufficient coverage to keep a user within HRPD system.

It should be noted that currently the HSGW only supports PMIP protocol (S2a interface) whereas the E-UTRAN access may be using either GTP or PMIP protocol and as such, the GRE keys must be provided to the HSGW even in case of GTP-based EPC for E-UTRAN access.

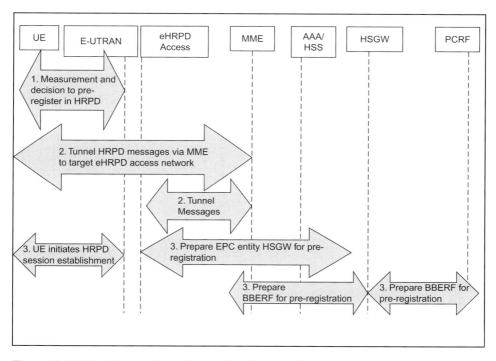

Figure 12.4.3.1.1 *Preparation and pre-registration overview.*

Figure 12.4.3.1.1 outlines the steps for the EUTRAN to HRPD access handover.

1. Some of the basic requirements on the UE and the E-UTRAN when supporting handover from E-UTRAN to HRPD networks are that the E-UTRAN provides the HRPD information (e.g. neighbouring cell information, CDMA timing and HRPD pre-registration control information) on the broadcast channel for UE measurement purposes. HRPD system information may also be provided to the UE by means of dedicated signalling. E-UTRAN is responsible for configuration and activation of HRPD measurements towards the UEs. The UE performs measurements in active mode when directed by the E-UTRAN network using information provided via the dedicated radio

signalling. Note that for idle mode, the UE performs cell reselection proce-dures in addition to being pre-registered in the target system prior to perform idle mode optimized mobility.

2. Once the UE decides to perform pre-registration to HRPD, it needs to tunnel the HRPD messages over E-UTRAN radio. The HRPD messages are encap-sulated in the appropriate uplink messages for pre-registration or for handover signalling and downlink messages for other HRPD messages. The handover signalling is given higher priority and the RAT type and other identifying information is also provided towards HRPD network for interpretation of the messages correctly. In order for the MME to select the correct target HRPD system, where the messages for that UE should be tunnelled to and also assist the HRPD network with the collected appropriate radio-related information, each eNodeB cell is associated with a HRPD Sector ID (also known as ref-erence cell ID). This Sector ID is provided to the MME during the message transfer over S1-MME. MME then uses this information to find the appropri-ate target HRPD entity and tunnels the messages over S101 towards that entity.

3. Based on the trigger from E-UTRAN radio network, the UE initiates estab-lishment of new session in eHRPD network. This process causes the HSGW to be connected with the HRPD access network and based on the informa-tion provided by the UE via the EPC, the HSGW also initiates the process of establishing connection with PCRF for non-primary BBERF connection in order to provide functions like bearer binding, QoS rules provisioning as the BBERF function is located in the HSGW for this access. At this point, HSGW has the latest bearer information, APN, PDN GW addresses, etc. for that user. HSGW acquires the information about the already allocated PDN GWs from HSS/AAA. When the source E-UTRAN makes the decision to trigger the UT to handover to target HRPD access, the UE starts sending the appropriate preparation messages to the HRPD access network via the tun-nels over E-UTRAN and MME.

Figure 12.4.3.1.2 outlines the steps for the EUTRAN to HRPD access handover completion phase:

1. Based on the measurement information, source E-UTRAN instructs the UE to perform handover to the HRPD access network, UE has already pre-registered to the HRPD network prior to any active mode optimized handover.

2. The UE initiates the procedures to establish traffic channel connection towards the HRPD access network over the E-UTRAN access, E-UTRAN forwards the message to the MME which then forwards the message to HRPD node via the already established S101 tunnel and also adds additional information such as uplink GRE keys, APNs and PDN GW addresses associated with each APN.

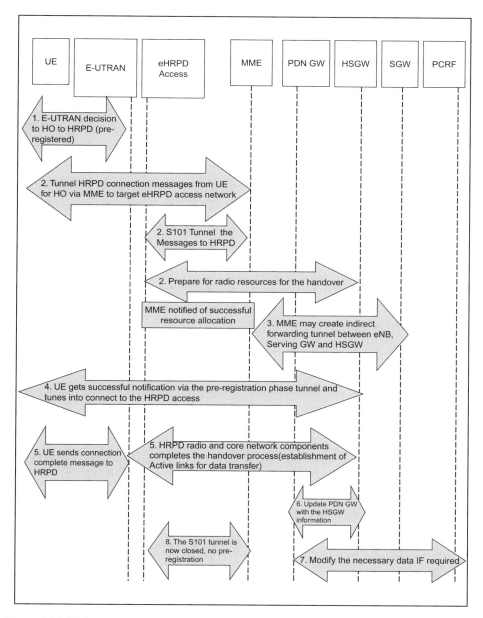

Figure 12.4.3.1.2 *Active mode handover from E-UTRAN to HRPD in the EPC.*

3. HRPD access network then establishes the necessary radio resources and requests HSGW for appropriate links and information in order to establish the connection between HRPD and HSGW. Once this process is completed, the MME is notified and then the MME may establish an indirect forwarding link between Serving GW and HSGW by sending the Serving GW the

necessary HSGW information. In this manner, there is a data forwarding path from eNodeB to Serving GW to HSGW.

4. UE is informed of successful resource allocation completion and the UE tunes to the HRPD access network. The UE no longer communicates via E-UTRAN access.

5. UE then sends a confirmation message directly to the HRPD network, at this time the UE has completely moved into HRPD access network. HRPD access network informs HSGW of UE's arrival and being prepared to receive/send data.

6. HSGW now establishes link with PDN GW where the PDN GW shall now forward data towards HSGW and not Serving GW.

7. PDN GW now interacts with PCRF function to receive any modified data for the new access.

8. The S10 tunnel is terminated for that UE between HRPD and MME access.

In case of multiple PDN connections existing for that UE, the HSGW must update the appropriate PDN GWs for each PDN connection individually.

The release of E-UTRAN resources is consistent with E-UTRAN Inter RAT handover scenario; note that the PDN GW can initiate resource release towards Serving GW anytime after it had successfully established the link with HSGW. Once MME receives confirmation from HRPD network about successful completion of handover, it initiates the release of UE context towards E-UTRAN.

MME may have timers to trigger the Serving GW resources for the E-UTRAN bearers and the indirect forwarding tunnels to be released as in Inter RAT handover case. When MME does trigger this release, it also indicates to Serving GW not to release resources towards the PDN GW.

In case of handover from HRPD to E-UTRAN access, the E-UTRAN access does not play any role in directing or assisting the UE. Due to possible impacts of HRPD radio access part, it is unclear during the writing of this book, if the HRPD RAN would assist in any way for this handover or it will be up to the UE to decide when to perform the handover based on measurements available to it. The steps are very similar to the E-UTRAN to HRPD handover with some differences due to the access network type (Figure 12.4.3.1.3).

The procedure is briefly described in the following steps.

1. The decision to handover to E-UTRAN is made by the UE. The UE is in active mode in HRPD network connected via EPC. Note that the UE may

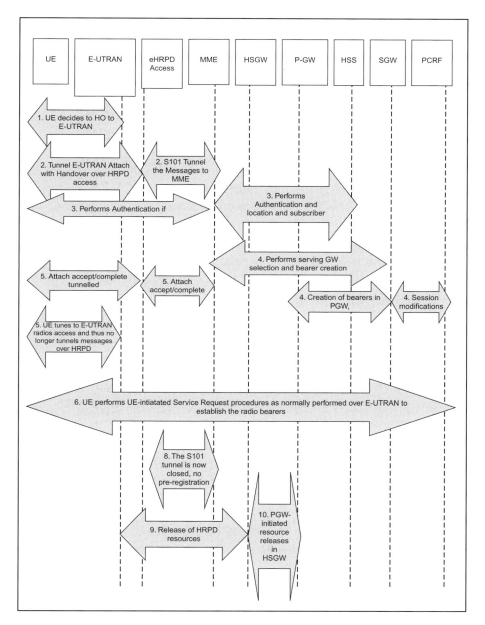

Figure 12.4.3.1.3 *HRPD to E-UTRAN active mode handover.*

choose to leave HRPD and execute E-UTRAN attach procedure directly over E-UTRAN, in such case the Attach type is not set to Handover.

2. UE sends a NAS message of Attach to E-UTRAN with indication set to Handover which is tunnelled over HRPD access network and then forwarded to MME via the pre-established S101 tunnel. The HRPD access node

determines the MME and the TAI using the Sector ID to MME mapping function located in HRPD network.

3. Based on E-UTRAN attach process, MME needs to make the appropriate decision of whether to perform authentication or not. MME also performs location update procedure in order to update the HSS with user information as well as retrieve the user's subscription data.

4. As in normal E-UTRAN attach procedure, the Serving GW is selected for that UE and bearer creation process for default bearer is initiated with an indication of handover. Serving GW triggers PDN GW to create the necessary information for the default bearer establishment. PDN GW performs this task and also triggers associated, dedicated bearer establishment procedures.

5. Once MME is informed of the bearer establishment procedure completion in the GWs, MME completes the attach procedure towards the UE via the HRPD access network and associated established tunnels. UE then completes the attach procedure over the HRPD access.

6. Once the attach procedure is completed, the UE moves to E-UTRAN access.

7. UE then performs necessary radio connection establishment via the service request procedure which triggers the establishment of radio links as well as the S1-user plane set up and then start the UE-initiated bearer establishment procedures to complete the process. For any bearers that could not be established during this process, the UE removes the resources for them.

8. Once the bearer establishment procedure is complete, the MME informs HRPD access via S101 tunnel that handover is complete.

9. HRPD access network initiates its resources release procedure.

10. PDN GW triggers the resources to be released towards HSGW anytime once it has established the Serving GW and PCRF relationship for the E-UTRAN access.

If there are multiple PDN connections to be established, the UE must initiate them via UE-initiated PDN connectivity process as described in Section 6.3.

What is evident is that by pre-establishing some of the connections within the target access network anytime prior to the actual handover itself, the execution of the actual handover process is reduced significantly during E-UTRAN to HRPD scenario. Even though the benefits may not be as significant for the reverse direction, for the consistency and simplicity of the network behaviour, the process has been kept consistent.

Even though this section focuses on active mode handover, it may benefit the readers to be exposed to the idle mode handover process between E-UTRAN and HRPD.

In case of E-UTRAN to HRPD direction, triggered by cell reselection process, the idle UE based on triggers internal or from E-UTRAN access, selects the HRPD access network where the UE is in dormant state due to either pre-registration or previous attachment to that access. Then the UE follows the HRPD procedures to connect to the access and in the core network, the HSGW establishes appropriate bearers towards the PDN GW and PCRF when applicable. PDN GW then triggers the release of the resources in the E-UTRAN access.

In case of HRPD to E-UTRAN direction, when the UE determines to perform an attach procedure to E-UTRAN access, the process it follows is the same as the active mode handover process until it tunes to the E-UTRAN access. Then the UE performs a TA update procedure over the E-UTRAN access. The resource release process is the same as in the reverse direction, triggered by PDN GW.

Basic non-optimized handover with non-3GPP access

When considering non-optimized handover between and within non-3GPP accesses using EPC, the main requirement is to be able to preserve the IP address(es) and maintain the IP connectivity/service continuity when handover is occurring. Since the handover decision is determined and executed from the UE, and there is no coordination between the access networks, this procedure is considered non-optimized compared to the optimized handover procedure between E-UTRAN and HRPD access.

Due to the number of protocol options as well as capability to support both host-based and network-based mobility, the handover procedure can get quite complex in terms of selecting the right combination to ensure that IP address preservation and session continuity is possible. As the reader may be by now already familiar with the IP mobility management selection process as described already in previous sections of this book, this plays a significant role in the handover process as well.

In case of handover between 3GPP access and non-3GPP access and between non-3GPP accesses, IPMS function performs the decision of how the IP connectivity shall be performed (i.e. preservation possible or not). In case of network-based mobility (i.e. PMIP), the PMIP protocol supports these functions and the UE capability is either indicated by the UE explicitly at handover or based on pre-configured information.

In case of host-based mobility, the decision is slightly differently taken. The decision may take place if the network is aware of the UE capability to support DSMIPv6 or MIPv4. This information may be acquired by the target non-3GPP access from the HSS/AAA (e.g. in case of DSMIPv6, the UE performed S2c bootstrap before moving to the target non-3GPP access). If the IP mobility management protocol selected is DSMIPv6, the non-3GPP access network provides the UE with a new IP address, local to the access network. In an untrusted non-3GPP access, the terminal also has to set up an IPSec tunnel with the ePDG. In these cases, in order to get IP address preservation for session continuity, the UE has to use DSMIPv6 over S2c reference point. The local IP address, allocated either by the trusted non-3GPP access network or the ePDG, is then used as a Care-of Address for DSMIPv6 within EPS. If the IP mobility management protocol selected is MIPv4, the address provided to the UE by the trusted non-3GPP access network is a FACoA and IP address preservation is performed over S2a using MIPv4 FACoA procedures. Note that MIPv4 is not supported when an ePDG is used. A basic handover flow when using DSMIPv6 is provided below. If the reader is interested in further details on the handover flows using host-based mobility protocols, then we recommend further detailed reading of 3GPP specifications, TS23.402 and TS24.303. Further details on the basic operation of DSMIPv6 are also described in Section 11.3.

As can be understood from IP Mobility Selection function, the protocol choice made by the UE and the protocol choice supported by the network requires to be in sync in order to perform handover, the UE indicates at the attach procedure that the attach function is being performed due to handover and it may also indicate its preference of host- or network-based mobility and the protocol choice for host-based mobility case. Note that in case of network-based mobility in 3GPP, the UE is unaffected by the network choice of protocol (GTP or PMIP). If UE does not indicate any preference, then PMIPv6 is the selected protocol and based on the PMIP protocol principles, there are two ways of making the decision of preserving the IP and maintain session continuity or not. An operator may configure its local policies at PDN GW whether to preserve the IP address based on a timer which allows the existing IP address to be maintained IF and only if the source/old access system tears down its connection before the timer expires and then assigns a new IP prefix, or assigns a new IP prefix immediately and thus no IP address preservation is performed.

In addition, the UE may also use information made available to it from ANDSF regarding its home operator's preference and other policies to assist it to select the preferred access network during handover.

Now let us focus on the scenario that has generated wider participation and opened operators' interest during the development and standardization of SAE (i.e. S2a interface using PMIPv6). This is also the default system behaviour if the operators choose not to configure any specific IPMS process in the UE and in the EPC network. Note that multiple PDN connectivity with different mobility protocols for the same UE is not supported, that means, for example, it is not possible for a UE that is connected to single or multiple PDNs over a 3GPP access to perform a handover to a non-3GPP access and then use different mobility protocols for the PDNs connected over non-3GPP access. We consider the flow in Figure 12.4.3.1.3 to cover the most likely scenario to be seen in the deployed network, where a handover is triggered from a trusted non-3GPP

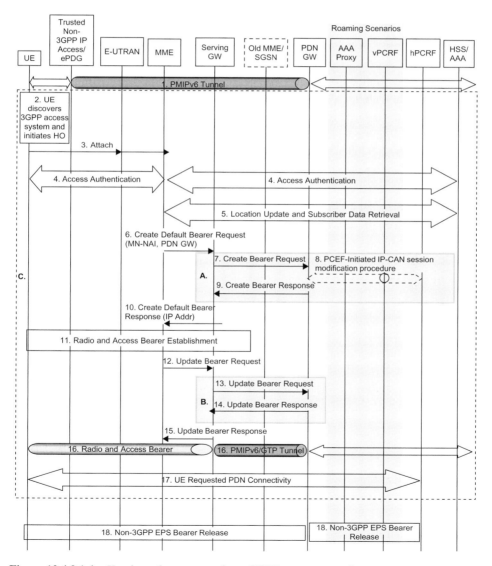

Figure 12.4.3.1.4 *Handover from a trusted non-3GPP access network.*

access network with S2a supporting PMIP to GTP-based (S5/S8 interface) EPC network (Figure 12.4.3.1.4).

The handover steps can be described as follows:

- *Steps 1–3*: The UE is connected to a non-3GPP access trusted by the operator and in active mode, the UE detects the E-UTRAN access and determines to handover (or rather transfer the sessions, as it is more a transfer of ongoing active sessions than 'handover' in the true meaning of this word as used in the existing mobile systems) to E-UTRAN from its current serving access system based on policies and other information available to the UE. Note that the UE is connected to an EPS network and has now moved to E-UTRAN access to perform an attach. The UE sends an Attach Request to the MME over the E-UTRAN access with attach type indicating 'Handover attach'. This attach procedure is handled as in case of normal attach in E-UTRAN and the UE should also include one of its APNs in this message.
- *Steps 4–6*: The MME may perform authentication of the UE as per E-UTRAN access via HSS. Once the authentication is successful, the MME continues as for an E-UTRAN access where it may perform location update procedure and subscriber data retrieval from the HSS. The MME receives information on the PDNs the UE is connected to over the non-3GPP access in the subscriber data obtained from the HSS. The MME selects an APN, either the default APN or the APN provided by the UE. Since the attach type sent by UE is 'Handover', the MME selects the PDN GW provided as part of the subscription data from the HSS. The MME then continues to select a Serving GW. The MME sends a Create Default Bearer Request to the selected Serving GW and includes the PDN GW address and handover indication.
- *Steps 7–10*: The Serving GW sends a Create Default Bearer Request with handover indication to PDN GW causing the PDN GW not to switch the tunnel from non-3GPP IP access to 3GPP access system at this point. PDN GW also interacts with PCRF to obtain the rules for the network IP-CAN and PDN connection for all established bearers due to handover. Due to the handover, the PDN GW stores the new PCC rules for E-UTRAN access as well as maintains the old PCC rules for the trusted or untrusted non-3GPP IP access and still applies the old PCC rules for charging. PDN GW returns the UE's IP address/prefix assigned for the non-3GPP access it is handing over from. This information is then passed onto MME indicating successful bearer establishment and set up of S5 tunnel establishment. Additional dedicated bearers may also be established during this process by PDN GW as in case of normal attach.
- *Step 11*: Radio and access bearers are established for E-UTRAN access as in the normal attach case.

- *Steps 12–15*: The MME sends an Update Bearer Request (eNodeB address, eNodeB TEID, handover indication) message to the Serving GW. Based on the presence of the handover indication, the Serving GW sends an Update Bearer Request message to the PDN GW to prompt the PDN GW to tunnel packets from non-3GPP IP access to 3GPP access system and immediately start routing packets to the Serving GW for the default and any dedicated EPS bearers established. PDN GW can now route the packets to E-UTRAN access and stop data transfer to non-3GPP access.
- *Step 16*: The UE sends and receives data at this point via the E-UTRAN system.
- *Step 17*: For connectivity to any remaining PDNs from the old non-3GPP access, the UE establishes connectivity to each PDN, by executing the UE requested PDN connectivity procedure.
- *Step 18*: The PDN GW initiates resource allocation deactivation procedure in the non-3GPP IP access.

In case the E-UTRAN access is using PMIP-based EPC, the difference in signalling sequences is illustrated by the steps (A) and (B) where dynamic PCC interaction based on off-path policy control related signalling are executed as seen in general where the Serving GW interacts with PCRF. In the case of PMIP-based S5/S8, instead of a Create Bearer Request and Update Bearer Request, the PBU/PBA is sent from the Serving GW to the PDN GW.

In case of multiple PDN connections need to be established, the UE-initiated PDN connectivity procedures are executed either in sequence or in parallel in order to establish these additional PDN connections.

In case the handover is performed towards a 2G/3G 3GPP access network, the procedures are very similar in the sense that the attach and PDP context activation procedures are executed according to the specific 3GPP access itself with the handover indication in the PDP context activation procedure which allows the Serving GW and PDN GW to appropriately preserve the IP address/prefix and handle the sessions like E-UTRAN case.

Next, we will illustrate a handover scenario from a 3GPP access to a non-3GPP access where S2c (i.e. DSMIPv6) is used in the target non-3GPP access (see Figure 12.4.3.1.4). The session starts in a 3GPP access (e.g. E-UTRAN) where either GTP or PMIP is used on the S5/S8 reference point. The session is then handed over a non-3GPP access. The IP mobility mode selection is made resulting in that DSMIPv6 is used. The terminal thus receives a local IP address in the target non-3GPP access. If the access is treated as untrusted, the terminal in addition sets up an IPSec tunnel towards the ePDG. In that case, the terminal also receives another local IP address from the ePDG. The terminal then invokes DSMIPv6 to the PDN GW in order to maintain the IP session. Note that the same PDN GW has to be used for the target access as in source access in order to maintain IP session continuity.

The procedure is shown in Figure 12.4.3.1.5 and is briefly described in the following steps.

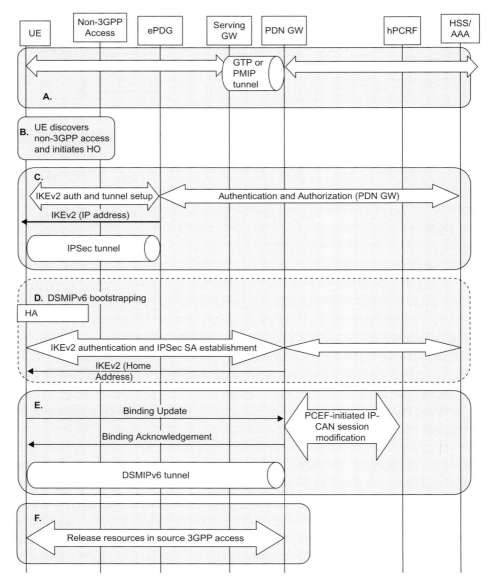

Figure 12.4.3.1.5   *Handover from 3GPP access to untrusted non-3GPP access.*

A. The terminal is attached over a 3GPP access, for example E-UTRAN.
B. The terminal discovers a non-3GPP access, for example WLAN, and decides to hand over the session.
C. If the decision is that this is an untrusted non-3GPP access, the terminal needs to establish an IPSec tunnel towards an ePDG. In this case, the terminal discovers an IP address of a suitable ePDG using DNS and initiates an IKEv2 procedure to authenticate and set up an IPSec SA. If successful, this results in that an IPSec tunnel can be established between UE and ePDG. The ePDG also allocates a local IP address and delivers it to the UE. The plane is from now on tunnelled in the IPSec tunnel, including the DSMIPv6 signalling in later steps.
D. If not done already, the terminal performs DSMIPv6 bootstrapping. This includes discovering a suitable PDN GW (acting as Home Agent) and performing the IKEv2 procedure with that PDN GW to set up an IPSec Security Association for DSMIPv6. The PDN GW returns the same IP address as was used in the source 3GPP access.
E. Then the terminal sends a Binding Update to the PDN GW to perform the actual user plane path switch from source to target access. The PDN GW informs the PCRF about the new access type and replies with a Binding Acknowledgement. A bidirectional IP-in-IP tunnel is now set up between UE and PDN GW. The terminal can continue its IP sessions using the same IP address as was used in the source 3GPP access. Note that the Binding Update, Binding Acknowledgement and the DSMIPv6-tunnelled user plane are all transported within the IPSec tunnel between terminal and ePDG.
F. The PDN GW informs the source 3GPP access that the terminal has handed over to a non-3GPP access. The resources in the 3GPP access can now be released.

The message flow above illustrates a handover from a 3GPP access to an untrusted non-3GPP access. In case of handover from a 3GPP access to a trusted non-3GPP accesses using DSMIPv6 the procedure is similar with a few key differences:

• The tunnel setup with ePDG in step B is not preformed. Instead, step B is replaced by an access authentication in the trusted non-3GPP access. The non-3GPP access also establishes a gateway control session with the PCRF via Gxx as part of step B.
• When the path switch has taken place in step E, the PCRF may provide the trusted non-3GPP access with updated QoS rules via the Gxx reference point.

12.5 Bearer and QoS-related procedures

12.5.1 General

As have been described in Section 6.2, the 3GPP accesses use the concept of a 'bearer' to manage the QoS between the UE and the PDN GW. For E-UTRAN, the EPS bearer extends between the UE and the PDN GW (for GTP-based system) and the UE and Serving GW (for PMIP-based system). For GERAN and UTRAN, the bearers are implemented as PDP contexts between the UE and the SGSN. When GERAN and UTRAN are connected to the EPS, the PDP contexts get mapped to EPS bearers between SGSN and the PDN GW or Serving GW.

Each EPS bearer is associated with a well-defined QoS class described by the QCI as well as packet filters that determine which IP flows gets transported over that particular bearers. Certain bearers also have associated GBR and MBR parameters.

The bearers can be dynamically established, modified or removed depending on the needs of applications being used by the UE. In E-UTRAN, the dedicated bearer procedures are always initiated by the network. The UE may send a request for certain resources (QCI, bit rates, packet filters) to the network and this request can result in bearer operations being initiated by the network. In GERAN/UTRAN, the PDP context procedures are initiated by the network or by the UE.

Other accesses not defined by 3GPP, such as HRPD or WiMAX, may have similar procedures for allocating resources in the access network. The terminology may be different but the purposes of the procedures are roughly the same; that is, to set up a (logical) transmission path between the UE and the network to transport traffic with certain QoS requirements. The mechanisms and procedures used between the UE and the access network are specific to each access and will not be described in detail in this book. Using the PCC architecture, EPS is able to interwork with such access-specific procedures.

In this section, we will describe some of the dynamic procedures available in EPS to handle bearers for 3GPP accesses and how to interwork with the access-specific procedures available in other accesses.

12.5.2 Bearer procedures for E-UTRAN

12.5.2.1 Dedicated bearer activation

The dedicated bearer activation procedure is used when the network decides that a new dedicated bearer needs to be established. The procedure is initiated by the PDN GW (when using GTP-based S5/S8) or the Serving GW (when using PMIP-based S5/S8). The trigger is typically that the PDN GW/Serving GW has received new PCC/QoS rules from the PCRF that require a new dedicated bearer to be established. The reason for why the PCRF provided new PCC/QoS rules can, for example, be due to an Rx interaction, or that the UE has sent a UE-initiated resource request that was provided to the PCRF.

The procedure is shown in Figure 12.5.2.1.1 and is briefly described in the following steps.

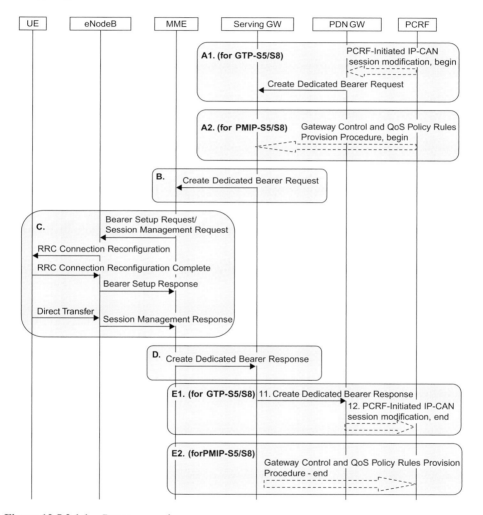

Figure 12.5.2.1.1   *Bearer procedures.*

A. The PCRF makes a policy decision because it received session information from an AF over the Rx interface or because it received a request for resources from the access network.
   - For a GTP-based system (A1), the PCRF sends the new PCC rules to the PDN GW. Based on the received PCC rules, the PDN GW decides to activate a new dedicated bearer and sends a Create Dedicated Bearer Request to the Serving GW.

- For a PMIP-based system (A2), the PCRF sends the new QoS rules directly to the Serving GW. Based on the received QoS rules, the Serving GW decides to activate a new dedicated bearer.

B. The Serving GW sends a Create Dedicated Bearer Request to the MME.

C. The MME sends a command to the eNodeB to initiate the appropriate E-UTRAN procedures to establish an appropriate radio bearer. The appropriate reconfiguration of the RRC connection between UE and eNodeB is performed.

D. The MME acknowledges the bearer activation to the Serving GW by sending a Create Dedicated Bearer Response (EPS Bearer Identity, S1-TEID) message.

E. The Serving GW acknowledges the dedicated bearer setup.

- For a GTP-based system (E1), the Serving GW sends a Create Dedicated Bearer Response to the PDN GW. The PDN GW sends an acknowledgement to the PCRF.

- For a PMIP-based system (E2), the Serving GW sends an acknowledgement directly to the PCRF.

12.5.2.2 UE-initiated resource request, modification and release

As described in Section 8.1, there are two concepts for how QoS is allocated in the NW, either triggered by the network or triggered by an explicit request from the UE. The procedure described in this section supports the latter scenario where, for example, an application in the UE would like to have premium QoS and triggers the E-UTRAN interface in the UE to make a corresponding request to the network. The procedure can also be used when the UE wants to modify or release a previously granted resource. If accepted by the network, the request invokes either the dedicated bearer activation procedure, the dedicated bearer modification procedure or a dedicated bearer is deactivated. The flow diagram for this procedure is illustrated in Figure 12.5.2.2.1.

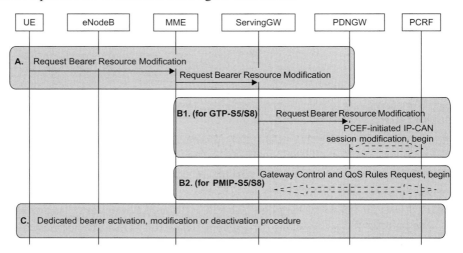

Figure 12.5.2.2.1 *UE-initiated resource request, modification and release.*

The procedure is briefly described in the following steps.

A. Based on, for example, application request, the UE sends a request for resource modification to the network. The UE may include packet filter information for the packet flows that are to be added, modified or deleted as well as associated QoS information (requested QCI and potentially a requested GBR). The MME forwards the request to the Serving GW.
B. For GTP-based system (B1), the request is forwarded to the PDN GW that in turn informs the PCRF about the request. For a PMIP-based system, the Serving GW informs the PCRF about the request, without involving the PDN GW.
C. The PCRF makes a policy decision and provides the policy decision to the PDN GW and potentially the Serving GW. Depending on the new, modified or deleted PCC/QoS rules, the PDN GW (for GTP-based S5/S8) or the Serving GW (for PMIP-based S5/S8) invokes the appropriate EPS bearer procedures.

12.5.3 Bearer procedures for GERAN/UTRAN

12.5.3.1 UE-initiated secondary PDP context establishment

When the UE is camped on GERAN or UTRAN, it uses the secondary PDP context activation procedure in order to establish a new PDP context for the same PDN connection. The procedure is triggered by the UE for similar reasons as the UE-initiated bearer resource modification procedure over E-UTRAN (see above).

The S4-SGSN maps between the PDP context procedures used over GERAN/UTRAN and the EPS bearer procedures used towards the Serving GW and PDN GW. As can be seen below, the S4-SGSN maps from the secondary PDP context activation/modification procedure to a UE-initiated bearer resource modification procedure over S4.

The procedure is shown in Figure 12.5.3.1.1 and is briefly described in the following steps.

A. Based on, for example, application request, the UE sends a secondary PDP context activation request. The S4-SGSN maps this into a UE-initiated bearer resource modification request and sends it to the Serving GW.
B. For GTP-based system (B1), the request is forwarded to the PDN GW that in turn informs the PCRF about the request. For a PMIP-based system, the Serving GW informs the PCRF about the request, without involving the PDN GW. The PCRF makes a policy decision and replies to the PDN GW or the Serving GW.
C. Depending on the new, modified or deleted PCC/QoS rules, the PDN GW (for GTP-based S5/S8) or the Serving GW (for PMIP-based S5/S8) invokes the appropriate EPS bearer procedures. In this case, the UE has requested a new PDP context; this means that the PDN GW (or Serving GW in PMIP

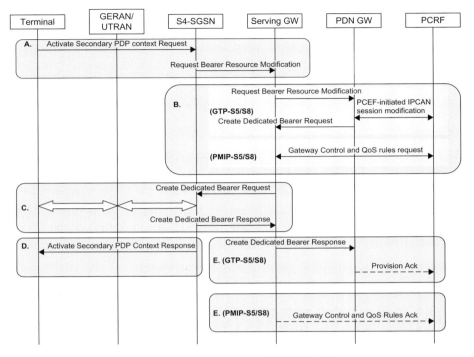

Figure 12.5.3.1.1 *Secondary PDP context activation (UE-initiated).*

case) must initiate the establishment of a new dedicated EPS bearer in order to maintain the 1:1 mapping between PDP contexts and EPS bearers. When the S4-SGSN receives the Create Dedicated Bearer Request, it initiates the appropriate GERAN or UTRAN procedures. The S4-SGSN also completes the procedure towards the Serving GW.

D. The S4-SGSN responds to the UE with an Activate Secondary PDP Context Response.

E. The Serving GW completes the dedicated bearer procedure.

12.5.3.2 Network requested secondary PDP context activation

The network requested secondary PDP context activation procedures for GERAN/UTRAN correspond to the network-initiated dedicated bearer procedures for E-UTRAN. This means that it is the network that decides to establish a new PDP context. There is however one key difference between the GERAN/UTRAN and the E-UTRAN procedures. With E-UTRAN, the network-initiated bearer procedures are the 'native' bearer procedures. A UE-initiated bearer resource modification request over E-UTRAN is not a bearer procedure as such, but may trigger a network-initiated bearer procedure. With GERAN/UTRAN, it is the other way around. The secondary PDP context procedures are UE-initiated as illustrated in the previous section. As shown in this section, the network may trigger such a procedure by sending a request to the UE for activating a secondary PDP context.

Similar to the UE-initiated procedures in the previous section, it is the SGSN that maps between PDP context procedures and EPS bearer procedures.

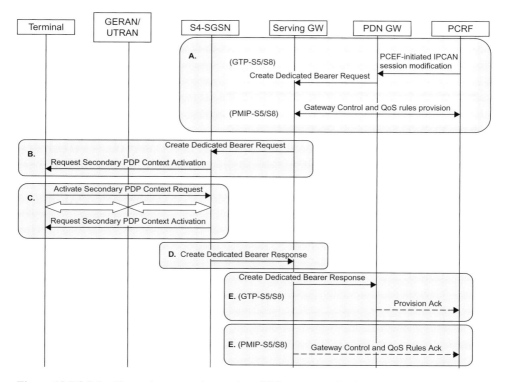

Figure 12.5.3.2.1 *Network requested secondary PDP context activation*

The procedure is shown in Figure 12.5.3.2.1 and is briefly described in the following steps.

A. Based on trigger from the PCRF, the PDN GW (for GTP-S5/S8) or Serving GW (for PMIP-S5/S8) decides to establish a new dedicated bearer. In case of GTP-S5/S8, the PDN GW sends a Create Dedicated Bearer Request message to the Serving GW. In case of PMIP-S5/S8, the PCRF interacts directly with the Serving GW.
B. The Serving GW sends the Create Dedicated Bearer Request message to the S4-SGSN. The S4-SGSN maps this into request fro the UE to activate a secondary PDP context.

C. The UE initiates the secondary PDP context activation procedure. The appropriate RAN procedures are performed and the S4-SGSN responds to the UE.
D. The S4-SGSN also responds to the Serving GW when the dedicated bearer (i.e. the secondary PDP context) has been activated.
E. For GTP-based S5/S8, the Serving GW forwards the response to the PDN GW that in turn may inform the PCRF. For PMIP-based S5/S8, the Serving GW responds to the PCRF directly.

12.5.4   Interworking with dynamic QoS mechanisms in other accesses

12.5.4.1   Network-initiated resource provision

The QoS procedures in the non-3GPP access network depend on the particular access technology being used. EPS defines generic procedures that can be used to interwork with different non-3GPP accesses. In this section, we illustrate a network-initiated QoS reservation procedure, similar to the network-initiated dedicated bearer and PDP context activation procedures illustrated above. The interface between the UE and the access network is however not described since it varies between the accesses (Figure 12.5.4.1.1).

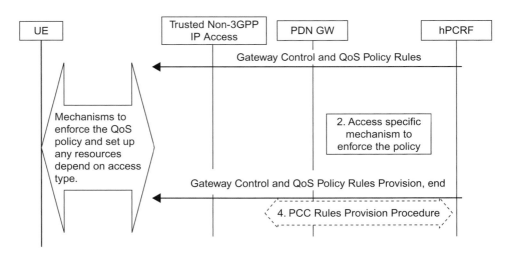

Figure 12.5.4.1.1   *Network-initiated resource provision.*

12.6   Single radio voice call continuity

As described in Chapter 5, the SRVCC is a solution that addresses the problem that there may not be full coverage for MMTel VoIP services used on E-UTRAN.

SRVCC solves this problem by offering a mechanism where the UE performs a coordinated radio level handover in combination with a change from IMS VoIP to circuit-switched voice using IMS procedures for service continuity.

The details of SRVCC and the related procedures spans outside the EPS into IMS and the CS core network. This chapter will outline the SRVCC procedure and the impact on the EPC network on a high level without going too far into the details of other parts of the system. The 3GPP TS 23.216 [23.216] elaborates more on the SRVCC procedure and impacts the EPC while the 3GPP TS 23.237 [23.237] details how the IMS handles service continuity.

The solutions for SRVCC towards GERAN/UTRAN and SRVCC towards CDMA are not exactly the same due to the differences in the interworking of EPS with CDMA and GERAN/UTRAN, respectively.

12.6.1   SRVCC from E-UTRAN to GERAN or UTRAN

SRVCC impacts the eNodeB, the MME and the MSC Server and requires service continuity support in the IMS. A new interface, Sv, has been specified between the MME and the MSC to execute the actual handover. Prerequisites for SRVCC are that both the UE and the EPS are SRVCC capable and that the session is anchored in a Service Centralization and Continuity Application Server (SCC AS) in the IMS. To illustrate the SRVCC functionality, we will examine an example of a call between two UE's where UE-A will experience an SRVCC. A simplified picture of the call path between two UE's before SRVCC is performed is shown in Figure 12.6.1.1.

In the example, the UE-A and UE-B are using IMS voice telephony to communicate. Only the most relevant nodes are shown. The IMS signalling traverses the IMS control nodes and is anchored in an SCC AS in UE-A's network. The IMS voice packets are sent end-to-end between the UEs using the EPS IP connectivity through eNodeB and Serving/PDN GW (the details on how the IMS voice packets are sent to UE-B are note relevant for this discussion).

When UE-A moves in the network, it may reach a point where there the LTE coverage will diminish and an SRVCC handover must be performed to GERAN CS or UTRAN CS. When the LTE signal strength decreases, the UE will measure on neighbour cell to discover suitable HO candidates and report the measurements to the eNodeB. Based on the bearer QCI and configured knowledge of to which cells SRVCC is necessary, the eNodeB can trigger an SR VCC HO if PS handover is not possible.

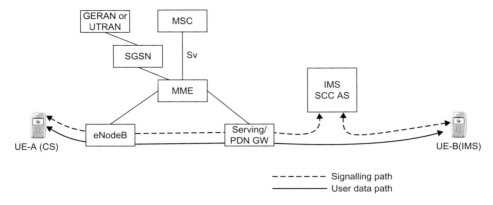

Figure 12.6.1.1   *Before SRVCC execution.*

The MME is informed by the eNodeB that SRVCC HO is needed via the hand-over request from eNodeB with the indication that this is for SRVCC handling. The MME then triggers the SRVCC procedure with the MSC Server via the Sv interface (the address of the MSC is provisioned to the MME). The MME also passes a special routing number that allows the MSC Server to establish a session with the IMS SCC AS. The MSC Server initiates the session trans-fer procedure to IMS and coordinates it with the CS handover procedure to the target cell. The MSC Server sends a response to the MME, which includes the necessary CS HO command information for the UE to access the target UTRAN/GERAN cell. The MME sends a handover command to the eNodeB that passes it to the UE. As a next step, the UE tunes into the GERAN/UTRAN cell and in principle experience a CS HO in the target cell where the UE will receive a CS telephony bearer. After the HO, the UE-A will be connected with CS telephony to the MSC Server. The MSC Server interworks the CS teleph-ony used by UE-A to the IMS MMTel voice service used by UE-B as shown in Figure 12.6.1.2.

IMS voice telephony can also be supported on WCDMA/HSPA. Hence, a simi-lar solution for SRVCC from IMS voice telephony to GERAN or UTRAN CS telephony is also specified in 3GPP. The only differences are that the SGSN takes the role of the MME and UTRAN performs the same functionality as the eNodeB in E-UTRAN.

For more details on the SR VCC solutions see 3GPP TS 23.216 [23.216] and TS 23.237 [23.237].

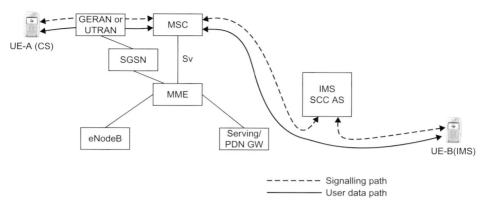

Figure 12.6.1.2   *After SRVCC handover.*

12.6.2   *SRVCC from E-UTRAN to 1xRTT*

The architecture for SRVCC between IMS voice telephony on E-UTRAN to CS telephony on 1xRTT is shown in Figure 12.6.2.1.

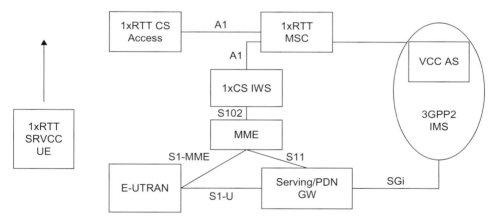

Figure 12.6.2.1   *SRVCC to 1xRTT architecture.*

The architecture for SRVCC between IMS voice telephony on E-UTRAN to CS telephony on 1xRTT looks similar to the one for SRVCC to GERAN and UTRAN, but there are a few key differences. Instead of an Sv interface between the MME and the interface to trigger handovers, there is a different type of mechanism employed. The UE communicates directly with the 1xRTT MSC using NAS message tunnelling. The MME is in principle just a signalling relay and the interworking function tunnels the signalling messages and interworks them towards the 1xRTT MSC. This allows the UE to communicate with the

1xRTT MSC to trigger the service continuity procedure in IMS and to prepare the access before performing the handover on the radio layer. For details on the SRVCC solution for 1xRTT, refer to 3GPP TS 23.216 [23.216] and 3GPP2 TS X.S0042-0 [X.S0042].

12.7 CS fallback

The main idea behind CS fallback is to allow UEs to camp on LTE and utilize the LTE for data services but reuse the GSM, WCDMA or CDMA network for circuit-switched voice services. The CS fallback feature was described briefly in Chapter 5.

To support CS fallback, there is a set of modifications of existing procedures and also some additional CS fallback specific procedures added to EPS. This section intends to illustrate the main principles of the CS fallback procedures by outlining an example of a CS fallback call.

There are special additions to the attach and TA update procedures which activates an interface, called SGs, between the MME and the MSC. This interface is used by the MSC to send paging messages for CS calls to the UE camping on LTE.

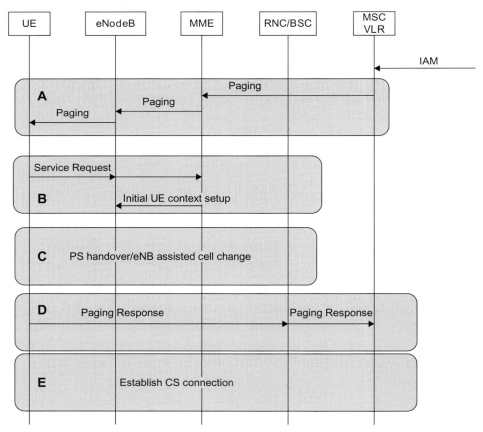

Figure 12.7.1 *Mobile terminating call in idle mode.*

The procedure is shown in Figure 12.7.1 and is briefly described in the following steps.

A. The MSC receives an incoming voice call and sends a CS page to the MME over a SGs interface. The MME uses the TMSI (or IMSI) received from the MSC to find the S-TMSI (which is used as the paging address on the LTE radio interface). The MME forwards the paging to the eNodeB in the TAs where the UE is registered. The eNodeBs perform the paging procedures in all the cells in the indicated TAs. The paging message includes a special CS indicator that informs the UE that the incoming paging is for a terminating CS call.

B. At the reception of the paging message, the UE performs a service request procedure which establishes the RRC connection and sends the Service Request to the MME. The Service Request message includes a special CS fallback indicator that informs the MME that the CS fallback is needed. This triggers the MME to activate the bearer context in the eNodeB with an indication to perform fallback to GERAN or UTRAN.

C. The eNodeB selects a suitable target cell, possibly by triggering the UE to send measurements on neighbour cells, and initiates a handover or cell change procedure. The selection between handover or cell change procedure is based on the target cell capabilities and is configured in the eNodeB.

D. After handover/cell change, the UE detects the new cell and establishes a radio connection and sends a page response to the MSC, via the target RAN.

E. When the page response arrives at the MSC, a normal mobile terminated call setup continues and the CS call is activated towards the UE.

The CS fallback specifications cover all necessary procedures to support fallback to GSM, WCDMA and 1xRTT for mobile originated and mobile terminating calls in both idle and active mode. The procedures for CS fallback are specified in 3GPP TS 23.272 [23.272].

Even though CS fallback was primarily introduced to support voice calls it also supports other CS services. In particular the support for SMS has been specified in such a way that it does not require the UE to switch to another radio interface. The UE can remain on LTE and still send and receive SMS. The SMS messages are tunnelled between the UE and the MSC through the MME NAS signalling and the SGs interface.

Part V
Conclusion and Future of EPS

13

Conclusions and looking ahead

It is clear that the SAE work and the development of the EPC specifications is a major achievement carried out by 3GPP and its partners, involving the global mobile industry community. This is due to the fact that the SAE work has been targeting a significantly broader scope than previous 3GPP releases, extending the functionality of the 3GPP packet core architecture to also encompass interworking with access technologies standardized outside of 3GPP, as well as provide an evolved packet-only core for the next generation of mobile broadband access technology-LTE. This in turn created significant excitement and interest from a vast number of contributing companies, ranging from mobile telecom operators to telecom equipment and handset vendors as well as research institutions.

Inclusion of CDMA interworking capabilities in the 3GPP architecture is naturally a major breakthrough, paving the way for network deployments of LTE and EPC that can be shared across an even wider operator community, as well as closer cooperation between 3GPP and 3GPP2. Global uptake of a single common technology means large volumes of handsets and network equipment, a highly competitive market leading to focus on cost efficient solutions and lots of attention from service and application developers. A global technology also means excellent roaming possibilities in that users can access and utilize services in a large number of countries while using their own personal mobile device.

Before closing this book, let us spend a few moments looking into the near future, even though the deployment of LTE and EPC networks are still in its early phase. Work on EPC will continue with the next phase, and there are a number of areas which can be exploited in terms of adding new features to the architecture further.

In near term, 3GPP is addressing a number of areas and features for Rel-9. The focus of this work is on solutions and architecture for handling specific telephony requirements (such as support for emergency access over EPC) and for small base stations for deployment in offices or homes (the so-called femto base stations or Home NodeB/eNodeB's). There are a number of other suggestions for enhancements and additions, including, for example, enhanced broadcasting over LTE and WCDMA (eMBMS), as well as more advanced mechanisms for supporting simultaneous connections over multiple radio access networks.

The impact from possible deployment of small base stations in the private homes of end-users is one focus area for the continued EPS evolution. These small base stations, referred to as femto or home base stations, would interconnect to the mobile network over fixed broadband connections that are assumed to be present in the homes of end-users. This solution could potentially offer operators cost efficient solutions for increasing network coverage and capacity while enabling new service offerings to the users.

Potential future deployments of telephony services based on IMS requires that a number of legal requirements are fulfilled, somewhat differing between countries and associated licence requirements for mobile network operators. These requirements include support for emergency calls, for example, over networks where the user in the emergency situation is not a subscriber. Emergency calls also often require that the geographical position of the user making the call is made available to the applicable authorities with sufficient accuracy. These requirements will also impact the evolution of EPC.

Broadcasting would allow for a more efficient resource usage in the case that many receivers shall receive the same content. This can be solved through transmitting the contents only once, and not once per user. This is already supported as part of the WCDMA/HSPA MBMS specifications, where MBMS is short for Multimedia Broadcast/Multicast Service. The corresponding feature in Rel-9 for LTE and WCDMA is called evolved MBMS (eMBMS).

One additional important activity that will influence EPC in the future is the work on evolving the LTE access towards capabilities matching the ITU definitions of 'IMT-advanced'. This work was started in 3GPP in 2008, and although the basic assumption is that the core network architecture and procedures shall remain as defined in Rel-9, the possibility that the core network will in fact be impacted cannot be ruled out at this early stages of development phase. The evolution of LTE can be assumed to be included earliest in 3GPP Rel-10 specifications.

Target requirements on LTE meeting IMT-advanced criteria can be found in 3GPP TR 36.913 [36.913] and include:

• Building on and evolving Rel-8 LTE including maintaining support for all relevant LTE requirements.

- Meeting or exceeding the ITU-R IMT-advanced requirements.
- A downlink peak data rate of 1 Gbit/s and an uplink peak data rate of 500 Mbit/s.
- Transition time from idle to connected mode less than 50 ms.
- Transition from dormant to active state in connected mode less than 10 ms.
- Reduced user plane latency compared to Rel-8 LTE.
- Downlink peak spectrum efficiency of 30 bps/Hz and uplink peak spectrum efficiency of 15 bps/Hz, assuming antenna configurations of 8×8 or less for DL and 4×4 or less for UL.
- Improved VoIP capacity compared to Rel-8 LTE.
- Support mobility for speeds up to 350 km/h or 500 km/h depending on the frequency band.
- Coverage requirements as for LTE Rel-8.
- Backwards compatibility in that Rel-8 LTE terminals can be supported and that new terminals supporting evolved LTE can work in a Rel-8 LTE network.
- Support for additional frequency bands and wider bandwidths (up to 100 MHz) compared to Rel-8 LTE.
- Support for handover to legacy mobile networks with capabilities and performance the same or better than for LTE Rel-8.
- Support for network sharing.
- Cost efficient solutions in general, addressing, for example, power consumption, backhaul costs, operation and maintenance, etc.

It remains to be seen to what extent these LTE requirements will also impact EPC. It should be noted that actual network deployments including Rel-10 based evolved LTE capabilities are several years away.

In addition to the features discussed above, there are also some smaller improvements being studied currently, focusing on more alignment of non-3GPP accesses with the 3GPP services.

Looking beyond pure mobile networks, the area of common solutions for fixed and mobile access has seen a lot of attention from large operators with both mobile and fixed business interests. While 3GPP Rel-8 specifies generic mechanisms for interconnection between EPC and just about any other network based on Mobile IP/Proxy Mobile IP mechanisms, there is a potential of further harmonization of, for instance, subscriber management and policy control across fixed and mobile access technologies. One common way of controlling subscriber credentials, access rights and service usage should prove valuable to operators wanting to provide a generic service offering over both fixed and mobile accesses. Such a harmonization would most likely require a close cooperation between 3GPP and the Broadband Forum (BBF) to arrive at common solutions.

13.1 Concluding words

Going forward from the EPC solution and specifications as of 3GPP Rel-8, there are obviously several areas that may be exploited and developed in the future. The authors of this book are convinced of the necessity that the decisions on the next steps to take should be based on strong commercial aspects, ensuring that the 3GPP focus remains on features and functions of interest to the global community of network operators, consumers and enterprise customers. With the specification work of EPC, 3GPP has provided an excellent platform for future core network evolution. It will for sure continue to be an interesting journey.

References

[Beming, 2007]   Beming, P., Frid, L., Hall, G., Malm, P., Noren, T., Olsson, M., Rune, G., October, 2007. LTE-SAE architecture and performance. Ericsson Review 3. http://www.ericsson.com/ericsson/corpinfo/publications/review/2007_03/files/5_LTE_SAE.pdf

[Blanchet, 2006]   Blanchet, M., December, 2005. Migrating to IPv6: A Practical Guide to Implementing IPv6 in Mobile and fixed Networks. John Wiley & Sons, p. 418. ISBN-10: 0471498920/ISBN-13: 9780471498926.

[Brenner, 2008]   Brenner, M., Unmehopa, M., April 4, 2008. The Open Mobile Alliance: Delivering Service Enablers for Next-Generation Applications. John Wiley & Sons, p. 530. ISBN-10: 0470519185/ISBN-13: 978-0470519189.

[Camarillo, 2008]   Camarillo G., Garcia-Martin M.-A., November, 2008. 3G IP Multimedia Subsystem (IMS). ISBN-10: 0470516623.

[Dahlman, 2008]   Dahlman, E., Parkvall, S., Sköld, J., Beming, P., August, 2008. 3G Evolution: HSPA and LTE for Mobile Broadband. Elsevier, p. 648. ISBN-10: 012372533X/ISBN-13: 978-0-12-374538-5.

[Hagen, 2006]   'IPv6 Essentials, second edition', Hagen, Silvia, O'Reilly (May 2006), ISBN-10: 0-596-10058-2/ISBN-13: 9780596100582, pp. 436.

[Li, 2006]   Li, Q., Jinmei, T., Shima, K., January, 2006. IPv6 Core Protocols Implementation. In: The Morgan Kaufmann Series in Networking. Morgan Kaufmann Publishers. ISBN-10: 0124477518/ISBN-13: 9780124477513.

[Mulligan, 2009]   Open API Standardisation and the NGN Platform. IEEE Communications Magazine, May 2009.

3GPP Technical Specifications

[21.905]   3GPP TS 21.905 Vocabulary for 3GPP Specifications

[22.011]   3GPP TS 22.011 Service accessibility

[22.168]   3GPP TS 22.168 Earthquake and Tsunami Warning System (ETWS) requirements

[22.278]   3GPP TS 22.278 Service requirements for the Evolved Packet System (EPS)

[23.003]   3GPP TS 23.003 Numbering, addressing and identification

[23.060]   3GPP TS 23.060 General Packet Radio Service (GPRS); Service description; Stage 2

[23.122]   3GPP TS 23.122 Non-Access-Stratum (NAS) functions related to Mobile Station (MS) in idle mode

[23.203]   3GPP TS 23.203 Policy and charging control architecture

[23.216]   3GPP TS 23.216 Single Radio Voice Call Continuity (SRVCC); Stage 2

[23.272]   3GPP TS 23.272 Circuit Switched (CS) fallback in Evolved Packet System (EPS); Stage 2

[23.401]   3GPP TS 23.401 General Packet Radio Service (GPRS) enhancements for Evolved Universal Terrestrial Radio Access Network (E-UTRAN) access

[23.402]   3GPP TS 23.402 Architecture enhancements for non-3GPP accesses

[23.882]   3GPP TR 23.882 3GPP system architecture evolution (SAE): Report on technical options and conclusions

[24.007]   3GPP TS 24.007 Mobile radio interface signalling layer 3; General aspects

[24.301]   3GPP TS 24.301 Non-Access-Stratum (NAS) protocol for Evolved Packet System (EPS); Stage 3

[24.302]   3GPP TS 24.302 Access to the Evolved Packet Core (EPC) via non-3GPP access networks; Stage 3

[24.303]   3GPP TS 24.303 Mobility management based on Dual-Stack Mobile IPv6; Stage 3

[24.304]   3GPP TS 24.304 Mobility management based on Mobile IPv4; User Equipment (UE) – foreign agent interface; Stage 3

[24.312]   3GPP TS 24.312 Access Network Discovery and Selection Function (ANDSF) Management Object (MO)

[25.913]   3GPP TS 25.913 Requirements for Evolved UTRA (E-UTRA) and Evolved UTRAN (E-UTRAN)

[29.060]   3GPP TS 29.060 General Packet Radio Service (GPRS); GPRS Tunnelling Protocol (GTP) across the Gn and Gp interface

[29.061]   3GPP TS 29.061 Interworking between the Public Land Mobile Network (PLMN) supporting packet based services and Packet Data Networks (PDN)

[29.118]   3GPP TS 29.118 Mobility Management Entity (MME) – Visitor Location Register (VLR) SGs interface specification

[29.168]   3GPP TS 29.168 Cell Broadcast Centre interfaces with the Evolved Packet Core; Stage 3

[29.213]   3GPP TS 29.213 Policy and charging control signalling flows and Quality of Service (QoS) parameter mapping

[29.214]   3GPP TS 29.214 Policy and charging control over Rx reference point

[29.215]   3GPP TS 29.215 Policy and Charging Control (PCC) over S9 reference point

[29.230]   3GPP TS 29.230 Diameter applications; 3GPP specific codes and identifiers

[29.272]   3GPP TS 29.272 Evolved Packet System (EPS); Mobility Management Entity (MME) and Serving GPRS Support Node (SGSN) related interfaces based on Diameter protocol

[29.273]   3GPP TS 29.273 Evolved Packet System (EPS); 3GPP EPS AAA interfaces

[29.274]   3GPP TS 29.274 3GPP Evolved Packet System (EPS); Evolved General Packet Radio Service (GPRS) Tunnelling Protocol for Control plane (GTPv2-C); Stage 3

[29.275]   3GPP TS 29.275 Proxy Mobile IPv6 (PMIPv6) based Mobility and
            Tunnelling protocols; Stage 3

[29.276]   3GPP TS 29.276 Optimized Handover Procedures and Protocols
            between EUTRAN Access and cdma2000 HRPD Access

[29.281]   3GPP TS 29.281 General Packet Radio System (GPRS) Tunnelling
            Protocol User Plane (GTPv1-U)

[29.303]   3GPP TS 29.303 Domain Name System Procedures

[33.106]   3GPP TS 33.106 Lawful Interception Requirements

[33.210]   3GPP TS 33.210 3G security; Network Domain Security (NDS); IP
            network layer security

[33.401]   3GPP TS 33.401 3GPP System Architecture Evolution (SAE);
            Security architecture

[33.402]   3GPP TS 33.402 3GPP System Architecture Evolution (SAE);
            Security aspects of non-3GPP accesses

[36.101]   3GPP TS 36.101 Evolved Universal Terrestrial Radio Access (E-
            UTRA); User Equipment (UE) radio transmission and reception

[36.300]   3GPP TS 36.300 Evolved Universal Terrestrial Radio Access
            (E-UTRA) and Evolved Universal Terrestrial Radio Access Network
            (E-UTRAN); Overall description; Stage 2

[36.304]   3GPP TS 36.304 Evolved Universal Terrestrial Radio Access
            (E-UTRA); User Equipment (UE) procedures in idle mode

[36.306]   3GPP TS 36.306 Evolved Universal Terrestrial Radio Access
            (E-UTRA); User Equipment (UE) radio access capabilities

[36.321]   3GPP TS 36.321 Evolved Universal Terrestrial Radio Access
            (E-UTRA); Medium Access Control (MAC) protocol specification

[36.322]   3GPP TS 36.322 Evolved Universal Terrestrial Radio Access
            (E-UTRA); Radio Link Control (RLC) protocol specification

[36.323]   3GPP TS 36.323 Evolved Universal Terrestrial Radio Access
            (E-UTRA); Packet Data Convergence Protocol (PDCP) specification

[36.331]   3GPP TS 36.331 Evolved Universal Terrestrial Radio Access
            (E-UTRA); Radio Resource Control (RRC); Protocol specification

[36.401]   3GPP TS 36.401 Evolved Universal Terrestrial Radio Access
            Network (E-UTRAN); Architecture description

[36.410]   3GPP TS 36.410 Evolved Universal Terrestrial Radio Access Network (E-UTRAN); S1 layer 1 general aspects and principles

[36.411]   3GPP TS 36.411 Evolved Universal Terrestrial Radio Access Network (E-UTRAN); S1 layer 1

[36.412]   3GPP TS 36.412 Evolved Universal Terrestrial Radio Access Network (E-UTRAN); S1 signalling transport

[36.413]   3GPP TS 36.413 Evolved Universal Terrestrial Radio Access (E-UTRA); S1 Application Protocol (S1AP)

[36.414]   3GPP TS 36.414 Evolved Universal Terrestrial Radio Access Network (E-UTRAN); S1 data transport

[36.420]   3GPP TS 36.420 Evolved Universal Terrestrial Radio Access Network (E-UTRAN); X2 general aspects and principles

[36.421]   3GPP TS 36.421 Evolved Universal Terrestrial Radio Access Network (E-UTRAN); X2 layer 1

[36.422]   3GPP TS 36.422 Evolved Universal Terrestrial Radio Access Network (E-UTRAN); X2 signalling transport

[36.423]   3GPP TS 36.423 Evolved Universal Terrestrial Radio Access Network (E-UTRAN); X2 Application Protocol (X2AP)

[36.424]   3GPP TS 36.424 Evolved Universal Terrestrial Radio Access Network (E-UTRAN); X2 data transport

[36.913]   3GPP TS 36.913 Requirements for further advancements for Evolved Universal Terrestrial Radio Access (E-UTRA) (LTE-Advanced)

3GPP2 Specifications

[X.S0042]   3GPP2 X.S0042-0 Voice Call Continuity between IMS and Circuit Switched System

IETF RFCs

[768]   IETF RFC 768; User Datagram Protocol

[793]   IETF RFC 793; Transmission Control Protocol

[1035]   RFC 1035, Domain Names – Implementation and Specification

[2003]   IETF RFC 2003; IP Encapsulation within IP

[2181]   RFC 2181, Clarifications to the DNS Specification

[2401]   IETF RFC 2401; Security Architecture for the Internet Protocol

[2402]   IETF RFC 2402; IP Authentication Header

[2406]   IETF RFC 2406; IP Encapsulating Security Payload (ESP)

[2407]   IETF RFC 2407; The Internet IP Security Domain of Interpretation
              for ISAKMP

[2408]   IETF RFC 2408; Internet Security Association and Key Management
              Protocol (ISAKMP)

[2409]   IETF RFC 2409; The Internet Key Exchange (IKE)

[2473]   IETF RFC 2473; Generic Packet Tunnelling in IPv6 Specification

[2606]   RFC 2606; Reserved Top Level DNS Names

[2784]   IETF RFC 2784; Generic Routing Encapsulation (GRE)

[2890]   IETF RFC 2890; Key and Sequence Number Extensions to GRE

[2960]   IETF RFC 2960; Stream Control Transmission Protocol

[3309]   IETF RFC 3309; Stream Control Transmission Protocol (SCTP)
              Checksum Change

[3344]   IETF RFC 3344; IP Mobility Support for IPv4

[3748]   IETF RFC 3748; Extensible Authentication Protocol (EAP)

[3775]   IETF RFC 3775; Mobility Support in IPv6

[3776]   IETF RFC 3776; Using IPsec to Protect Mobile IPv6 Signalling
              Between Mobile Nodes and Home Agents

[3588]   IETF RFC 3588; Diameter Base Protocol

[3958]   RFC 3958; Domain-Based Application Service Location Using
              SRV RRs and the Dynamic Delegation Discovery Service (DDDS)

[4005]   IETF RFC 4005; Diameter Network Access Server Application

[4006]   IETF RFC 4006; Diameter Credit-Control Application

[4072]   IETF RFC 4072; Diameter Extensible Authentication Protocol (EAP) Application

[4186]   IETF RFC 4186; Extensible Authentication Protocol Method for Global System for Mobile Communications (GSM) Subscriber Identity Modules (EAP-SIM)

[4187]   IETF RFC 4187; Extensible Authentication Protocol Method for 3rd Generation Authentication and Key Agreement (EAP-AKA)

[4285]   IETF RFC 4285; Authentication Protocol for Mobile IPv6

[4301]   IETF RFC 4301; Security Architecture for the Internet Protocol

[4302]   IETF RFC 4302; IP Authentication Header

[4303]   IETF RFC 4303; IP Encapsulating Security Payload (ESP)

[4306]   IETF RFC 4306; Internet Key Exchange (IKEv2) Protocol

[4555]   IETF RFC 4555; IKEv2 Mobility and Multihoming Protocol (MOBIKE)

[4877]   IETF RFC 4877; Mobile IPv6 Operation with IKEv2 and the Revised IPsec Architecture

[4960]   IETF RFC 4960; Stream Control Transmission Protocol

[5094]   IETF RFC 5094; Mobile IPv6 Vendor Specific Option

[5213]   IETF RFC 5213; Proxy Mobile IPv6

[5216]   IETF RFC 5216; The EAP-TLS Authentication Protocol

[5447]   IETF RFC 5447; Diameter Mobile IPv6: Support for Network Access Server to Diameter Server Interaction

[5448]   IETF RFC 5448; Improved Extensible Authentication Protocol Method for 3rd Generation Authentication and Key Agreement (EAP-AKA)

[5555]   IETF RFC 5555; Mobile IPv6 Support for Dual Stack Hosts and Routers

IETF Internet Drafts

[draft-ietf-mext-binding-revocation]	IETF Internet Draft, Binding Revocation for IPv6 Mobility (draft-ietf-mext-binding-revocation)
[draft-ietf-netlmm-grekey-option]	IETF Internet-Draft, GRE Key Option for Proxy Mobile IPv6, (draft-ietf-netlmm-grekey-option)
[draft-ietf-netlmm-pmip6-ipv4-support]	IETF Internet-Draft, IPv4 Support for Proxy Mobile IPv6, (draft-ietf-netlmm-pmip6-ipv4-support)
[draft-ietf-dime-pmip6]	IETF Internet Draft, Diameter Proxy Mobile IPv6: Mobile Access Gateway and Local Mobility Anchor Interaction with Diameter Server

Note: At the time of preparing this book, several of the Internet Drafts listed above are close to becoming approved RFCs. The interested reader should consult the IETF web page (www.ietf.org) for the latest status.

ITU Recommendations

[I.112]   ITU-T Recommendation I.112: I.112 INTEGRATED SERVICES DIGITAL NETWORK (ISDN), GENERAL STRUCTURE, VOCABULARY OF TERMS FOR ISDNs

Appendix A
Standards bodies associated with EPS

Third Generation Partnership Project (3GPP)

As we mentioned in Chapter 1, GSM was originally developed as a European standard within ETSI. 3GPP was established in 1998 in order to unite a number of different regional standardization bodies for the creation of a global cellular standard. These regional bodies are referred to as Organizational Partners. As of 2009, the partners were ETSI (Europe), ARIB (Japan), ATIS (US), CCSA (China), TTA (Korea) and TTC (Japan).

Originally, 3GPP was to produce specifications and reports for a 3G Mobile System based on evolved GSM core and radio networks. It rapidly evolved, however, to also have responsibility for the development of the GSM technologies, such as GPRS and EDGE.

More recently, 3GPP has led the work on a set of common IMS specifications and, in parallel with SAE, the Long Term Evolution of the radio network, also known as LTE. The core network evolution, SAE, is naturally closely related to the LTE work item.

Structure of 3GPP

3GPP is organized into several different Technical Specification Groups (TSGs). These TSGs are responsible for the technical work and production of the specifications. There are four TSGs, split across the different areas that 3GPP works with:

TSG GERAN: responsible for the specification of the radio access part of GSM/EDGE.
TSG RAN: responsible for the definition of the functions, requirements and interfaces of the UTRA/E-UTRA (WCDMA and LTE) radio networks covering both FDD and TDD variants.

TSG SA: responsible for the overall architecture and service capabilities of systems based on 3GPP specifications, and also for co-ordination between the different TSGs.

TSG CT: responsible for specifying terminal interfaces and capabilities. Also responsible for specifying the core network protocols of 3GPP systems.

Each of these TSGs has a number of Working Groups (WGs) associated with them; each WG is responsible for a certain number of tasks within the mandate of the TSG that they fall under. For example, WG SA1 is responsible for system requirements, while WG SA2 is responsible for the system architecture. WG CT4, meanwhile, handles the protocol definition for Basic Call Processing and protocols between nodes within the Network. The relationship between the TSGs and the different WGs is illustrated below.

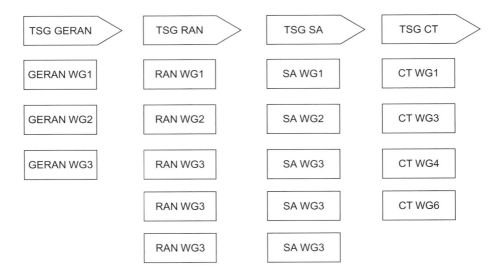

The overall management of 3GPP, for example, organisation and allocation of work, is handled by the highest decision-making body in 3GPP – the Project Coordination Group (PCG).

Each WG in 3GPP is responsible for producing Technical Specifications (TSs) and Technical Reports (TRs), that is, the actual technical documents that can be downloaded from the 3GPP website (www.3gpp.org), but they are not allowed to just create whatever specification they like and publish it; each TS and TR must go through an approval process within the TSG responsible for their particular WG. Once this is completed, the specifications can then be taken to regional organizations (such as ETSI for Europe) to be approved as formal standards, or the ITU for use in their set of standards.

Each TS and TR is referred to be a set of digits; 'xx.yyy', where xx refers to the so-called series number, while yyy refers to the particular specification within that series. As an example, 23.401 indicates that it is a system architecture document (23 series), while 401 refers to the actual specification, in this case: GPRS enhancements for E-UTRAN access.

LTE specifications are handled in RAN WG1, RAN WG2, RAN WG3, RAN WG4 and RAN WG5. SAE specifications, meanwhile, are handled in SA WG1, SA WG2, SA WG3, CT WG1, CT WG3 and CT WG4. Each of these WGs is responsible for a different section of the SAE work and will be covered in more detail in subsequent chapters.

Stages in 3GPP standardization

When 3GPP develops a new standard or amends an existing standard, the work proceeds in three logical phases; stage 1, stage 2 and stage 3. It has been adopted by 3GPP and is also used by several other standardization organizations. It can be noted that work on the three stages often takes place simultaneously, or at least with significant temporal overlap, in order to make the standardization work time efficient.

During stage 1 the service requirements are specified, that is, the functions that the system as a whole is intended to support. In the next step, stage 2, the architectural requirements are specified, taking the stage 1 requirements into account. This means that the different logical network entities and the reference points between the network entities are defined. The purpose and functions of each network entity and each reference point are also specified. The procedures, that is, the logical message flow between network entities, are defined; including what information is transferred across the reference points. In the third stage, stage 3, the actual protocols are defined based on the architectural work done during stage 2. Each message is specified in detail and the message content such as parameter formats, information element structure and so on, is defined. The stage 3 work also has the very important task of making sure that error cases are handled appropriately by the system. This includes, for example, defining relevant result codes in response messages.

Tracking down the right specification in 3GPP

All 3GPP specifications are freely available online for anyone to read. In order to find the right specification, it is generally easiest to search for it by its specification number, for example, 23.401, or via the working group responsible for its development, for example, SA2.

An important thing to remember when searching for specifications is to ensure that the release number is correct. 3GPP uses a parallel release mechanism in order to ensure that any developer has a stable platform to work upon. Once a release is frozen, it means that no more functional changes may be made to the specifications; they may, however, be changed if errors are found or for maintenance purposes.

Different WGs often work towards different releases at the same time. For example, the requirements specifications developed in SA1 are often frozen at a much earlier date than for the System Architecture documents created in SA2. So, SA1 may be working on Rel-9 requirements, while SA2 is working on Rel-8 architecture documents. This is quite natural; in order to ensure that SA2 has a stable set of requirements to work upon, they need to be frozen before work can commence.

A full list of different specifications and the working groups responsible for them at the time of writing is available in Appendix A.

Internet Engineering Task Force (IETF)

The IETF, in comparison to 3GPP, is more loosely organized. The IETF is comprised of individuals, rather than companies with participants from all different areas of the industry; it does not take membership fees and as a result anyone can participate. It takes care of IP-related protocols and has developed most of the protocols in use on the Internet. It handles only protocols, however, and does not define the network architectures that combine the different protocols together. Nor does it define the functionality of nodes on a network.

Structure of the IETF

The IETF is split into different areas: Applications Area, General Area, Internet Area, Operations and Management Area, Real-time Applications and Infrastructure Area, Routing Area, Security Area and the Transport Area. Each of these areas has several WGs under its directorship and each WG has a particular technical subject that it works on and produces a set of documents for. The WGs are therefore referred to as 'Area Directorates'. WGs are created for specific purposes and after their documents are complete, they are either disbanded or 'rechartered' with a new set of deliverable documents. As a result, the active WGs in the IETF change; the latest list of WGs is, at the time of writing, available at: http://www.ietf.org/html.charters/wg-dir.html.

WGs are assigned a unique acronym which identifies the task that they are working on, for example, mip4 relates to Mobility for IPv4, or sipping refers to the group that handles the SIP, the protocol that forms a key component of the IMS.

In a similar fashion to 3GPP, the IETF has an overall group that forms the technical management team of the IETF; this is called the Internet Engineering Steering Group (IESG). Each Area Directorate has one or two directors who join the IETF chairman in the IESG. It is the IESG's responsibility to review all specifications that the WGs produce and also to decide on the overall technical direction that the IETF will take, that is, what areas the IETF should work on.

A high-level view of the IETF is illustrated below:

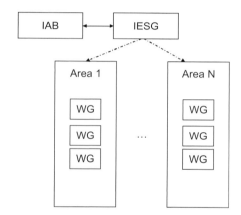

Open Mobile Alliance (OMA)

The OMA was created in 2002 comprising over 200 companies, including wireless vendors, IT companies, mobile operators and application and content providers. OMA is intended to be the focal point for the development of mobile service enabler specifications, supporting the creation of interoperable end-to-end mobile services.

OMA specifications are independent of the underlying network architectures. Within the 3GPP specifications for EPS, references to OMA are made for several different reasons, for example Device Management (DM).

More information about OMA can be found on their website: http://www.open-mobilealliance.org

Appendix B
SAE/EPC specifications

The EPS is specified in a set of new specifications as well as updates of existing GSM and UMTS specifications. Some of the specifications listed below are hence introduced for EPS in Rel-8 while others have existed for GSM and UMTS and have just been updated to cater for EPS.

All specs are Rel-8 if not explicitly mentioned otherwise.

SA1

TS 22.011	Service accessibility	
TS 22.168	Earthquake and Tsunami Warning System (ETWS) requirements	
TS 22.220	Service requirements for Home Node B (HNB) and Home eNode B (HeNB)	
TS 22.268	Public Warning System (PWS) requirements	Rel-9
TS 22.278	Service requirements for the EPS	

SA2

Click on spec number for details.

TS 23.002	Network architecture
TS 23.060	GPRS; Service description; Stage 2
TS 23.203	PCC architecture
TS 23.216	Single Radio Voice Call Continuity (SRVCC); Stage 2
TS 23.221	Architectural requirements
TS 23.236	Intra-domain connection of Radio Access Network (RAN) nodes to multiple Core Network (CN) nodes
TS 23.272	Circuit-Switched (CS) fallback in EPS; Stage 2
TS 23.401	GPRS enhancements for E-UTRAN access
TS 23.402	Architecture enhancements for non-3GPP accesses

SA3

Click on spec number for details.

TS 33.401	3GPP SAE; Security architecture
TS 33.402	3GPP SAE; Security aspects of non-3GPP accesses

CT1

TS 23.003	Numbering, addressing and identification
TS 23.122	Non-Access-Stratum (NAS) functions related to Mobile Station (MS) in idle mode
TS 24.301	NAS protocol for EPS; Stage 3
TS 24.302	Access to the EPC via non-3GPP access networks; Stage 3
TS 24.303	Mobility management based on Dual-stack Mobile IPv6; Stage 3
TS 24.304	Mobility management based on MIPv4; User Equipment (UE)-foreign agent interface; Stage 3
TS 24.312	Access Network Discovery and Selection Function (ANDSF) Management Object (MO)

CT3 and CT4

TS 29.002	Mobile Application Part (MAP) specification	?
TS 29.060	GPRS; GTP across the Gn and Gp interfaces	
TS 29.061	Interworking between the Public L and Mobile Network (PLMN) supporting packet based services and PDN	
TS 29.118	Mobility Management Entity (MME)-Visitor Location Register (VLR) SGs interface specification	
TS 29.168	Cell Broadcast Centre interfaces with the EPC; Stage 3	
TS 29.210	Charging rule provisioning over Gx interface	?
TS 29.211	Rx interface and Rx/Gx signalling flows	?
TS 29.212	PCC over Gx reference point	?
TS 29.213	PCC signalling flows and QoS parameter mapping	?
TS 29.214	PCC over Rx reference point	
TS 29.215	PCC over S9 reference point	
TS 29.230	Diameter applications; 3GPP specific codes and identifiers	
TS 29.272	EPS; MME and SGSN-related interfaces based on Diameter protocol	
TS 29.273	EPS; 3GPP EPS AAA interfaces	
TS 29.274	3GPP EPS; Evolved GPRS Tunnelling Protocol for Control plane (GTPv2-C); Stage 3	
TS 29.275	PMIPv6-based Mobility and Tunnelling protocols; Stage 3	

(Continued)

TS 29.276	Optimized Handover Procedures and Protocols between E-UTRAN Access and cdma2000 HRPD Access
TS 29.277	Optimized Handover Procedures and Protocols between E-UTRAN Access and 1xRTT Access
TS 29.279	MIPv4-based Mobility protocols; Stage 3
TS 29.280	EPS; 3GPP Sv interface (MME to MSC, and SGSN to MSC) for SRVCC
TS 29.281	GPRS Tunnelling Protocol User Plane (GTPv1-U)
TS 29.303	Domain Name System Procedures; Stage 3
TS 29.304	MIPv4-based Mobility protocols; Stage 3
TS 29.305	Interworking Function (IWF) between MAP-based and Diameter-based interfaces

E-UTRAN specs, WG RAN1-5:

The E-UTRAN is specified in a separate specification series, the 36.xxx series

Click on spec number for details.

TS 36.101	E-UTRA; UE radio transmission and reception
TS 36.104	E-UTRA; Base Station (BS) radio transmission and reception
TS 36.106	E-UTRA; FDD repeater radio transmission and reception
TS 36.113	E-UTRA; BS and repeater Electromagnetic Compatibility (EMC)
TS 36.124	E-UTRA; EMC requirements for mobile terminals and ancillary equipment
TS 36.133	E-UTRA; Requirements for support of radio resource management
TS 36.141	E-UTRA; BS conformance testing
TS 36.143	E-UTRA; Repeater conformance testing
TS 36.201	E-UTRA; LTE physical layer; General description
TS 36.211	E-UTRA; Physical channels and modulation
TS 36.212	E-UTRA; Multiplexing and channel coding
TS 36.213	E-UTRA; Physical layer procedures
TS 36.214	E-UTRA; Physical layer measurements
TS 36.300	E-UTRA and E-UTRAN; Overall description; Stage 2
TS 36.302	E-UTRA; Services provided by the physical layer
TS 36.304	E-UTRA; UE procedures in idle mode
TS 36.306	E-UTRA; UE radio access capabilities
TS 36.314	E-UTRAN; Layer 2 measurements
TS 36.321	E-UTRA; Medium Access Control (MAC) protocol specification
TS 36.322	E-UTRA; Radio Link Control (RLC) protocol specification
TS 36.323	E-UTRA; Packet Data Convergence Protocol (PDCP) specification
TS 36.331	E-UTRA; Radio Resource Control (RRC); Protocol specification

(Continued)

TS 36.401	E-UTRAN; Architecture description
TS 36.410	E-UTRAN; S1 layer 1 general aspects and principles
TS 36.411	E-UTRAN; S1 layer 1
TS 36.412	E-UTRAN; S1 signalling transport
TS 36.413	E-UTRA; S1 Application Protocol (S1AP)
TS 36.414	E-UTRA; S1 data transport
TS 36.420	E-UTRAN; X2 general aspects and principles
TS 36.421	E-UTRAN; X2 layer 1
TS 36.422	E-UTRAN; X2 signalling transport
TS 36.423	E-UTRAN; X2 Application Protocol (X2AP)
TS 36.424	E-UTRAN; X2 data transport
TS 36.440	E-UTRAN; General aspects and principles for interfaces supporting Multimedia Broadcast Multicast Service (MBMS) within E-UTRAN
TS 36.441	E-UTRAN; Layer 1 for interfaces supporting MBMS within E-UTRAN
TS 36.442	E-UTRAN; Signalling Transport for interfaces supporting MBMS within E-UTRAN
TS 36.443	E-UTRAN; M2 Application Protocol (M2AP)
TS 36.444	E-UTRAN; M3 Application Protocol (M3AP)
TS 36.445	E-UTRAN; M1 Data Transport
TS 36.446	E-UTRAN; M1 User Plane protocol
TS 36.508	E-UTRA and EPC; Common test environments for UE conformance testing
TS 36.509	E-UTRA; Special conformance testing function for UE
TS 36.521-1	E-UTRA; UE conformance specification; Radio transmission and reception; Part 1: conformance testing
TS 36.521-2	E-UTRA; UE conformance specification; Radio transmission and reception; Part 2: Implementation Conformance Statement (ICS)
TS 36.521-3	E-UTRA; UE conformance specification; Radio transmission and reception; Part 3: Radio Resource Management conformance testing
TS 36.523-1	E-UTRA and E-UTRAN; UE conformance specification; Part 1: Protocol conformance specification
TS 36.523-2	E-UTRA and E-UTRAN; UE conformance specification; Part 2: ICS
TS 36.523-3	E-UTRA and E-UTRAN; UE conformance specification; Part 3: Abstract Test Suites (ATS)
TR 36.801	E-UTRA; Measurement requirements
TR 36.803	E-UTRA; UE radio transmission and reception
TR 36.804	E-UTRA; BS radio transmission and reception
TR 36.814	Further advancements for E-UTRA Physical layer aspects
TR 36.902	E-UTRAN; Self-configuring and self-optimizing network (SON) use cases and solutions
TS 36.903	E-UTRA; Derivation of test tolerances for multi-cell Radio Resource Management (RRM) conformance tests
TR 36.913	Requirements for further advancements for E-UTRA (LTE-Advanced)

(Continued)

TR 36.938	E-UTRAN; Improved network-controlled mobility between E-UTRAN and 3GPP2/mobile WiMAX radio technologies
TR 36.942	E-UTRA; Radio Frequency (RF) system scenarios
TR 36.956	E-UTRA; Repeater planning guidelines and system analysis

IETF documents

RFCs

- RFC 768; User Datagram Protocol
- RFC 793; Transmission Control Protocol
- RFC 2003; IP Encapsulation within IP
- RFC 2401; Security Architecture for the Internet Protocol
- RFC 2402; IP Authentication Header
- RFC 2406; IP Encapsulating Security Payload (ESP)
- RFC 2407; The Internet IP Security Domain of Interpretation for ISAKMP
- RFC 2408; Internet Security Association and Key Management Protocol (ISAKMP)
- RFC 2409; The Internet Key Exchange (IKE)
- RFC 2473; Generic Packet Tunnelling in IPv6 Specification
- RFC 2784; Generic Routing Encapsulation (GRE)
- RFC 2890; Key and Sequence Number Extensions to GRE
- RFC 2960; Stream Control Transmission Protocol (SCTP)
- RFC 3309; SCTP Checksum Change
- RFC 3344; IP Mobility Support for IPv4
- RFC 3748; Extensible Authentication Protocol (EAP)
- RFC 3775; Mobility Support in IPv6
- RFC 3776; Using IPsec to Protect MIPv6 Signalling between Mobile Nodes and Home Agents
- RFC 3588; Diameter Base Protocol
- RFC 4005; Diameter Network Access Server Application
- RFC 4006; Diameter Credit-Control Application
- RFC 4072; Diameter EAP Application
- RFC 4186; EAP Method for Global System for Mobile Communications (GSM) Subscriber Identity Modules (EAP-SIM)
- RFC 4187; EAP Method for Third Generation Authentication and Key Agreement (EAP-AKA)
- RFC 4285; Authentication Protocol for MIPv6
- RFC 4301; Security Architecture for the Internet Protocol
- RFC 4302; IP Authentication Header
- RFC 4303; IP Encapsulating Security Payload (ESP)

- RFC 4306; Internet Key Exchange (IKEv2) Protocol
- RFC 4555; IKEv2 Mobility and Multihoming Protocol (MOBIKE)
- RFC 4877; MIPv6 Operation with IKEv2 and the Revised IPsec Architecture
- RFC 4960; SCTP
- RFC 5213; PMIPv6
- RFC 5216; The EAP-TLS Authentication Protocol

Internet drafts

- MIPv6 Support for Dual Stack Hosts and Routers (DSMIPv6) (draft-ietf-mext-nemo-v4traversal)
- GRE Key Option for PMIPv6 (draft-ietf-netlmm-grekey-option)
- IPv4 Support for PMIPv6 (draft-ietf-netlmm-pmip6-ipv4-support)
- Improved EAP Method for Third Generation Authentication and Key Agreement (EAP-AKA') (draft-arkko-eap-aka-kdf)
- Diameter PMIPv6: Support For Mobile Access Gateway and Local Mobility Anchor to Diameter Server Interaction (draft-korhonen-dime-pmip6)
- Diameter MIPv6: Support for Network Access Server to Diameter Server Interaction (draft-ietf-dime-mip6-integrated)

Appendix C
Mobile broadband application development

The following is a list of different initiatives ongoing in the mobile telecommunications industry in relation to mobile broadband applications. It is not exhaustive by any means, but is meant only as a starting point.

AT&T devCentral

devCentral is AT&T's official resource for wireless development. AT&T provides a whole range of different tools for wireless developers.

For more information, see: http://developer.att.com/

Vodafone Betavine

Betavine is Vodafone's open community for mobile application developers, covering Operating Systems, widgets and a whole host of other tools.

For more information, see: http://www.betavine.net/bvportal/home.html

Ericsson Developer Connection and Ericsson Labs

Ericsson Developer Connection assists developers creating applications that incorporate telecommunication network capabilities, such as location-based services, charging, messaging and presence. For more information, see: http://www.ericsson.com/developer/

Ericsson Labs is built upon the concepts of Open Innovation; Ericsson provides users and developers the opportunity to interact and discuss new ideas and prototypes. For more information, see: http://labs.ericsson.net/

Google Android

Google Android is an open source platform for mobile application development, containing code examples and many other resources for developers.

For more information, see: http://developer.android.com/

iPhone DevCenter

This is Apple's website dedicated to development of applications on the iPhone.

For more information, see: http://developer.apple.com/iphone/

Forum Nokia

Forum Nokia is a one-stop shop for mobile developers on Nokia devices, with everything from development advice, design ideas and support for taking applications to market.

For more information, see: http://www.forum.nokia.com/

Sony Ericsson Developer World

Sony Ericsson Developer World provides the documentation, tools, training, technical and go-to-market support for developers on Sony Ericsson mobile devices.

For more information, see: http://developer.sonyericsson.com

Appendix D
Abbreviations

0–9

2G	2nd Generation
3G	3rd Generation
3GPP	Third Generation Partnership Project
3GPP2	Third Generation Partnership Project 2

A

AAA	Authentication, Authorization and Accounting
ABMF	Account Balance Management Function
AF	Application Function
AH	Authentication Header
AKA	Authentication and Key Agreement
AMBR	Aggregate Maximum Bit Rate
AN	Access Network
ANDSF	Access Network Discovery and Selection Function
AP	Application Protocol
API	Application Programming Interface
APN	Access Point Name
APN-NI	APN Network Identifier
APN-OI	APN Operator Identifier
ARIB	Association of Radio Industries and Businesses (Japan)
ARP	Allocation and Retention Priority
ARQ	Automatic Repeat ReQuest
AS	Application Server
ASME	Access Security Management Entity
ATIS	Alliance for Telecommunications Industry Solutions
ATM	Asynchronous Transfer Mode
AuC	Authentication Centre
AUTN	Authentication Token
AV	Authentication Vector
AVP	Attribute Value Pair

B

BA	Binding Acknowledgement
BBERF	Bearer Binding and Event Reporting Function
BBF	Bearer Binding Function; Broadband Forum
BGCF	Breakout Gateway Control Function
BRA	Binding Revocation Acknowledgement

BRI   Binding Revocation Indication
BS   Base Station
BSC   Base Station Controller
BS ID   Base Station Identity
BSS   Base Station Subsystem
BSSID   Basic Service Set Identifier
BU   Binding Update

C

CAMEL   Customized Application for Mobile network Enhanced Logic
CAP   CAMEL Application Part
CBC   Cell Broadcast Centre
CCSA   China Communications Standards Association
CDF   Charging Data Function
CDMA   Code Division Multiple Access
CDR   Charging Data Records
CGF   Charging Gateway Function
CGI   Cell Global Identity
CHAP   Challenge Handshake Authentication Protocol
CK   Cipher Key
CN   Core Network; Correspondent Node
CoA   Care-of Address
CS   Circuit-Switched
CSCF   Call Session Control Function
CSFB   Circuit Switched Fall Back
CTF   Charging Trigger Function

D

DAD   Duplicate Address Detection
DCCA   Diameter Credit Control Application
DHCP   Dynamic Host Configuration Protocol
DL   DownLink
DM   Device Management
DNS   Domain Name System
DPI   Deep Packet Inspection
DRA   Diameter Routing Agent
DRX   Discontinuous Reception
DSCP   DiffServ Code Point
DSL   Digital Subscriber Line
DSMIPv6   Dual Stack Mobile IPv6
DTF   Domain Transfer Function
DTX   Discontinuous Transmission

E

EAP   Extensible Authentication Protocol
ECM   EPS Connection Management
EDGE   Enhanced Data rates for GSM Evolution

eHRPD	Evolved High Rate Packet Data
EIR	Equipment Identity Register
eMBMS	Evolved Multicast Broadcast Multimedia Service
EMM	EPS Mobility Management
eNB	E-UTRAN NodeB
EPC	Evolved Packet Core
ePDG	Evolved Packet Data Gateway
EPS	Evolved Packet System
E-RAB	E-UTRAN Radio Access Bearer
ESM	EPS Session Management
ESP	Encapsulated Security Payload
ETSI	European Telecommunications Standards Institute
ETWS	Earthquake and Tsunami Warning System
E-UTRAN	Evolved Universal Terrestrial Radio Access Network
EV-DO	Evolution - Data Only

F

FA	Foreign Agent
FDD	Frequency Division Duplex
FEC	Forward Error Correction
FMC	Fixed Mobile Convergence
FQDN	Fully Qualified Domain Names

G

GBR	Guaranteed Bit Rate
GERAN	GSM EDGE Radio Access Network
GGSN	Gateway GPRS Support Node
GPRS	General Packet Radio Service
GRE	Generic Routing Encapsulation
GRX	GPRS Roaming eXchange
GSM	Global System for Mobile communications
GSMA	GSM Association
GSN	GPRS Support Node
GTP	GPRS Tunnelling Protocol
GTP-C	GPRS Tunnelling Protocol for Control Plane
GTP-U	GPRS Tunnelling Protocol for User Plane
GUMMEI	Globally Unique MME Identifier
GUTI	Globally Unique Temporary Identifier
GW	Gateway

H

HA	Home Agent
HLR	Home Location Register
HO	Handover
HoA	Home Address
HOM	Higher Order Modulation
H-PCRF	Home PCRF

HPLMN	Home Public Land Mobile Network
HRPD	High Rate Packet Data
HSDPA	High Speed Downlink Packet Access
HSGW	HRPD Serving Gateway
HSPA	High Speed Packet Access
HSS	Home Subscriber Server
HSUPA	High Speed Uplink Packet Access

I

IAB	Internet Architecture Board
IANA	Internet Assigned Numbers Authority
ICMP	Internet Control Message Protocol
I-CSCF	Interrogating-CSCF
ICV	Integrity Check Value
IEEE	Institute of Electrical and Electronics Engineers
IESG	Internet Engineering Steering Group
IETF	Internet Engineering Task Force
IK	Integrity key
IKEv1	Internet Key Exchange version 1
IKEv2	Internet Key Exchange version 2
IMEI	International Mobile Equipment Identity
IMS	IP Multimedia Subsystem
IMSI	International Mobile Subscriber Identity
IMT-2000	International Mobile Telecommunications 2000
IMT-Advanced	International Mobile Telecommunications-Advanced
IP	Internet Protocol
IP-CAN	IP Connectivity Access Network
IPMS	IP Mobility Mode Selection
IPSec	IP Security
IPX	IP Packet eXchange
I-RAT	Inter Radio Access Technology
ISAKMP	Internet Security Association and Key Management Protocol
ISDN	Integrated Services Digital Network
ISP	Internet Service Provider
ISR	Idle mode Signalling Reduction
ITU	International Telecommunication Union
ITU-R	ITU Radiocommunication Sector
ITU-T	ITU Telecommunication Sector
IWF	Interworking Function
I-WLAN	Interworking Wireless LAN

L

LA	Location Area
LAC	Location Area Code
LAN	Local Area Network
LBO	Local Breakout
LEA	Law Enforcement Agencies
LI	Lawful Intercept

LMA	Local Mobility Anchor
LTE	Long-Term Evolution

M

M2M	Machine-to-Machine
MAG	Mobile Access Gateway
MAP	Mobile Application Part
MBMS	Multimedia Broadcast Multicast Service
MCC	Mobile Country Code
MGCF	Media Gateway Control Function
MGW	Media Gateway
MH	Mobility Header
MID	Mobile Internet Device
MIMO	Multiple Input, Multiple Output
MIP	Mobile IP
MIPv4	Mobile IPv4
MIPv6	Mobile IPv6
MM	Mobility Management
MME	Mobility Management Entity
MMEC	MME Code
MMEGI	MME Group Identifier
MMEI	MME Identifier
MMS	Multimedia Messaging Service
MMTel	MultiMedia Telephony
MN	Mobile Node
MNC	Mobile Network Code
MO	Managed Object
MOBIKE	IKEv2 Mobility and Multi-homing Protocol
MPLS	Multi-Protocol Label Switching
MRFC	Media Resource Function Controller
MRFP	Media Resource Function Processor
MS	Mobile Station
MSC	Mobile Switching Centre
MSC-S	MSC Server
MSIN	Mobile Subscriber Identification Number
MSISDN	Mobile Subscriber ISDN Number

N

NAI	Network Access Identifier
NAP-ID	Network Access Provider Identity
NAPTR	Name Authority Pointer
NAS	Non-Access Stratum; Network Access Server
NAT	Network Address Translation
NB	NodeB
NDS	Network Domain Security
NID	Network Identification
NRI	Network Resource Identifier
NW	Network

O

OCF	Online Charging Function
OCS	Online Charging System
OFDM	Orthogonal Frequency Division Multiplexing
OFCS	Offline Charging System
OMA	Open Mobile Alliance

P

PBA	Proxy Binding Acknowledgement
PBU	Proxy Binding Update
PC	Personal Computer
PCC	Policy and Charging Control
PCEF	Policy and Charging Enforcement Function
PCO	Protocol Configuration Options
PCRF	Policy and Charging Rules Function
P-CSCF	Proxy-CSCF
PDCP	Packet Data Convergence Protocol
PDN	Packet Data Network
PDN GW	Packet Data Network Gateway
PDP	Packet Data Protocol
PDU	Protocol Data Unit
P-GW	PDN GW
PIN	Personal Identification Number
PKI	Public Key Infrastructure
PLMN	Public Land Mobile Network
PMIP	Proxy Mobile IP
PMM	Packet Mobility Management
PON	Passive Optical Networks
PPP	Point-to-Point Protocol
PS	Packet-Switched

Q

QAM	Quadrature Amplitude Modulation
QCI	QoS Class Identifier
QoS	Quality of Service

R

RA	Router Advertisement; Routing Area
RAB	Radio Access Bearer
RAC	Routing Area Code
RADIUS	Remote Authentication Dial In User Service
RAI	Routing Area Identity
RAN	Radio Access Network
RANAP	Radio Access Network Application Protocol
RAT	Radio Access Technology
RAU	Routing Area Update
Rel-8	Release 8
Rel-9	Release 9

Rel-99	Release 99
RF	Rating Function
RFC	Request For Comments
RLC	Radio Link Control
RNC	Radio Network Controller
RO	Route Optimization
RRC	Radio Resource Control
RRM	Radio Resource Management
RTSP	Real Time Streaming Protocol

S

SA	Security Association
SAE	System Architecture Evolution
SBC	Session Border Controller
SCC AS	Service Centralization and Continuity Application Server
S-CSCF	Serving-CSCF
SCTP	Stream Control Transmission Protocol
SDF	Service Data Flow
SDP	Session Description Protocol
SEG	Security Gateway
SGSN	Serving GPRS Support Node
S-GW	Serving GW
SID	System Identification
SIM	GSM Subscriber Identity Module
SIP	Session Initiation Protocol
SLA	Service Level Agreement
SMS	Short Message Service
SMS-C	Short Message Service Centre
SN	Serving Network
SN ID	Serving Network Identity
S-NAPTR	Straightforward-NAPTR
SPI	Security Parameters Index
SPR	Subscription Profile Repository
SQN	Sequence Number
SRNS	Serving Radio Network Subsystem
SRVCC	Single-Radio Voice Call Continuity
SS7	Signalling System No 7
SSID	Service Set Identifier

T

TA	Tracking Area
TAC	Tracking Area Code
TAI	Tracking Area Identity
TAP	Transferred Account Procedure
TAU	Tracking Area Update
TCP	Transmission Control Protocol
TDD	Time Division Duplex
TDMA	Time-Division Multiple Access

TEID	Tunnel End Point Identifier
TFT	Traffic Flow Template
TISPAN	Telecommunications and Internet converged Services and Protocols for Advanced Networking
TLS	Transport Layer Security
TMSI	Temporary Mobile Subscriber Identity
TOS	Type of Service
TR	Technical Report
TS	Technical Specification
TSG	Technical Specification Group
TTA	Telecommunication Technology Association of Korea
TTC	Telecommunication Technology Committee (Japan)

U

UDP	User Datagram Protocol
UE	User Equipment
UICC	Universal Integrated Circuit Card
UL	UpLink
UMTS	Universal Mobile Telecommunications System
USB	Universal Serial Bus
USIM	Universal Subscriber Identity Module
UTRAN	Universal Terrestrial Radio Access Network

V

VCC	Voice Call Continuity
VoIP	Voice over IP
V-PCRF	Visited PCRF
VPLMN	Visited Public Land Mobile Network
VPN	Virtual Private Network

W

WCDMA	Wideband Code Division Multiple Access
WG	Working Group
WiMAX	Worldwide Interoperability for Microwave Access
WLAN	Wireless Local Area Network

X

XRES	eXpected RESult

Index